VOLUME FIFTY

# ADVANCES IN CHEMICAL ENGINEERING

## Spatially Resolved Operando Measurements in Heterogeneous Catalytic Reactors

# ADVANCES IN CHEMICAL ENGINEERING

VOLUME FIFTY

# ADVANCES IN CHEMICAL ENGINEERING

## Spatially Resolved Operando Measurements in Heterogeneous Catalytic Reactors


Edited by

**ANTHONY G. DIXON**

*Department of Chemical Engineering,*
*Worcester Polytechnic Institute*

**OLAF DEUTSCHMANN**

*Institute for Chemical Technology and Polymer Chemistry,*
*Karlsruhe Institute of Technology*



ELSEVIER

ACADEMIC PRESS

An imprint of Elsevier

Academic Press is an imprint of Elsevier
50 Hampshire Street, 5th Floor, Cambridge, MA 02139, United States
525 B Street, Suite 1800, San Diego, CA 92101-4495, United States
The Boulevard, Langford Lane, Kidlington, Oxford OX5 1GB, United Kingdom
125 London Wall, London, EC2Y 5AS, United Kingdom

First edition 2017



ISBN: 978-0-12-812589-2
ISSN: 0065-2377

For information on all Academic Press publications
visit our website at https://www.elsevier.com/books-and-journals



*Publisher:* Zoe Kruze
*Acquisition Editor:* Sam Mahfoudh
*Editorial Project Manager:* Shellie Bryant
*Production Project Manager:* Vignesh Tamil
*Cover Designer:* Greg Harris

Typeset by SPi Global, India

# CONTENTS

# CONTRIBUTORS

**Farid Aiouache**
Lancaster University, Lancaster, United Kingdom

**Alessandra Beretta**
LCCP—Laboratory of Catalysis and Catalytic Processes, Dipartimento di Energia, Politecnico di Milano, Milano, Italy

**Andrea Carrera**
LCCP—Laboratory of Catalysis and Catalytic Processes, Dipartimento di Energia, Politecnico di Milano, Milano, Italy

**Jae-Soon Choi**
Oak Ridge National Laboratory, Knoxville, TN, United States

**Alessandro Donazzi**
LCCP—Laboratory of Catalysis and Catalytic Processes, Dipartimento di Energia, Politecnico di Milano, Milano, Italy

**Alexandre Goguet**
School of Chemistry and Chemical Engineering, Queen's University Belfast, Belfast, United Kingdom

**Gianpiero Groppi**
LCCP—Laboratory of Catalysis and Catalytic Processes, Dipartimento di Energia, Politecnico di Milano, Milano, Italy

**Michael P. Harold**
University of Houston, Houston, TX, United States

**Dan Luss**
University of Houston, Houston, TX, United States

**Kevin Morgan**
School of Chemistry and Chemical Engineering, Queen's University Belfast, Belfast, United Kingdom

**Hoang Nguyen**
University of Houston, Houston, TX, United States

**William P. Partridge Jr.**
Oak Ridge National Laboratory, Knoxville, TN, United States

**Po-Yu Peng**
University of Houston, Houston, TX, United States

**Caomhán Stewart**
School of Chemistry and Chemical Engineering, Queen's University Belfast, Belfast, United Kingdom

**Jamal Touitou**
School of Physics, Trinity College Dublin, College Green, Dublin 2, Ireland

# PREFACE

Heterogeneous catalysis and experimental investigation are essential components of the continuing search for better utilization of raw materials and energy, with reduced impact on the environment. *Operando* measurement techniques in heterogeneous chemical reactors have progressed rapidly due to developments in laboratory capabilities, and are used extensively for model validation and insight into the development of catalytic reactors and processes. This thematic volume on "Spatially resolved operando measurements in heterogeneous catalytic reactors" presents reviews of the state of the art of *operando* measurements in various catalytic reactors and processes.

There have been many reviews and chapters written on the use of *operando* probes to investigate the catalytic surface in detail, at the atomic scale. The intentions of the editors for this volume were that it should

- focus on spatial information within catalytic reactors, rather than on catalyst surfaces;
- cover selected topics from electromagnetic probe methods (e.g., MRI, near-IR tomography, NMR thermometry) and physical probe methods (e.g., Spaci-MS, IR Imaging, Spaci-IR, Optical Fiber Reflectometry, Optical Fiber Pyrometry, Spaci-FB); and
- focus on applications to fixed beds, trickle beds, fluidized beds, foams, honeycomb monoliths, or other heterogeneous reactors.

Although the nominal title of this volume emphasizes experimental and measurement techniques, some of the authors found it was essential to include discussions of modeling and simulation. This illustrates both the need for experimental validation of the modeling efforts, and the use of the modeling to interpret and understand the experimental results. As the power of computers has grown, enabling more detailed and realistic numerical models by CFD and other techniques, so too have the demands for more accurate experimental measurements by such methods as Spaci-MS, NIR tomography, and others. This is especially true in complex reactor geometries and/or multiphase processes.

Chapter 1 presents an overview from the group that originated the spatially resolved capillary-inlet mass spectrometry technique (Spaci-MS). Partridge and Choi at Oak Ridge National Laboratory have provided an insightful overview of that technique and its many descendants. The chapter discusses the spatial resolution of concentration profiles inside monolithic catalytic converters for automobile applications and gives much practical advice for using this technique. Aside from the major focus of the chapter on lean $NO_x$ traps and SCR catalysts, the authors also include applications of Spaci-MS in oxidation catalysts and particulate filters.

The second chapter, from the collaboration of Harold and Luss at the University of Houston, discusses coherent optical frequency domain reflectometry for measurement of spatiotemporal temperatures inside catalytic reactors. The chapter also covers extensions of this method to couple with Spaci-MS to add concentration measurements. This method is applied to derive spatiotemporal profiles in exhaust gas after treatment systems. In particular, the dynamics in the storage behavior in diesel oxidation catalyst is elucidated.

Continuing the theme of extending the highly popular Spaci-MS technology, the third chapter by Goguet and Morgan and their colleagues at Queen's University, Belfast, provides a look at recent developments in using Spaci techniques for packed beds of powdered catalysts. The authors provide an illuminating history of several generations of the method and its extensions to improve both measurement accuracy and capabilities by combination with other techniques. They conclude with a look forward to future developments and applications of this method.

Chapter 4 from Beretta et al. at Politecnico di Milano focuses on the in situ probing of temperature and concentration profiles in high temperature and short contact time partial oxidation processes. The authors review their own and other work on the use of spatially resolved sampling techniques to improve the understanding of the interplay between homogeneous and heterogeneous chemistry, and how to balance the two.

The final contribution, from Aiouache at Lancaster University, United Kingdom, presents a comprehensive description of the near-infrared tomography technique in both 2D and 3D versions. The author and coworkers developed the applications of this method while at Queen's University, Belfast, and he has continued his work in this area. The method is introduced and illustrated with examples in packed-bed adsorbers, dispersion, and catalyst activity profiling in gas–solid packed beds, including deactivation modeling.

Our thanks are due to the authors and reviewers for all their hard work, to the editorial team at Elsevier, and to Series Editor Guy B. Marin for the invitation to act as Guest Editors for this volume of *Advances in Chemical Engineering*.

ANTHONY G. DIXON
Professor
Department of Chemical Engineering
Worcester Polytechnic Institute

OLAF DEUTSCHMANN
Professor
Institute for Chemical Technology and Polymer Chemistry
Karlsruhe Institute of Technology

CHAPTER ONE

# Understanding the Performance of Automotive Catalysts via Spatial Resolution of Reactions Inside Honeycomb Monoliths

**William P. Partridge Jr.[1], Jae-Soon Choi**
Oak Ridge National Laboratory, Knoxville, TN, United States
[1]Corresponding author: e-mail address: partridgewp@ornl.gov

## Contents



## Abstract

By directly resolving spatial and temporal species distributions within operating honeycomb monolith catalysts, spatially resolved capillary inlet mass spectrometry (SpaciMS) provides a uniquely enabling perspective for advancing automotive catalysis. Specifically, the ability to follow the spatiotemporal evolution of reactions throughout the catalyst is a significant advantage over inlet-and-effluent-limited analysis. Intracatalyst resolution elucidates numerous catalyst details including the network and sequence

*Advances in Chemical Engineering*, Volume 50
ISSN 0065-2377
https://doi.org/10.1016/bs.ache.2017.09.002

of reactions, clarifying reaction pathways; the relative rates of different reactions and impacts of operating conditions and catalyst state; and reaction dynamics and intermediate species that exist only within the catalyst. These details provide a better understanding of how the catalyst functions and have basic and practical benefits; e.g., catalyst system design; strategies for on-road catalyst state assessment, control, and on-board diagnostics; and creating robust and accurate predictive catalyst models. Moreover, such spatiotemporally distributed data provide for critical model assessment, and identification of improvement opportunities that might not be apparent from effluent assessment; i.e., while an incorrectly formulated model may provide correct effluent predictions, one that can accurately predict the spatiotemporal evolution of reactions along the catalyst channels will be more robust, accurate, and reliable. In such ways, intracatalyst diagnostics comprehensively enable improved design and development tools, and faster and lower-cost development of more efficient and durable automotive catalyst systems. Beyond these direct contributions, SpaciMS has spawned and been applied to enable other analytical techniques for resolving transient distributed intracatalyst performance. This chapter focuses on SpaciMS applications and associated catalyst insights and improvements, with specific sections related to lean $NO_x$ traps, selective catalytic reduction catalysts, oxidation catalysts, and particulate filters. The objective is to promote broader use and development of intracatalyst analytical methods, and thereby expand the insights resulting from this detailed perspective for advancing automotive catalyst technologies.

## 1. INTRODUCTION

Spatially resolved capillary inlet mass spectrometry (SpaciMS) is an analytical technique that provides spatially resolved measurements of transient species concentrations within confined-space chemical reactors under operating conditions; although in application these reactors have been primarily honeycomb or channelized automotive catalyst, it has been applied to study numerous other reactor systems. Such intracatalyst measurements elucidate evolution of the catalyst chemistry, the network, and the sequence of the various catalyst reactions, how the various reaction steps interact, and how these are impacted by temperature, reaction feed, aging, etc. This new direct perspective enabled by SpaciMS and similar diagnostics has provided a considerable leap in advancing catalyst understanding and technologies relative to integral analysis where only catalyst-inlet and -effluent measurements are available. The result is better models, faster and lower-cost development of advanced efficiency engine catalyst systems, and these systems arriving in the market faster with the associated societal, economic, and environmental benefits.

SpaciMS was originally developed in c.1997, in a United States Department of Energy (US DOE) funded partnership between Oak Ridge National Laboratory (ORNL) and Cummins Inc. to resolve fast lean $NO_x$ trap (LNT) catalyst effluent transients that were too fast to resolve with conventional analyzers common to automotive emissions characterization analyzers; e.g., chemiluminescence ($NO_x$), paramagnetic ($O_2$), nondispersive infrared (CO, $CO_2$, other), and flame-ionization detection (hydrocarbons). In those early days, still focused on catalyst effluent analysis, SpaciMS was known as the fast-MS and mobile-MS highlighting its temporal resolution and ability to transcend neat laboratory environments and be taken on the road for applications in corporate development facilities to real world, nonneat systems; i.e., unfiltered and undiluted exhaust from diesel engine catalyst systems. At the time, it was fairly unconventional to imagine sampling undiluted engine exhaust into a mass spectrometer (MS), much less routinely transporting such an instrument in the back of a van crosscountry for field applications. Indeed, it was while observing LNT effluent transients on just such a diesel engine catalyst system application campaign at the Cummins Technical Center in October 1999 (Partridge et al., 2000) that the team realized how such measurements could be made within and along the catalyst channels to spatiotemporally resolve how the integral effluent transients evolve. To reflect the new capability, team member Neal Currier of Cummins coined the more descriptive and creative name and acronym SpaciMS.

Beyond this initial concept and early intracatalyst applications, significant additional effort was required to advance the hardware, analysis, and basic understanding of how to interpret complex spatiotemporal data. Indeed, effectively communicating concepts based on data that is varying in both the spatial and the temporal domains to a broad audience remains one of the more complicated aspects of SpaciMS and similar diagnostics. Two-dimensional plots in either the spatial or the temporal domain can communicate detail, but often fall short of effectively capturing the full multidimensional complexity. Three-dimensional plots at select times or locations have similar limitations, and can be similarly too complicated for effective communication on presentation timescales. Simulation movies, on the other hand, while very effective for quickly conveying general spatiotemporal concepts, are less effective for communicating specific details such as gradients and relationships between species and functions. Nevertheless, the automotive catalysis community has embraced intracatalyst analysis and significantly advanced in effectively communicating the associated spatiotemporal data to the point that such discussions are now common to related conferences, meetings, and literature. Parallel analytical improvements were significant and included improved capillary and instrument heating for better temporal resolution, multicapillary inlet options for switching between sampling locations without capillary repositioning, enhanced capillary positioning control for better spatial resolution, new mass filters and operation strategies for better sensitivity, strategies for optimizing temporal resolution and range of measured species, and improved data analysis for more accurate measurement of a greater number of species.

Broader SpaciMS implementation outside the original ORNL–Cummins partnership drove further technique development, catalyst insights, and commercial development. Early adopters included internal engine- and catalyst system development applications at Cummins, Goguet's team at Queen's University Belfast (Sá et al., 2010), Epling's team at University of Waterloo (Irani et al., 2009a), and Dearth's team at Ford (Crocker et al., 2010). These and other researchers using SpaciMS added their unique configurations, interpretations, and contributions, thus furthering the technique. Beyond catalyst applications, SpaciMS was applied to engine-development studies related to cylinder-managed reductant generation (West et al., 2004), exhaust gas recirculation (EGR) uniformity (Partridge et al., 2002), nonthermal plasma reactors, PEM fuel cells (Partridge et al., 2006), and fuel reformer catalysts (Schwank and Tadd, 2010). Applications of SpaciMS to engine and catalyst development within Cummins and via their partnerships contributed to their new 2007 6.7*L* ISB engine and associated catalyst system. This LNT-based system was the first heavy duty diesel $NO_x$ catalyst system to market, and met 2010 emissions regulations 3 years early. Also notable is the QUB and Hiden Analytical partnership which developed a SpaciMS configuration with multicapillary inlet and integrated capillary translation, capable of providing an extensive and high-resolution three-dimensional map of transient species distributions throughout an operating catalyst; this configuration was ultimately commercialized as the Hiden SpaciMS (Hiden Analytical, SpaciMS, n.d.). Considering this development, application, and commercialization, the SpaciMS was recognized with an R&D100 Award in 2008 as being among the top 100 technologies commercialized that year (Hiden Analytical, ORNL, n.d.).

The main SpaciMS applications and contributions have been in advancing automotive catalyst technologies, and it has been a major force in providing direct evidence of intracatalyst transient reaction distributions and furthering these efforts. Its operando spatiotemporally resolved intracatalyst measurements enhance our understanding of how catalyst function and improves the accuracy and robustness of conceptual and detailed models via application of the resulting information for model development, validation and critical assessment, and determining kinetic parameters under operating conditions (vs integral and differential conditions required by conventional approaches). But probably more generally, SpaciMS was a major driver of highlighting the transient and spatially varying intracatalyst perspective to the broader research-and-development community, and supporting the now routine expectation for considering such intracatalyst reaction interplay. This is reflected in the literature, conference presentations, and discussions, and subsequent development of other analytical tools for resolving intracatalyst reaction distributions (Morgan et al., 2016); e.g., Phosphor Thermography (Choi et al., 2006), SpaciFTIR (Luo et al., 2011a), SpaciFB (Goguet et al., 2017; Touitou et al., 2012, 2013), and optical frequency domain reflectometry (OFDR) (Nguyen et al., 2013; Peng et al., 2017). Interestingly, a method similar to SpaciMS, but for resolving spatiotemporal reaction distributions within foam monolith reformer catalyst was independently developed in Schmidt's group at the University of Minnesota and continues to be advanced by its inventor Horn (Horn et al., 2006). Practically, SpaciMS and insights from like diagnostics promote improved process design and development tools; e.g., improved models, better analysis-led design, control software, and virtual sensors. Similarly, intracatalyst insights promote improved process control, by elucidating the relationship between measurement parameters and catalyst state or health, and what/where/when to measure parameters for implementing advanced control. In this way, SpaciMS enabled insights contribute to hastening the development and commercialization of better, less expensive, more efficient, and durable catalyst systems to the community.

This chapter focuses on SpaciMS applications and related insights for advancing automotive catalyst technologies. An initial section addresses some general SpaciMS configurations and analytical considerations relevant to the chapter, and application-specific considerations are discussed as appropriate in the corresponding sections. Following sections highlight SpaciMS applications to LNT, selective catalytic reduction (SCR), particulate filter, and oxidation catalysts. While the perspective is primarily focused on general insights enabled from intracatalyst measurements and SpaciMS specifically, we observe the relevance of this information to modeling and specific modeling applications. Moreover, non-SpaciMS methods for resolving intracatalyst reaction distributions are noted. Our goal is to highlight unique insights that have been shown or demonstrated by direct intrareactor spatiotemporally resolved measurements, and so motivate broader development and application of intrareactor diagnostics.

## 2. SpaciMS METHODS

While specific SpaciMS methods are described in the references discussed in the following sections and have been specifically discussed in a recent publication reviewing spatially resolved intracatalyst methods (Morgan et al., 2016), some additional minor discussion is relevant here. The intent is to provide a general summary of technique configurations and considerations relevant to SpaciMS applications and interpretations in the subsequent sections.

A general SpaciMS configuration is shown in Fig. 1 and involves extractive sampling from a specific location within a catalyst via a capillary to an MS for analysis. Generally, the instrument including the sampling capillary is heated to mitigate sample adsorption on instrument surfaces. Capillaries are generally uncoated (nonpassivated) fused-silica capillary tubes, with polyimide coating that burns off above c.300°C, and the sampling location is defined by the capillary tip location. For spatially resolving reaction distributions a single capillary can be translated, and/or sampling can be switched between different capillaries arrayed throughout a catalyst. Plug-like capillary flow causes species transients at the capillary tip sampling point to maintain their stacked orientation during transport through the capillary and appear as corresponding transients at the MS inlet. The capillary transit time causes some fixed offset between the transient and its measurement; this

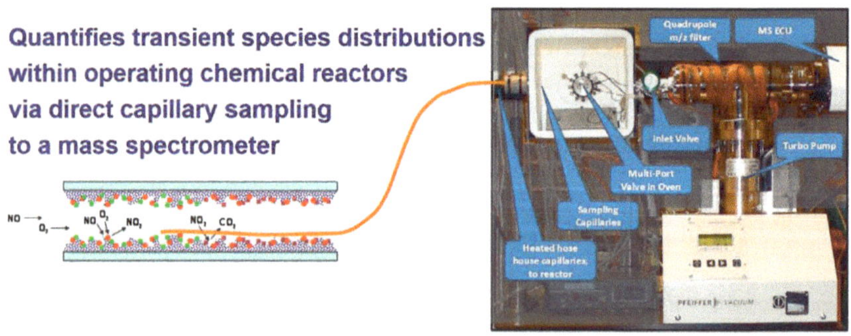


Fig. 1 Typical SpaciMS configuration. The schematic shows a typical catalyst monolith channel being probed by a sampling capillary, and the picture identifies typical SpaciMS components.

offset is generally constant, does not define the temporal resolution, and can be accounted for in postprocessing. The temporal resolution is defined by diffusion at the interface of adjacent sample units or plugs and will broaden step transients; sample adsorption on capillary walls or other instrument parts can also contribute to instrument response but is typically mitigated by instrument heating. Both capillary transit time and instrument response are influenced by the SpaciMS inlet configuration as well, which is discussed below. The broad mass-to-charge filtering capabilities of the MS permits similarly broad species measurements, and notably allows for measurement of diatomic species which many analytical tools (e.g., FTIR) are not sensitive to; specifically, direct nitrogen, hydrogen, and oxygen measurements enable corresponding balances to be assessed.

The SpaciMS inlet configuration will impact capillary size, sampling rate, temporal resolution, and potentially invasive nature; the two general types are direct and indirect sampling. For a direct sampling configuration, the entire capillary sample is injected into the MS. Direct sampling uses smaller capillaries (e.g., 50-μm ID, 180-μm OD, c.2-m long) and smaller sampling rates (c.10μL/min), and correspondingly has longer capillary transport times (10 s of seconds). The longer transport times allow more extensive diffusion between sample plugs, and thus generally results in lower temporal resolution. Direct sampling was used for the original SpaciMS and continues to be the primary configuration we use at ORNL. The broader community seems to use SpaciMS configurations with indirect sampling which is also known as differentially pumped or sampled. For an indirect sampling configuration, only a small portion of the capillary flow is sampled into the MS via an orifice, while the major capillary flow is exhausted. Indirect sampling uses larger capillaries and larger sampling rates leading to shorter capillary transport times and greater temporal resolution. Specific indirect sampling configurations vary, but a common setup we use at ORNL to configure a Fast-SpaciMS with indirect sampling includes a large-bore capillary (100-μm ID, 200-μm OD, c.2 m long) to achieve c.20 ccm sample flow, and an orifice tuned to achieve the desired MS pressure. These inlet options allow the SpaciMS configuration to be tailored to the specific constraints and priorities of a given experiment. For instance, indirect sampling is the natural choice for experiments requiring high temporal resolution, such as with fast three-way catalyst (TWC) modulation (i.e., controlling performance by active modulation of the air-to-fuel ratio), but a high cell density catalyst might require smaller capillaries like those suitable for direct sampling. Both the capillary size and sampling rate influence how effectively noninvasive the technique is, and thus the inlet configuration impacts this figure of merit. Because of the various inlet options, and the wide array of specific hardware that can be used to implement SpaciMS, it is important to understand these differences, report them carefully, and consider such details when comparing studies with different specific SpaciMS configurations.

It is important to consider the potential impact of the capillary on the catalyst reactions within the probed catalyst channel relative to that in the unprobed channels. While SpaciMS can be effectively noninvasive, certain specific configurations can change the catalyst performance, and might even be expected due to the experimental tradeoffs. The physical basis for capillary impact involves the capillary occluding the sampled channel flow and diverting flow to the nonprobed channels; this impact can be balanced to some degree by the capillary sampling rate, and the Epling group has described a method applicable to SpaciMS using indirect sampling for location-specific tuning of capillary flow to maintain constant channel flow as the capillary is moved to sample different intrachannel locations (Hou et al., 2012). Catalyst channel occlusion is impacted by capillary OD, sampling rate, catalyst-channel size, space velocity, and position of the capillary within the catalyst channel (Hettel et al., 2013; Nguyen et al., 2017), and experimental methods for assessing practical impact of a specific configuration have been shown (Morgan et al., 2016; Nguyen et al., 2017). Catalyst orientation can constrain the location of the sampling capillary within the channel and thus offers a simple strategy for enhancing the measurement's minimally invasive nature. Specifically, a horizontal reactor (i.e., horizontally oriented catalyst channels) with diagonally oriented catalyst channel walls, will cause flexible capillaries to naturally rest in the channel corner; this corner orientation has been shown to be the optimum geometry for mitigating possible occlusion impacts of an intrachannel probe (Hettel et al., 2013). Thus, there are numerous SpaciMS configurations, capillary and catalyst geometry, environment, and reactor considerations that can be varied to optimize the minimally invasive nature of SpaciMS within the required experimental tradeoffs. Nevertheless, even with experimental setups showing clear evidence of an invasive capillary impact (Easterling et al., 2012), SpaciMS results have been used to gain valuable insights for advancing catalyst technology. The salient point here is that numerous studies have shown that many SpaciMS setups are practically noninvasive (Hettel et al., 2013; Morgan et al., 2016; Nguyen et al., 2017; Sá et al., 2010).

While special catalyzed capillaries (Currier and Yezerets, 2007) have been developed to enhance SpaciMS applications, capillary contamination is generally undesirable and should be avoided and monitored. Indeed, small catalyst particles can be carried into the sample capillary, and such contamination can be exacerbated by pushing a capillary tip along the surface of a channel wall (Easterling et al., 2012). Such contamination should be greater with the higher sampling rates of indirect sampling configuration, but have been observed even with the microliter-scale rates of direct sampling. While indirect sampling might result in greater capillary contamination, the relative net impact might be somewhat compensated for by the shorter capillary transit time (higher intracapillary space velocity). Regardless, capillary contamination should be avoided. Simple methods for avoiding contamination are to pull rather than push the capillary, and purge the capillary during translation; i.e., push a gas (Ar, $N_2$, etc.) backward through the capillary causing flow out the tip. The SpaciMS in Fig. 1 is configured for this option where the common or nonselected outlet can be reversed from a backing pump to a pressurized gas source during capillary translation, and then back to pumping once the capillary is stationary in the desired location; the green-handled MS inlet valve can be closed during this process as an instrument safety precaution. In general, it is preferable to purge the capillary when not making SpaciMS measurements to avoid contamination. Capillary contamination can be readily monitored for by positioning the capillary tip upstream of the catalyst and checking for appropriate catalyst-activated reaction; e.g., oxidation, $NO_x$ conversion, etc.

## 3. LNT APPLICATIONS

Lean $NO_x$ trap catalysts are designed for $NO_x$ conversion from lean-burn engines with excess exhaust oxygen (e.g., diesel engines) and are also known as $NO_x$ storage and reduction (NSR) catalysts (Epling et al., 2004; Roy and Baiker, 2009). These catalysts contain alkali- or alkaline-earth (e.g., K, Ba) and precious metal (e.g., Pt) components on a high-surface area support (e.g., $Al_2O_3$). For practical automotive applications, this catalyst formulation is washcoated on a honeycomb monolith structure consisting of many typically square channels that are much longer than their cross-sectional area (e.g., 1.27 mm sides for 400 cells per square inch (cpsi), 150-mm long). Operation of an LNT is inherently cyclic or transient, where the exhaust is cycled between longer (minutes scale) lean phases and shorter (seconds scale) rich phases. During lean operation, exhaust $NO_x$ is stored on the alkali/alkaline-earth metal component as nitrites and/or nitrates. The result of this adsorption is that an evolving surface $NO_x$ distribution develops along the LNT channels during operation, with adsorption naturally occurring first near the inlet and progressing along the channels with further lean operation. Because of the finite $NO_x$-storage capacity and to avoid $NO_x$ emissions, a short rich exhaust pulse is implemented to regenerate the LNT before the surface $NO_x$ distribution evolves too far along the catalyst as to cause effluent $NO_x$ slip. While early rich-phase implementations involved diesel injection into the exhaust upstream of the LNT catalyst, modern systems use in-cylinder late-cycle fuel injection to create exhaust with low oxygen and an abundance of reducing agents (e.g., $H_2$, CO, and various hydrocarbons). The rich phase reduces the surface nitrites/nitrates, ideally with high nitrogen selectivity, and regenerates the LNT for additional $NO_x$ storage in the subsequent lean phase; such cyclic lean–rich operation allows for high $NO_x$ conversion from diesel and other lean-burn engine systems. During regeneration, the reductant front lags the bulk-gas velocity because of reaction with the surface $NO_x$, resulting in evolving and distributed reduction along the LNT channels. This cyclic operation and evolving intracatalyst spatiotemporally distributed nature, makes SpaciMS analysis ideally suited and beneficial to the study of LNT catalysts.

Many fundamental and practical insights regarding development of efficient LNT systems have resulted from SpaciMS measurements of intra-LNT transient species distributions. Most basically, these have provided direct measures of how the $NO_x$-storage and $NO_x$-reduction distributions evolve throughout the lean and rich phases, respectively, as well as through continuous cycling, and the regeneration effectiveness and unique process inherent to different reductants (e.g., $H_2$ vs CO). In addition to the basic storage and reduction functions, LNT catalysts have other functions including oxidation and water–gas shift reaction (WGSR), which can also have transient distributed performance. Moreover, sulfur poisoning of the LNT occurs in a transient distributed fashion, and can have varying impacts on the different catalyst functions. The result is that the LNT catalyst is a spatiotemporally complex device, whose transient distributed nature continues to change throughout its life. This section highlights such major insights resulting from SpaciMS studies of LNT catalysts.

## 3.1 $NO_x$ Storage and Regeneration

Figure 2T shows the transient $NO_x$ distribution throughout an entire 54-s/4-s lean/rich operation cycle of a $Pt/K/Al_2O_3$/cordierite LNT catalyst

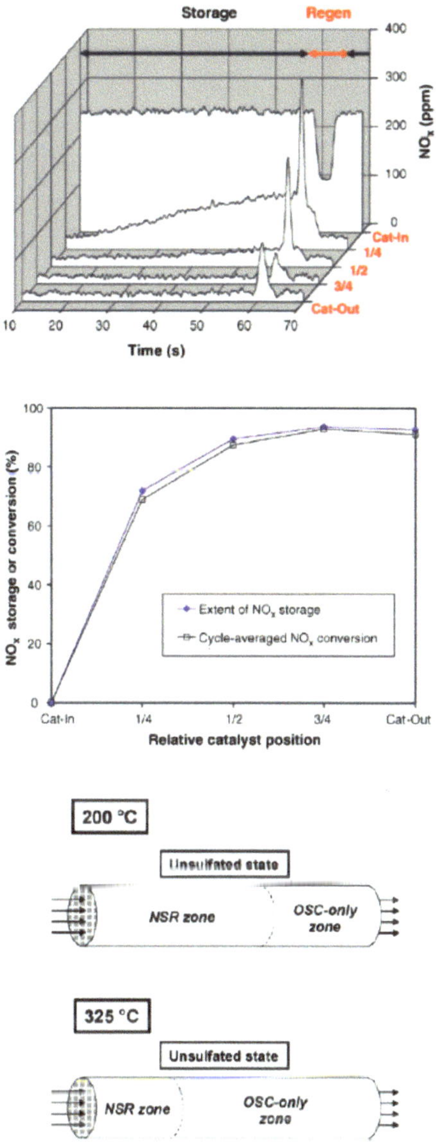


**Fig. 2** $NO_x$ removal performance of a 3-in. long Pt/K/$Al_2O_3$-washcoated monolith during 54-s storage/4-s regeneration cycling. Storage gas: 250 ppm NO, 8% $O_2$, 5% $H_2O$, and $N_2$ balance; regeneration gas: 4% CO, 1% $O_2$, 5% $H_2O$, and $N_2$ balance; catalyst mid-bed $T=430°C$ (*Top*: T) $NO_x$ breakthrough profiles at different axial locations obtained by SpaciMS (Choi et al., 2006). (*Middle*: M) Corresponding cycle-averaged $NO_x$-storage extent and $NO_x$ conversion (Choi et al., 2006). (*Bottom*: B) An LNT is axially divided between an NSR zone, where $NO_x$ is stored, and a downstream OSC-only zone (Choi et al., 2008a).

(Choi et al., 2006); in this example (which was obtained from a laboratory bench-top flow reactor using simulated exhaust gas streams), SpaciMS measurements were made at the LNT inlet, $1/4L$, $1/2L$, $3/4L$, and outlet location, where $L$ is the catalyst length. The inlet $NO_x$ is steady throughout the lean storage phase and drops during the rich regeneration phase; the nonzero measured $NO_x$ during regeneration is apparently due to the reactor and instrument response. At the $1/4L$ location and early storage times (e.g., $1/4L$ and 10s in Fig. 2T) very little $NO_x$ is measured, indicating practically complete $NO_x$ trapping in the front quarter of the catalyst. As the lean storage phase continues, measured $NO_x$ slip at the $1/4L$ location increases indicating progressive filling of the $NO_x$-storage capacity in this front quarter, and corresponding utilization of downstream $NO_x$-storage capacity. Similar progressive but much smaller $NO_x$ slip is indicated at the $1/2L$ location, with practically zero $NO_x$ measured throughout the storage phase at both the $3/4L$ and outlet locations; this indicates that $NO_x$ storage is confined to the front portion of the LNT, with the maximum $NO_x$ penetration depth being somewhere between the $1/2L$ and $3/4L$ locations. Figure 2M quantifies the net surface–$NO_x$ distribution during the lean phase which was determined by integrating the area above the transient curves during storage. This demonstrates how surface–$NO_x$ is spatially distributed along the LNT axis, with c.70% of the total storage-phase $NO_x$ being stored in the front $1/4L$ and progressively lower levels being stored in the subsequent downstream sections of the NSR zone. Based on the practically flat $NO_x$-storage curve in the back $1/2L$, it is clear that very little $NO_x$ is stored in the third $1/4L$ of the LNT. The deviation from 100% storage at the $3/4L$ and outlet locations in Fig. 2M, and the nonzero storage-phase $NO_x$ at these locations in Fig. 2L, is likely due to the finite uncertainties in the SpaciMS measurements; nevertheless, the results trends clearly indicate that the back $1/4L$ of the LNT, and likely the majority of the third $1/4L$ location, is not used for $NO_x$ storage during the lean phase.

To achieve high $NO_x$ conversion and avoid $NO_x$ slip, only a limited front portion of the LNT is used for $NO_x$ storage. Thus, the LNT can be visualized as being divided into NSR and OSC-only zones as shown in Fig. 2B (Choi et al., 2008a). Typically, oxygen storage capacity (OSC) exists throughout the catalyst, including in the front NSR zone. In the case of simple model LNTs such as $Pt/K/Al_2O_3$ shown in Fig. 2, the OSC function is not intentionally added (vs commercial catalysts for which OSC material such as $CeO_2$ is a common component), but as oxygen atoms stored on Pt or K consume an extra amount of reductants (although much less than $CeO_2$), the net effect is as if there were an OSC function on model catalysts as well. Surface $NO_x$ is progressively spatially distributed within the NSR zone, which spatially evolves through the storage phase, both of which are demonstrated in Fig. 2T and M; for that case, the interface between the NSR and OSC-only zones occurs in the third quarter of the LNT. The length of the NSR zone will vary with changes in net NSR efficiency, and variations in the lean-phase period. For instance, lower $NO_x$ capacity due to actual or effective availability of $NO_x$-storage sites will require a greater fraction of the LNT to trap a given level of exhaust $NO_x$ and proportionally extend the NSR-zone length; such lower capacity could be due to catalyst composition, aging/sulfation, incomplete regeneration, and temperature-dependent capacity. Clearly, for a fixed LNT length, the NSR and OSC-only zones vary together in such a way that lengthening of the NSR zone shortens that of the OSC-only zone. The NSR-zone length increases as the storage-phase period increases, and at sufficiently long storage times the NSR zone could occupy the entire LNT length while the OSC-only zone would correspondingly shrinks to nonexistence. Although fuel-penalty considerations favor maximizing the lean-to-rich phase timing, conversion requirements mandate the NSR zone to be less than the overall LNT length. Moreover, the OSC-only zone can contribute to secondary $NO_x$ adsorption (discussed below), overall $NO_x$ conversion and selectivity, and serve to reduce reductant slip by oxidizing residual reducing species. In general, high $NO_x$ conversion can be achieved as long as the NSR-zone length is less than that of the LNT.

Just as a spatiotemporal surface–$NO_x$ distribution develops through the lean storage phase, a reductant wave sweeps along the LNT axis reacting with the stored $NO_x$ during the rich regeneration phase. Figure 3 shows transient $NO_x$ and $H_2$ distributions during operation of a commercial LNT, and using $H_2$ as the rich-phase reducing agent (Choi et al., 2012). In this case, and as indicated by Fig. 3L, the NSR zone ends somewhere between $1/4L$ and $1/2L$, with the major $NO_x$ storage occurring in the front $1/4L$. Figure 3R details how the $H_2$ reducing front moves through the catalyst during regeneration; while the temporal response of the direct sampling SpaciMS setup temporally broadened the more step or top-hat shaped rich-phase pulse, the spatiotemporal nature of the reductant pulse, and its relation to other distributed catalyst functions are apparent. Quite naturally, it shows that $H_2$ is progressively depleted at the reductant pulse front as it travels through the LNT catalyst; this is apparent in Fig. 3R by the systematically shifting $H_2$ breakthrough times to later rich-phase times with increasing

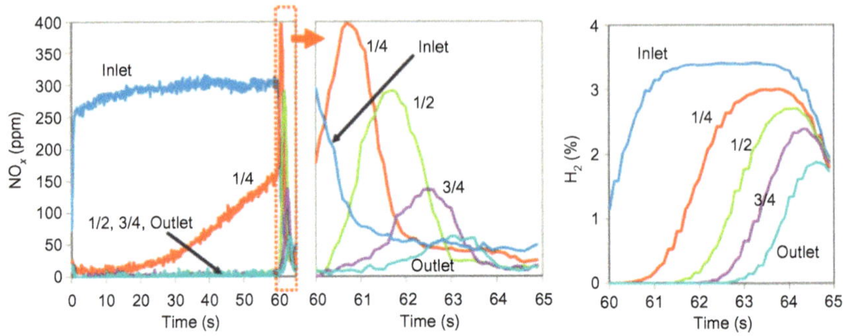


**Fig. 3** Spatiotemporal $NO_x$ and $H_2$ profiles during 65-s/5-s lean/rich cycling at 400°C of a commercial LNT catalyst with $H_2$ as the rich-phase reductant (Choi et al., 2012). (*Left*: L) Transient $NO_x$ distributions throughout the lean/rich cycle, with the rich phase indicated by the *red-dotted box*. (*Middle*: M) Transient $NO_x$ distributions during the rich phase. (*Right*: R) Transient $H_2$ distributions during the rich phase.

distance into the catalyst, and the common reductant tail or pulse end. Figure 3R also indicates that the $H_2$ consumption varies along the catalyst axis, and the local-zone $H_2$ consumption can be determined from the integral of the difference in the $H_2$ profiles at the bounding locations; i.e., $H_2$ consumption in the front $1/4L$ is indicated by the integral of the difference in the inlet and $1/4L$ profiles. Not surprisingly, the $H_2$-utilizaiton distribution is consistent with the distributions of the NSR and OSC-only zones. Specifically, $H_2$ consumption is the greatest in the NSR zone, where it reacts with both stored $NO_x$ and OSC, and the absolute greatest $H_2$ consumption is in the front $1/4L$ where $NO_x$ storage is the greatest. As the OSC-only zone includes the catalyst back $1/2L$, the $H_2$ gradient in that region is consistent with OSC reduction; notice that $H_2$ consumption is relatively uniform in this region suggesting spatially uniform OSC along the catalyst channel.

Figure 4 more clearly demonstrates differences in how the reductant pulse front propagates through regions with and without stored $NO_x$ (Choi et al., 2007). Figure 4L was measured during OSC cycling (i.e., without $NO_x$ in the feed gas), and thus $H_2$ consumption reflects uniformly distributed reaction with the LNT OSC function; this figure also clearly demonstrates the shifting breakthrough time and common reductant tail indicative of $H_2$ consumption at and propagation of the reductant front. Relatively uniform $H_2$ consumption in each catalyst quarter indicates similarly spatially uniform OSC and washcoat distributions. In contrast, Fig. 4R shows $H_2$ consumption distributions during NSR cycling where $NO_x$ storage occurred during the lean cycling phase; these are similar to Fig. 3R

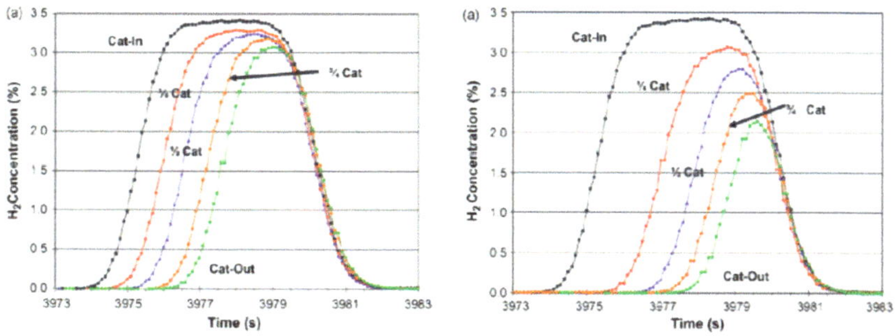


**Fig. 4** Spatiotemporal $H_2$ utilization profiles during OSC and NSR cycling of a commercial LNT catalyst at 325°C (Choi et al., 2007). (*Left*: L) OSC cycling without lean-phase $NO_x$ storage. (*Right*: R) NSR cycling with $NO_x$ storage during the lean phase.

curves, but are for operation at 325°C. The spatiotemporal $H_2$ consumption trends reflect the stored $NO_x$ distribution within the NSR zone which is contained in the front $1/2L$, and with the greatest $NO_x$ storage in the front $1/4L$. Hydrogen consumption in the downstream OSC-only zone with NSR cycling is similar to that without $NO_x$ storage (i.e., OSC cycling).

In addition to the spatial $H_2$ gradients indicated by the integrated difference in the bounding temporal profiles, the spatiotemporal $H_2$ profiles in Figs. 3R and 4 indicate variations in propagation velocity of the $H_2$ front. Because $H_2$ is being locally depleted from the inlet gas stream via reaction, the $H_2$ front moves slower than the gas velocity. The speed of the reductant front corresponds to the spatial variations in the breakthrough times. Figure 4L shows that with OSC cycling, the $H_2$ front advances at a relatively uniform pace through the catalyst; i.e., similar breakthrough-timing delay in each $1/4L$ of the catalyst. In contrast, Fig. 4R shows that the $H_2$ front advances more slowly through the NSR region and is the slowest in the front regions where surface–$NO_x$ is the greatest. The reductant front speed and advancement through an LNT catalyst was specifically studied by Clayton et al. using a SpaciMS corollary, where effluent measurements were made from catalyst of varying length (Clayton et al., 2008); they specifically studied the speed of both $H_2$- and intermediate-$NH_3$ (discussed below) reductant fronts. In general, the reductant fronts travel more slowly where consumption is greater and are slowest in regions where $NO_x$ storage is the greatest.

A $NO_x$ puff at early rich-phase times when the reductant front first encounters the zone of the highest $NO_x$ storage at the catalyst front is shown in Fig. 3M; i.e., the $NO_x$ pulse at $1/4L$ and early regen times. The SpaciMS measurements clearly indicate readsorption and corresponding $NO_x$ redistribution as that pulse travels along the catalyst axis. Significant readsorption is apparent in the NSR-zone back (i.e., between the 1/4*L* and 1/2*L* locations), and additional adsorption occurs in the OSC-only zone (back LNT 1/2*L*). By comparing the breakthrough timings of the $NO_x$ puff and reductant front in Fig. 3M and R, it is clear that the $NO_x$ puff advances in front of the reductant front and adsorbs on a downstream section that has not yet been reduced. This readsorption gives a second opportunity for $NO_x$ reduction when the reductant front arrives at later regen times, and thus increases overall $NO_x$ conversion. This demonstrates a benefit of the OSC-only zone, in that significant $NO_x$ puff readsorption occurs in that zone, and the potential for second-chance $NO_x$ reduction correspondingly decreases with the OSC-only zone length; e.g., much of the $NO_x$ puff would slip unconverted from the LNT if the NSR zone occupied the entire catalyst length. Such $NO_x$ readsorption and redistribution have been broadly observed in SpaciMS studies (Choi et al., 2006, 2012; Luo et al., 2011b) and influence both overall conversion and selectivity.

In cases where both $NO_x$ and reductants remain on the LNT surface at the end of the rich phase, a secondary $N_2O$ peak can occur at the rich–lean transition due to reaction of such residual species in a partially reduced environment. Nitrous oxide is particularly environmentally relevant in that it has a 114-year lifetime in the atmosphere, and 300-times greater greenhouse gas impact than does $CO_2$ (United States Environmental Protection Agency, n.d.). Readsorption and redistribution of $NO_x$ and incomplete regeneration can contribute to such residuals and thus impact selectivity. Such secondary $N_2O$ can be eliminated by using a sufficiently long regeneration time so as to completely eliminate residual surface $NO_x$, or by implementing a graded rich–lean transition (Mráček et al., 2016). SpaciMS was instrumental to understanding this complex process, developing a global-kinetics model capable of accurately predicting both primary and secondary $N_2O$ emissions (Bártová et al., 2014; Choi et al., 2012; Kočí et al., 2013; Mráček et al., 2015). Because this process and the related SpaciMS studies are detailed in a separate dedicated effort (Choi et al., 2017), they are only summarized here to note the contributions of SpaciMS to this catalytically and environmentally significant issue.

## 3.2 Regeneration With CO

SpaciMS has been instrumental in understanding LNT regeneration with CO, the role of the WGSR, and associated mechanistic perspectives (Choi et al., 2005, 2006, 2017). Because LNT catalysts can also be active for WGSR, it is possible that CO-based regeneration could occur via direct action or an intermediate-$H_2$ pathway. Other questions include the relative effectiveness and copresence impacts of CO and $H_2$; distributed utilization of CO for $O_2$ reduction, $NO_x$ reduction, and WGSR; the nature and impact of CO poisoning of PGM sites including intraregeneration transitions between "high-reaction-rate branch" and "low-reaction-rate branch" (Bär et al., 1992; Carlsson et al., 2004); and the role of PGM sites and impact of CO on LNT regeneration chemistry (Choi et al., 2005). The spatiotemporal trends afforded by intra-LNT SpaciMS measurements have provided clarity to such a broad range of LNT issues, and in that way contributed to advancing the technology.

Figure 5 shows SpaciMS measurements from a $Pt/K/Al_2O_3$-washcoated monolith LNT catalyst with varying surface- and gas-phase ($X_s$, $X_g$) composition (Fig. 5L, without $NO_x$ or $O_2$; Fig. 5M, with $O_{2,g}$ and $O_s$; and Fig. 5R, with $NO_{x,s}$ and $O_{2,g}$ and $O_s$), and regeneration using CO; surface oxygen refers to oxygen contained in and on the catalyst beyond that in the Blank condition, where the Blank condition contains no gas-phase oxygen as described in the caption of Fig. 5. For the conditions studied, the NSR zone was primarily in the front $1/2L$ of the LNT (not shown here). Varying the storage- and regeneration phase composition allows for the relative competition between CO utilization for stored $NO_x$ reduction, surface and gas-phase $O_2$ ($O_s$, $O_{2,g}$) reduction, and WGSR to be assessed. Figure 5Ll indicates that in the absence of surface- and gas-phase oxygen, the catalyst is active for $H_2$ generation via WGSR throughout regeneration; spatially, $H_2$ is produced throughout the catalyst with a peak effluent value of c.1.4%. The corresponding $H_2O$ measurements of Fig. 5Ll indicate proportional loss of gas-phase $H_2O$ to replenish surface $H_2O$ utilized for $H_2$ production. With competition from surface- and gas-phase oxygen, WGSR remains active throughout regeneration, but with certain degradation features. Figure 5Mu indicates that CO oxidation occurs throughout regeneration, and that $O_2$ is depleted within $1/4L$, producing a corresponding but asymmetric $CO_2$ profile; i.e., major $CO_2$ generation occurs in the front $1/4L$, and while the $O_2$-depletion transient would suggest a temporal top-hat like $CO_2$ profile, it in fact shows missing $CO_2$ at early regeneration times. The missing $CO_2$ is due to surface carbonate formation, which displaces surface $H_2O$, resulting in the $H_2O$ puff at early regeneration times; as major $CO_2$ production is in the front $1/4L$ where gas-phase $O_2$ is depleted, there is very little growth of the $H_2O$ puff beyond the $1/4L$ position. The reduction in $H_2$ concentration at $1/4L$ (compared to the case without $O_2$ in

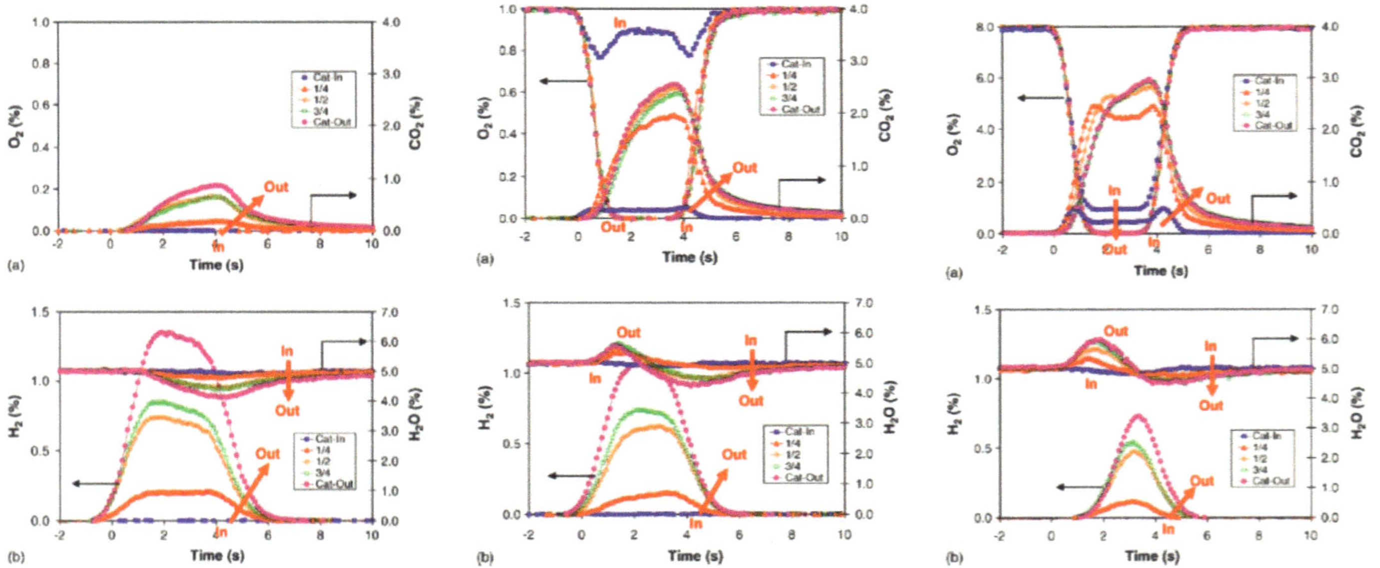


**Fig. 5** Spatially resolved (*Upper*: u) $O_2$, $CO_2$, (*Lower*: l) $H_2$ and $H_2O$ breakthrough profiles just before, during, and just after regeneration of a 3-in.-long Pt/K/$Al_2O_3$-washcoated monolith with varying storage-phase composition. Regeneration is via *X* + 4% CO + 5% $H_2O$ + $N_2$ balance, and the storage gas is *X* + 5% $H_2O$ + $N_2$ balance, where *X* is: (*Left*: L) blank, (*Middle*: M) 1% $O_2$, and (*Right*: R) 250 ppm NO + 8% $O_2$ for Neat, OSC, and NSR cycling, respectively (Choi et al., 2006).

the gas phase) reflects WGSR degradation due to competition between $H_2O$ and $O_2$ for available CO. Competition with both surface $NO_x$ and surface and gas-phase $O_2$ significantly degrades WGSR as shown in Fig. 5Rl. More extensive surface–$H_2O$ displacement is apparent throughout the NSR zone due to $CO_2$ formation at early regeneration times from both CO reaction with $O_2$ and surface $NO_x$ and is reflected in the spatiotemporal $H_2O$ puff and missing $CO_2$ at early regeneration times; i.e., the $H_2O$ puff is greater and grows progressively throughout the NSR zone. In this full-competition case, WGSR is delayed and degraded, with $H_2$ being apparent to a lesser extent and only at later regeneration times.

Besides providing insights to the nature of various LNT processes during regeneration, Fig. 5 suggests that CO acts to directly regenerate the catalyst as opposed to regeneration via an intermediate-WGSR pathway. The WGSR is relatively slow, and produced only c.33% CO conversion in the most favorable neat case without competition from $O_2$ and $NO_x$ (Fig. 5L), which would presumably degrade this neat case conversion. In contrast, CO reduction of $O_2$ and $NO_x$ was very fast occurring in the front $1/4L$ and $1/2L$, respectively, and completely (100%) depleting the CO at early regeneration times; i.e., even in the most WGSR-favorable case, there is not sufficient $H_2$ generation to account for the observed $O_2$ and $NO_x$ reduction. Furthermore, there is no correlation between the $H_2O$ puff and missing $H_2$ in the various cases that would suggest a reduction of $O_2$ and $NO_x$ by $H_2$; in this case the missing $H_2$ would be the difference between the neat case (Fig. 5L) and the two cases with combinations of $O_2$ and $NO_x$. Specifically, does the $H_2O$ puff occur from generation via an intermediate-WGSR route, or via displacement of surface $H_2O$ by carbonates resulting from a direct CO reduction route? Although CO was fully consumed in the front $1/2L$, leaving no WGSR potential in the third quarter, the $H_2O$ puff increase nevertheless beyond the $1/2L$ position; this is more likely due to downstream convection of $CO_2$ from a direct CO reduction than $H_2$ convection from an intermediate-WGSR process. While the $Pt/K/Al_2O_3$ LNT catalyst was active for WGSR, this reaction does not appear to be active during early regen times when CO is consumed by more favorable reaction with $O_2$ and stored $NO_x$; specifically, CO acts directly to regenerate the LNT vs an intermediate-WGSR route. Nevertheless, the WGSR might contribute to slower processes like desulfation, and as the basis for a catalyst state diagnostic as discussed in the section on LNT sulfation.

Figure 5 and the associated studies (Choi et al., 2005, 2006) provide additional examples of the broadly applicable insights available from operando intracatalyst SpaciMS measurements. At low temperatures, such as 200°C, CO conversion (inferred by $CO_2$-product profiles) inflections at rich-phase onset are observed indicating progressive slowing of initially rapid reaction as CO coverage builds up on the PGM (Choi et al., 2005); this progressive CO-PGM buildup degrades all reactions involving PGM including regeneration (not shown here). The conversion inflection at the rich-to-lean transition indicates the reverse process as incoming lean phase cleans the PGM via CO oxidation. This dynamic and reversible CO poisoning occurs through each regeneration event and has notable impacts on LNT performance relative to their modeling and development. The process is indicative of transition between a "high-reaction-rate branch" and a "low-reaction-rate branch" of the $CO + O_2$ reaction on $Pt/Al_2O_3$ as has been reported (Bär et al., 1992; Carlsson et al., 2004), and more generally, the various ($H_2$, CO, $NO_x$, $O_2$, $CO_2$) spatially distributed transients confirm the key role of PGM sites in release and reduction of stored $NO_x$ and that non-PGM-dependent mechanisms are of little importance to practically relevant automotive LNT applications (Choi et al., 2005). In addition to CO poisoning, varying reductant effectiveness ($H_2 > H_2 + CO > CO$) impacts NSR-zone length, as does anything that impacts storage and regeneration effectiveness, with the corresponding conversion and selectivity consequences as previously discussed.

## 3.3 Direct $H_2$ vs Indirect Intermediate-$NH_3$ Regeneration Pathways

Application of SpaciMS to demonstrate the indirect-$H_2$ pathway for LNT regeneration via intermediate $NH_3$ provides an excellent example of how an effluent perspective can obscure the actual intracatalyst reaction network and sequence, and how intracatalyst measurements can make these so obvious and clear. Because of this significance, we summarize this work here, while leaving more detailed discussion to our other work focusing on LNT regeneration and selectivity (Choi et al., 2017) and the references cited in this section. The typical effluent temporal sequence of $N_2 < H_2 < NH_3$ observed in Fig. 6L suggests separate fast and slow regeneration processes selective to $N_2$ and $NH_3$ products, respectively (Castoldi et al., 2004; Luo et al., 2011b; Nova et al., 2007); specifically, the earlier $N_2$ appearance suggests that it may be a unique product of a fast regeneration process possibly of surface–$NO_x$ sites proximal to PGM sites, and that $NH_3$, due to its later appearance, may be favored for a slower regeneration process, possibly of bulk and surface–$NO_x$ sites more distal to the PGM sites. Pihl et al. (2006) and

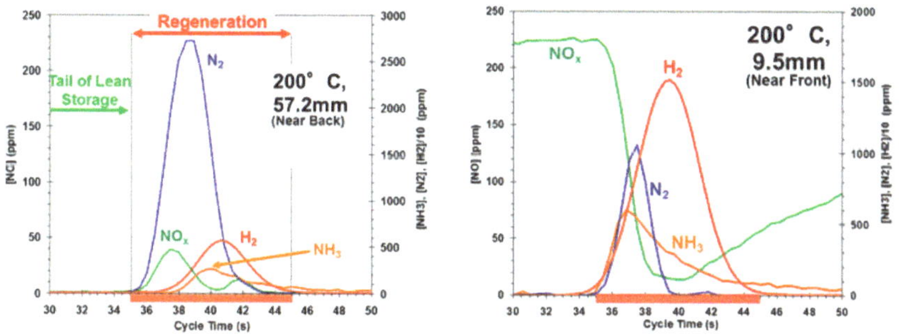


**Fig. 6** Species transients 57.2 mm (*Left*: L) and 9.5 mm (*Right*: R) along a 76-mm long Pt/Ba/$Al_2O_3$-washcoated monolith LNT catalyst during regeneration with $H_2$ (Partridge and Choi, 2009). The regeneration phase is highlighted by the *red bar* along the horizontal axis, and is bound by the end of the previous and start of the subsequent lean phases.

Cumaranatunge et al. (2007) were the first to propose that an alternate explanation may be possible whereby the later $NH_3$ appearance in the effluent could be due to its role in regeneration. In general, this perspective proposes that regeneration selectivity varies with the local reductant-to-$NO_x$ ratio, which varies across the reductant front, and spatiotemporally as the reductant front sweeps along the catalyst axis during regeneration (Pihl et al., 2006); specifically, the selectivity varies in the order of NO, $N_2O$, $N_2$, and $NH_3$ as the reductant-to-$NO_x$ ratio varies from low to high values, such as downstream and upstream of the reductant front, respectively. Experimental SpaciMS-based evidence of this alternate explanation first appeared in public presentations (Partridge et al., 2008; Partridge and Choi, 2008) and later in the literature (Clayton et al., 2008; Easterling et al., 2012; Partridge and Choi, 2009). Subsequent work further leveraged these insights and spatially resolved measurements to elucidate $N_2O$ selectivity (Bártová et al., 2014, 2015; Choi et al., 2012; Mráček et al., 2015), and develop improved LNT models (Clayton et al., 2008; Easterling et al., 2012; Kočí et al., 2013) and operation protocols (Mráček et al., 2016). The discussion here focuses primarily on related interpretation of SpaciMS measurements.

Figure 7T shows the nature of a reductant, where in this case $H_2$ is progressively consumed at the reductant front as it travels through the LNT during regeneration. As discussed in Section 3.1, this produces the characteristic pattern of progressively delayed onset time with increasing distance into the catalyst and a common tail, which is generally indicative of a reductant. The local $H_2$ consumption in each catalyst section is determined by the difference in the spatially bounding $H_2$ transients; e.g., $H_2$ consumption in the front

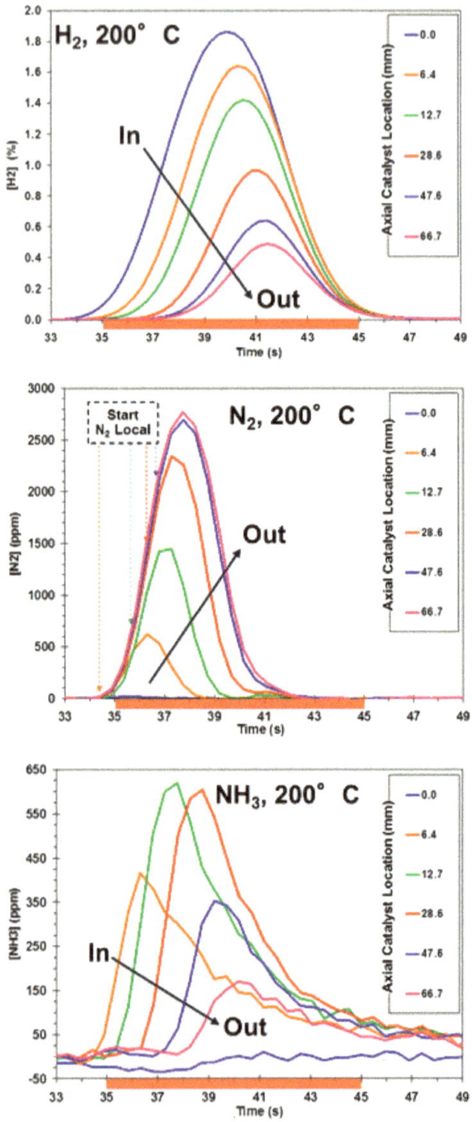


**Fig. 7** Intra-LNT spatially distributed $H_2$ (*Top*: T), $N_2$ (*Middle*: M), and $NH_3$ (*Bottom*: B) transients during regeneration of a Pt/Ba/$Al_2O_3$-washcoated monolith LNT catalyst with $H_2$ (Partridge and Choi, 2009).

6.4-mm LNT section is the difference in the blue and orange curves in Fig. 7T. The spatiotemporal $N_2$ profiles shown in Fig. 7M display the opposite pattern whereby the various spatially distributed $N_2$ transients have a common leading edge and a progressively delayed tail with increasing distance into the catalyst; this is the characteristic pattern of an inert product. Specifically, locally generated $N_2$ slips through the catalyst without further reaction, which creates the common leading edge of the family of $N_2$ transients. Like with $H_2$, the locally generated $N_2$ in each catalyst section is indicated by the difference in the spatially bounding curves; e.g., $N_2$ generated in the section bounded by 6.4 and 12.7 mm is equal to the difference in the corresponding green and orange curves in Fig. 7M. The $N_2$ observed at any point along the catalyst is the sum of that produced in and slipping from upstream catalyst sections, and that produced in the section of interest; e.g., the $N_2$ observed at 12.7 mm is the sum of that from the first two catalyst sections as defined in Fig. 7M. Figure 7B shows how $NH_3$ is both generated and consumed along the catalyst axis, and specifically how it displays the characteristic reductant pattern. The varying $NH_3$ onset time indicates that some portion of the locally generated $NH_3$, from $H_2$ reduction of stored $NO_x$ in that section, is consumed by $NH_3$-based $NO_x$ reduction in the adjacent downstream section; of course as the finite zones shown in Fig. 7B shrink to zero, the process becomes continuous. The varying $NH_3$ profile height along the catalyst indicates different zones in the catalyst where the balance between $NH_3$ generation and consumption differ; $NH_3$ generation is the greatest in the NSR-zone front where surface–$NO_x$ density is the greatest. This creates buildup, balanced, and deficit zones where the net $NH_3$ is spatially growing, constant, and depleting, respectively (Partridge and Choi, 2009). A less salient but important point is that due to reactivity differences, the $NH_3$ front may surpass the $H_2$ front, and be the first reductant to reach the downstream stored $NO_x$ that is yet to be reduced (Clayton et al., 2008). The temporally steep $NH_3$ leading edge is indicative of complete $NH_3$ consumption at the front in an $NH_3$-limited regeneration process creating a plug-$NH_3$ flow (Clayton et al., 2008; Easterling et al., 2012; Partridge and Choi, 2009).

With this improved understanding, it is now clear that the effluent $N_2 < H_2 < NH_3$ sequence is an integral effect, and not indicative of slow vs fast $NO_x$ sites. Figure 6R shows the various transients within and near the LNT front, and specifically indicates that during LNT regeneration, $NH_3$ is produced on similarly fast timescales as is $N_2$. However, whereas the indicated $N_2$ will be convected without further reaction along and out of the LNT to be immediately detected, some portion or all of the indicated $NH_3$ will be consumed in the downstream NSR and OSC-only zones, thus delaying its detection in the effluent. Figure 8T shows the buildup, balanced, and depletion $NH_3$ zones, and how the net $NH_3$ is growing in the

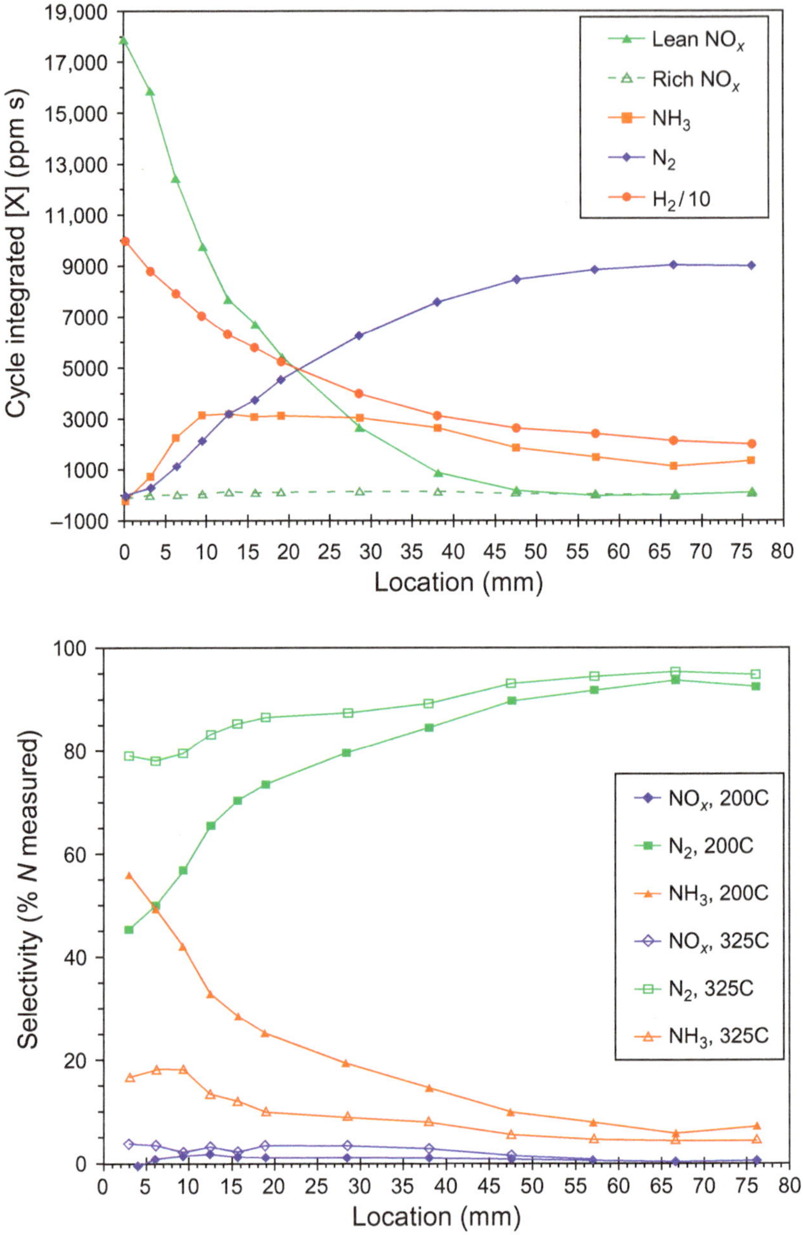


**Fig. 8** Species and selectivity distributions for a Pt/Ba/$Al_2O_3$-washcoated monolith LNT catalyst regenerated with $H_2$ (Partridge and Choi, 2009). (*Top*: T) Cycle-integrated species distributions at 200°C. (*Bottom*: B) Nitrogen selectivity distributions along the catalyst.

buildup zone where c.50% of the $NO_x$ is stored; net $NH_3$ falls in the NSR-zone end and downstream OSC-only zone. Despite the higher surface–$NO_x$ density at the LNT front, the high $NH_3$ selectivity shown in Fig. 8B indicates reaction at a low $NO_x/H_2$ condition; this indicates slow propagation of the $H_2$ front in this region, producing an increasing $NO_x/H_2$ ratio deeper into the catalyst more favorable to $N_2$ production.

## 3.4 Sulfation and Desulfation

Sulfation and desulfation (S and DeS) are spatially distributed processes, and the varying sensitivities of different LNT functions cause corresponding variations in the spatial dispersion of these functions as S and DeS progress. Understanding this spatiotemporally complex network of interactions and functions explains many aspects of LNT performance variation throughout S and DeS, including $NO_x$ conversion and selectivity, and their changes with temperature and catalyst formulation (Choi et al., 2007, 2008a, b; Luo et al., 2011b; Partridge et al., 2009). Figure 9 shows the general integral performance trends of an LNT catalyst with progressive sulfation. Specifically, while the initial S-induced $NO_x$-conversion degradation can be little to nil, it can increase dramatically above a specific sulfation level. In contrast, $NH_3$ selectivity increases relatively uniformly with progressive sulfation; and in further contrast, sulfation had little impact on $N_2O$ conversion. These diverse and different sulfation impacts are a consequence of the spatially distributed nature of LNT sulfation, and its specific action on different catalyst functions. This complex behavior becomes relatively simple and straightforward to understand via SpaciMS measurements, as discussed in this section.

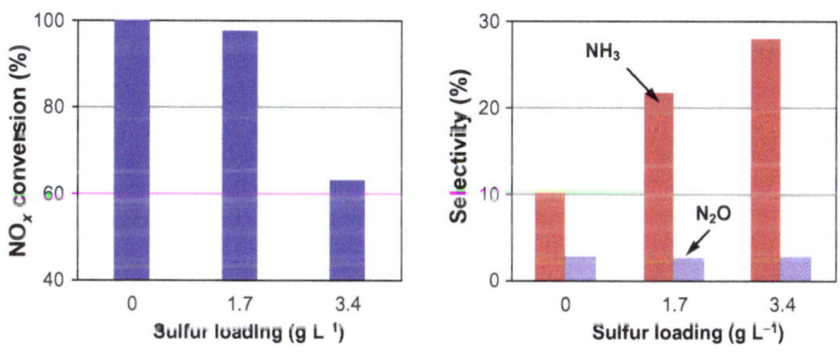


**Fig. 9** Variation in integral $NO_x$ conversion, and $NH_3$ and $N_2O$ selectivity of a commercial LNT catalyst with increasing sulfation (Choi et al., 2008b); 60 s/5 s lean/rich fast cycling at 325°C, with 3.4% $H_2$ in the regeneration phase.

Figure 10 shows the spatially distributed impact of sulfation on $NO_x$ storage, OSC and WGSR at three sulfation levels for a model $Pt/Ba/Al_2O_3$ catalyst; in this case CO is used as the reductant to investigate sulfation impact on the WGSR function, but the spatiotemporal sulfation impact and trends are generally similar to that when regenerating a commercial fully formulated LNT with $H_2$ (Choi et al., 2007, 2008a). In the sulfur-free baseline case, Fig. 10T, the NSR zone occupies the front c.$1/2L$ and is consistent with the conceptual model of Fig. 2B in which the catalyst is spatially divided between a front NSR zone and downstream OSC-only zone. The first (0.85 gS/L) and second (1.7 gS/L total) sulfation events progressively displace the NSR zone down the LNT axis by practically poisoning the front $1/4L$ and $1/2L$, respectively, for NSR. This demonstrates how the more stable Ba–S species produce a plug-like front-biased sulfation zone, effective NSR poisoning in the sulfated zone, and corresponding downstream displacement of the NSR zone. Practically, the sulfation front is not exactly step like, but has some finite gradient across which NSR is progressively degraded, and which in turn lengthens the NSR zone proportional to the width of this sulfation front gradient. Because only half of the unsulfated LNT length is required for complete $NO_x$ conversion under the reaction conditions used, and the initial sulfation zone occupies only the front $1/4L$–$1/2L$, high integral $NO_x$ conversion is sustained; i.e., in this case the NSR zone has been displaced to the back LNT $3/4L$–$1/2L$, and the OSC-only zone has shrunk to zero. On the contrary, the second sulfation poisons the front $1/2L$–$3/4L$ leaving an insufficient length for complete $NO_x$ conversion; i.e., the NSR zone has been pushed out the LNT back. In general, high integral conversion can be sustained with progressive sulfation provided the length of the displaced NSR zone does not reach the catalyst exit. However, sulfation beyond the point where the NSR-zone leading edge reaches the LNT exit, and the OSC-only zone shrinks to zero, will cause the onset of dramatic $NO_x$-conversion degradation.

Figure 10M shows that OSC is much less sensitive to sulfation than is NSR; our other SpaciMS studies have shown similar sensitivity for Ce-containing LNT formulations with active and much greater OSC (Choi et al., 2007, 2008a). In the sulfur-free case, OSC is relatively uniformly distributed along the LNT axis reflecting the correspondingly uniform washcoat distribution. While progressive sulfation causes corresponding OSC degradation in the sulfated zone, the impact is minor, and notably, OSC remains active in the sulfated zone. In agreement with the NSR impact,

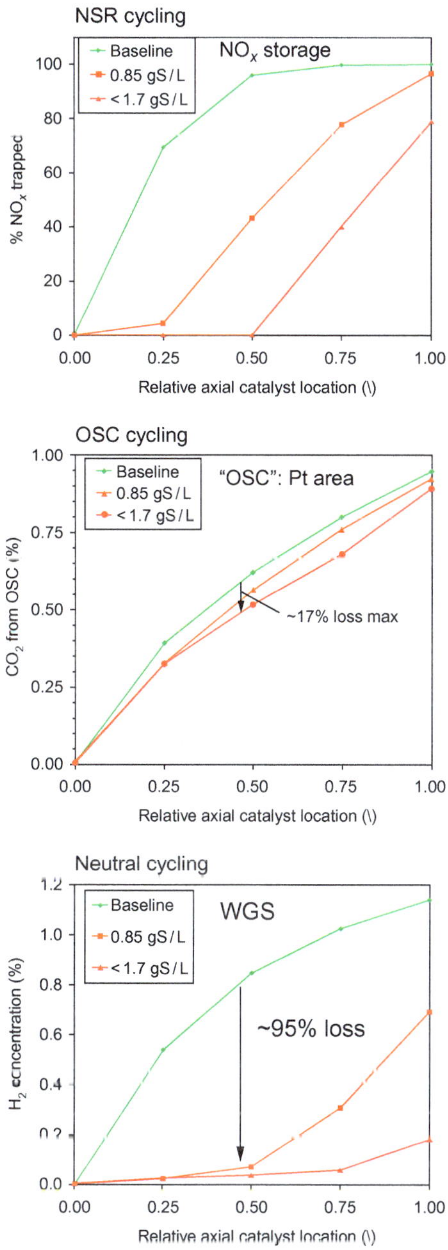


**Fig. 10** Spatially distributed impact of sulfation on $NO_x$ storage, OSC, and WGSR of a model $Pt/Ba/Al_2O_3$ LNT catalyst; the sample has no OSC-specific component such as $CeO_2$. Fast cycling (60 s/5 s lean/rich) at 325°C, and 2% CO + 5% $H_2O$ + $N_2$ balance during regeneration; the lean phase consists of 5% $H_2O$ + $N_2$ balance for neutral cycling (*Bottom*: B), and adds 10% $O_2$ for OSC cycling (*Middle*: M), and 300 ppm NO + 10% $O_2$ for NSR cycling (*Top*: T) (Partridge et al., 2009).

the initial sulfation degrades OSC in the front 1/4*L*; the proportional (parallel) downstream OSC distributions for the sulfur-free and 0.86 gS/L cases indicate practically zero sulfur impact on OSC in the back 3/4*L*. The second sulfation event shows progression of the same process. The salient point is that whereas NSR is effectively poisoned in the sulfated zone, OSC remains active, and significant in the front-biased sulfated zone.

Based on multiple SpaciMS studies of progressive sulfation of model and fully formulated (Ce-containing) LNT catalysts, we developed a conceptual model of the distributed impact of sulfation shown in Fig. 11. In the sulfur-free case, an NSR zone is established in the catalyst front, and a downstream OSC-only zone fills the remaining LNT length; both NSR and OSC are active in the NSR zone. Reductants including $NH_3$ escaping the NSR zone are oxidized in the OSC-only zone; $NH_3$ oxidation occurs with high $N_2$ selectivity, and does not contributed to $N_2O$ (Choi et al., 2012). Initial sulfation creates a sulfated zone at the LNT front within which NSR is inactive, resulting in downstream displacement of the NSR zone, and a corresponding shortening of the OSC-only zone. High integral $NO_x$ conversion will be sustained as long as sufficient LNT length exists such that the NSR zone is contained within the catalyst, such as in the 1.7 gS/L case in

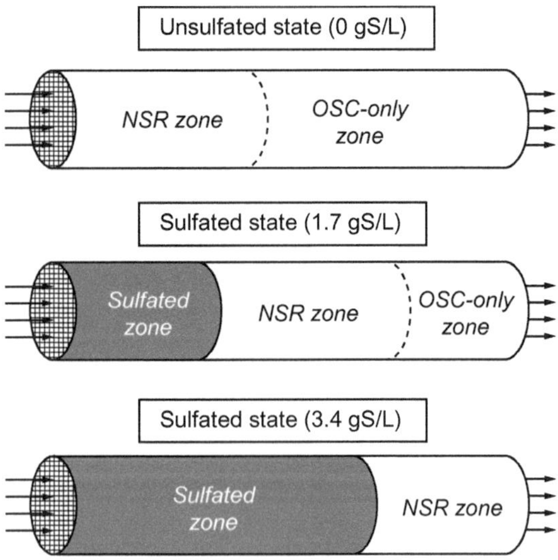


**Fig. 11** Schematic of conceptual model explaining spatially distributed impact of progressive sulfation of an LNT catalyst (Choi et al., 2008b); fully formulated commercial LNT catalyst, 60 s/5 s lean/rich fast cycling at 325°C, with 3.4% $H_2$ in the regeneration phase.

Fig. 11; and thus the high $NO_x$ conversion for the corresponding case in Fig. 9L. However, even while the displaced NSR zone is completely contained within the catalyst with progressive sulfation, the capacity to oxidize reductants in the OSC-only zone is proportionally diminished with shortening of the OSC-only zone; thus, $NH_3$ selectivity will progressively increase as continued sulfation shrinks the OSC-only zone to zero as shown in Fig. 9R. Continued sulfation beyond the critical level where the OSC-only zone becomes nonexistence (e.g., 3.4 gS/L case in Fig. 11), will push the NSR zone out the LNT exit, leaving insufficient NSR-zone length, and resulting in a dramatic loss in $NO_x$ conversion as shown in Fig. 9L. Beyond this critical sulfation level, slip of the primary reductant should increase and $NH_3$ selectivity should remain relatively unchanged as the NSR zone progressively shrinks with zero downstream OSC-only zone. The capacity for secondary $NO_x$ conversion and selectivity associated with $NO_x$ redistribution as discussed in Section 3.1, will vary with the OSC-only zone length. The relative insensitivity of $N_2O$ selectivity with sulfation shown in Fig. 9R indicates its primary formation in the NSR zone at the reductant front, and subsequent reaction-free slip through the varying length OSC-only zone.

While the specific sulfated-LNT performance varies with operating conditions, the conceptual model of sulfation impacts holds and explains the variations. For instance, the longer lower temperature NSR zone (e.g., 200°C case in Fig. 2B) results in higher $NH_3$ selectivity due to the lower capacity for its oxidation in the shorter OSC-only zone, and has a lower critical sulfation level where NSR zone first contacts the LNT exit plane and beyond which dramatic $NO_x$-conversion loss occurs (Choi et al., 2008a). In general, the same will occur in a proportional manner for anything that increases the NSR-zone length; e.g., less effective reductants, temperature. Thus, gradual transition to more effective reductants with progressive sulfation, could be a strategy for avoiding the negative impacts of sulfation, and delaying costly and potentially irreversible damage of desulfation, by shortening the NSR zone; e.g., increasing $H_2$ reductant-pool concentration with increasing sulfation. Similarly, less effective reductant pools might be possible in early LNT life applications and have corresponding fuel economy and engine durability benefits; e.g., via the consequences of cylinder-managed reductant generation via late-cycle fuel injection. The conceptual model provides a framework for balancing such design tradeoffs in addition to LNT sizing and associated capital and operational costs.

The WGS and reverse WGS reactions have been extensively studied for Pt/Ce systems, and a common theme is the importance of the metal–support interface and activation on Pt (Goguet et al., 2004; Jacobs and Davis, 2007; Luo and Gorte, 2003). We have shown spatiotemporal WGS activity on $Pt/K/Al_2O_3$, $Pt/Ba/Al_2O_3$, and Ce-containing LNT catalysts (Choi et al., 2005, 2006; Ji et al., 2008; Partridge et al., 2009). Figure 10B shows the uniquely high sensitivity of the LNT WGS function to sulfation. Specifically, the LNT suffers a maximum c.90%–95% loss in WGS activity relative to the sulfur-free baseline condition in the front $1/2L$ and $3/4L$ for the two sulfur exposures, 0.85 and 1.7 gS/L, respectively. This WGS-sulfated zone front leads the NSR-sulfated zone front by $1/4L$ and reflects the greater sensitivity of WGS to sulfation relative to that of NSR; degraded but finite NSR activity exists in the WGS-sulfated zone. Among the WGS, NSR, and OSC functions, WGS provides the most sensitive indicator of the absolute sulfation front location. This sensitivity and the importance of the metal–support interface to the WGS reaction suggest a further conceptual model for LNT sulfation, pictorially represented in Fig. 12. In the sulfur-free state, which initially exists throughout the catalyst, all functions are active starting at the inlet face. The first sulfur exposure dramatically impacts WGS, likely

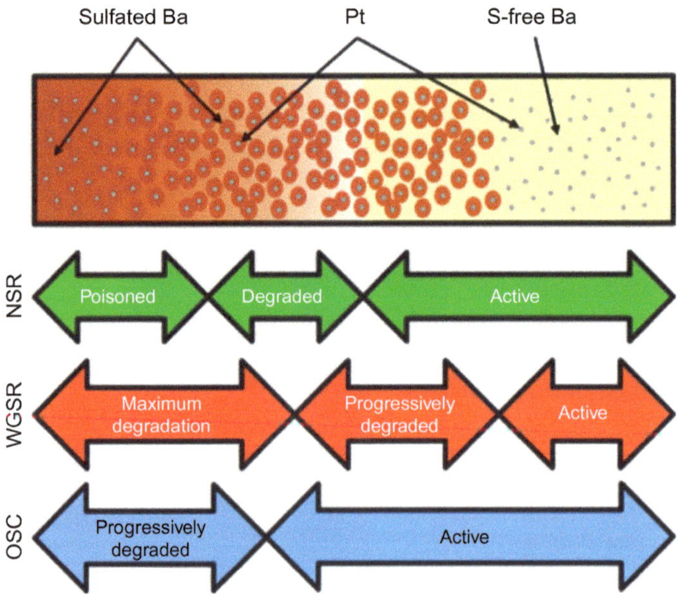


**Fig. 12** Conceptual model describing the function-specific spatially distributed impact of progressive sulfation on LNT NSR, WGSR, and OSC functions (Partridge et al., 2009).

by changing the Pt–support interface structure; the spatial extent of WGS degradation is indicative of the absolute sulfation front, and in this region OSC and NSR activity persists likely due to the ability to access sulfur-free Ba, Ce, etc., via spillover, surface diffusion, and gas-phase transport. The WGS sulfur sensitivity is suitably high that within the WGS-sulfated zone it is not further degraded by progressive sulfation; i.e., the WGS sulfation impact is practically switch or digital like, and progressive sulfation spatially advances the WGS-sulfated zone front. As sulfation progresses, sulfur-free Ba becomes increasingly difficult to access, and NSR is progressively degraded. At a critical sulfation level, sulfur-free Ba is apparently practically unavailable, and NSR is locally poisoned; further sulfation spatially advances the NSR-sulfated front. Because of the lower binding energy of sulfur to Ce (Choi et al., 2008a), OSC remains active albeit degraded to a significantly lesser extent in the NSR-sulfated and longer WSR-sulfated zones. In this way, the WGS, NSR, and OSC functions, which were all initially active from the LNT front in the sulfur-free state, become spatially dispersed with sulfation. Similarly, the spatial activity gradients at the sulfated fronts are the greatest for WGS, intermediate for NSR, and very distributed and least plug-like for OSC. This OSC activity in the NSR-poisoned zone can lead to the strange observation of enhanced $NO_x$ conversion in a sulfated state; i.e., by increasing $NO_2$ fraction entering the downstream-displaced NSR zone and correspondingly enhancing $NO_x$ storage. Understanding the extreme WGS sulfur sensitivity has led to control LNT strategies for improved durability and fuel economy as discussed in the next subsection.

An on-board diagnostics (OBD) patent demonstrates how basic information such as the spatiotemporally varying impact of sulfation on various catalyst functions can be used to improve catalyst efficiency, durability, and operating costs (Yezerets et al., n.d.). The patent describes how the varying sensitivity of exhaust'oxygen (Lambda, UEGO) sensors to different reductants can be used to assess an LNT's $NO_x$ reduction capacity, and in turn as the basis for OBD. In a specific example, UEGO sensors are sensitive to $H_2$, and changes in catalyst state that impact the WGSR will correspondingly change $H_2$ generation across an LNT and the sensor signals. From the conceptual model discussed above, and via the unique sensitivity of WGSR to sulfation, it is apparent that such sensor could be implemented to follow the S and DeS processes. Since DeS can degrade LNT performance and has significant fuel penalty, it is desirable to minimize the frequency and duration of DeS events. Although the patent is specifically focused on OBD, its concepts could be applied to implement an optimized DeS strategy that only occurred when and of the duration required for a specific application; compared to a prescheduled DeS frequency and duration common for all applications, such a customized control strategy could result in significant efficiency, durability, and cost benefits. This demonstrates a specific practical insight and benefit of intracatalyst spatiotemporally resolved knowledge of catalyst performance, and the rich basic and practical benefits of SpaciMS.

Complementing the sulfation studies described above, the Epling group has spatiotemporally studied DeS of Ba-based LNT samples with and without active Ce-based OSC via differentially sampled SpaciMS (Luo et al., 2011b). Figure 13T shows spatiotemporal progression of the $SO_2$ front, and how after 15 min of sulfation somewhat less than the front 2.4-cm is heavily sulfated. The LNT samples were studied in two sulfation states corresponding to 30- and 60-min $SO_2$ exposure as indicated in Fig. 13T; about half the catalyst was heavily sulfated after 30 min, while the full catalyst was heavily sulfated after 60 min. Desulfation was performed at 600°C under continuously rich conditions composed of 0.2% $H_2$, 5% $H_2O$, 5% $CO_2$, and $N_2$ balance; while this is sufficient to desorb sulfur weakly adsorbed to alumina, ceria, and well-dispersed barium, it is significantly less than the 750–800°C required to DeS the more strongly basic bulk-like barium sites. During DeS, $SO_2$ is promptly released in the front 3.6 cm but delayed by about 0.5 min at the 4.8 cm location as is shown in Fig. 13M. This delay is significant compared to the c.0.12 s gas residence time and indicates delay in the reductant front arriving at the 4.8-cm location due to upstream consumption via DeS and OSC reduction; advancement of the reductant front is much slower with 60-min sulfation, and the delays correspondingly longer (not shown here). Onset of $H_2S$ release is delayed relative to that of $SO_2$ as shown in Fig. 13B and is consistent of its formation in the richer conditions behind the reductant front.

Figure 13M and B indicates readsorption of $SO_2$ released from the front sections in the back catalyst sections. There is a clear loss of $SO_2$ from the 3.0–3.6–4.8-cm locations, where the catalyst was mildly sulfated for the 30-min exposure. In fact, they noted that by summing the $SO_2$ and $H_2S$ profiles, the total sulfur increased in the front 3.0 cm, and decreased in the downstream sections and particularly beyond 3.6 cm (Luo et al., 2011b). As these downstream sections were very mildly sulfated, $SO_2$ readsorption likely takes place on strongly basic sulfur-free bulk-like barium sites. Just as downstream $NO_x$ readsorption can influence conversion and selectivity, $SO_2$ readsorption can increase $H_2S$ emissions by increasing the probability of adsorbed sulfur experience the more deeply rich conditions required for

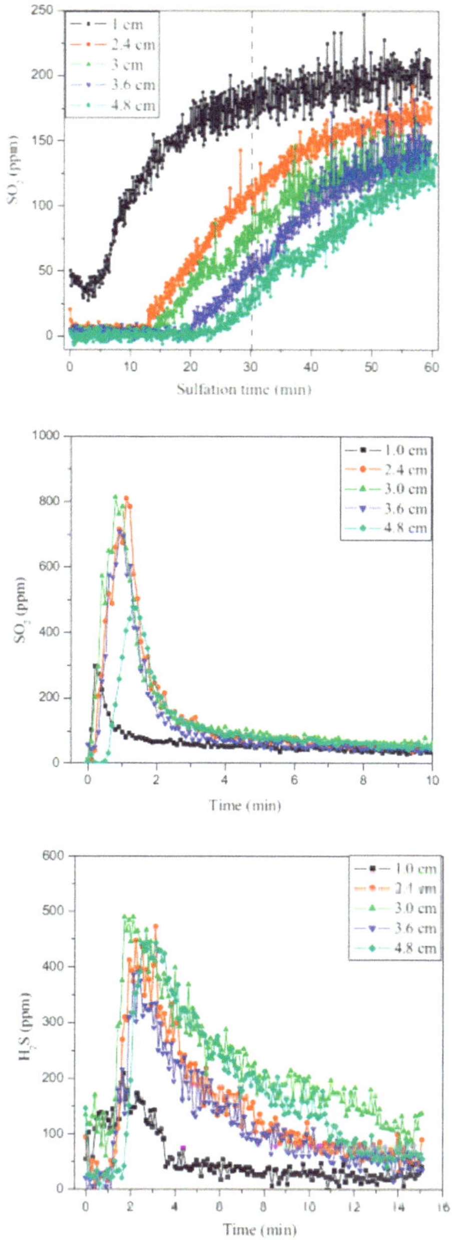


**Fig. 13** (*Top*: T) Spatiotemporal $SO_2$ profiles for sulfation of a 5-cm long, OSC-containing commercial LNT sample by exposure to 200 ppm $SO_2$, 10% $O_2$, 5% $H_2O$, 5% $CO_2$, and $N_2$ balance at 300°C. Spatiotemporal $SO_2$ (*Middle*: M) and $H_2S$ (*Bottom*: B) profiles during DeS of the 30-min sulfated sample under continuously rich 0.2% $H_2$, 5% $H_2O$, 5% $CO_2$, and $N_2$ balance at 600°C (Luo et al., 2011b).

its formation. No readsorption was observed for 60-min sulfation as heavy sulfation existed along the entire catalyst length, and downstream sulfur-free barium sites were not available to readsorb $SO_2$ released from upstream sites. This demonstrates spatial redistribution of weakly adsorbed sulfur in the more heavily sulfated front catalyst sections, to more stable surface'sulfur species adsorbed on originally sulfur-free bulk-like barium sites in the downstream catalyst regions. Such readsorption will only occur when stable downstream barium sites are available and weakly bound sulfur is available to desorb; thus, no significant readsorption in the heavily sulfated condition. While the ceria-free model catalyst showed $SO_2$ desorption, there was no readsorption in the back catalyst sections, even for very mildly 15-min sulfated conditions; this indicated ceria's role in promoting readsorption in the commercial catalyst by supplying oxygen necessary for formation of the downstream stable surface sulfur species. This SpaciMS-based work provided basic and practical insights related understanding DeS and the role of different catalyst formulation components.

## 3.5 LNT Modeling Using Spatiotemporal Data

SpaciMS promotes improved catalyst understanding and modeling by providing direct evidence of intracatalyst processes not apparent in the effluent, and measurement of functional and overall performance over a range of locally distributed conditions. The local intracatalyst performance is influenced by the overall environmental conditions, catalyst state, and the evolving interactions of the reaction network up to that point in the catalyst. These spatiotemporal variations can be complex, and direct operando spatiotemporally resolved measurements under application conditions provide a critical basis for assessing understanding and specific models of the catalyst reaction network. In this section, we highlight some SpaciMS contributions to development of automotive LNT catalyst models.

The SpaciMS and Spaci-like works investigating intra-LNT formation and utilization of intermediate $NH_3$ during regeneration provide clear examples of the benefit of intracatalyst measurements. As discussed in Section 3.3, the dramatically different $N_2$ and $NH_3$ effluent sequence initially obscured the understanding of LNT regeneration chemistry. While it was careful and clever analysis of integral experiments (Cumaranatunge et al., 2007; Pihl et al., 2006) that initially proposed the intermediate-$NH_3$ pathway, it was SpaciMS and Spaci-like studies that provided direct demonstration of that process (Bhatia et al., 2009; Clayton et al., 2008;

Partridge et al., 2008; Partridge and Choi, 2008, 2009); interestingly, these various studies were independently taking place at about the same time. In addition to details already discussed in Section 3.3, these studies provided insights promoting LNT modeling. Our studies showed parallel direct-$H_2$ and intermediate-$NH_3$ regeneration pathways, with partitioning between the practically equivalently effective branches varying with operation temperature (Partridge and Choi, 2009). Work by the Harold group provided much more extensive modeling-related insights based on Spaci-like measurements and analysis (Clayton et al., 2008). Specifically, they proposed a five-path reaction network based on spatiotemporally resolved measurements during LNT regeneration. They demonstrated that intracatalyst performance can be reductant limited despite the overall stoichiometry, and that $H_2$ and $NH_3$ breakthrough occur when $H_2$ supply is no longer the rate-limiting process. These observations provided supporting evidence of a previously proposed process where $NO_x$ diffusion to PGM sites was the reaction bottleneck in certain conditions. Moreover, in addition to showing parallel direct-$H_2$ and intermediate-$NH_3$ regeneration paths, their measurements demonstrated changes in the limiting process through regeneration; i.e., being initially $H_2$ feed rate limited, to $NO_x$ supply limited.

In other examples, SpaciMS has provided insights that are not apparent or only indirectly apparent from effluent observations. This includes the conceptual models of sulfation discussed in Section 3.4, and the corresponding progressive changes in the spatial extents and spatial dispersion of the various zones. Particularly, how OSC activity in the NSR-sulfated zone, can lead to the counterintuitive enhanced $NO_x$ conversion by increasing $NO_2$ feed to the NSR-active zone. While this and the more basic consequences of changes in NSR and OSC-only zone length with sulfation may be apparent through very extensive, careful, and clever effluent analysis, SpaciMS analysis accelerates establishment of these understandings, and correspondingly hastens catalyst modeling and other advances. As Koci and Harold have independently mentioned, SpaciMS is not required for model development, but it expedites the process and can clarify many reaction networks which might not be readily apparent from effluent analysis (Harold, 2017; Kočí, 2017); more specifically, SpaciMS captures the all-important intracatalyst thermal coupling impacts that cannot be replicated via effluent measurements with varying flowrate (Harold, 2017). While new or clarified insights from SpaciMS analysis can be understood from conventional perspectives and may seem obvious once demonstrated, other observations are simply not apparent from effluent analysis under typical reaction conditions.

In other applications, SpaciMS is used in combination with modeling to assess and develop modeling concepts, and to provide spatiotemporal data for critical assessment of model performance. In one example, we applied SpaciMS to better understand $N_2O$ selectivity during LNT regeneration (Bártová et al., 2014; Kočí et al., 2013; Mráček et al., 2015). This work focused on assessing various proposed mechanisms for secondary $N_2O$ emission occurring at the rich–lean transition. It built on a spatiotemporal LNT global kinetic model, which had been developed based on effluent measurements, and used selectivity functions dependent on the local PGM site state to implement the spatial functionality (Kočí et al., 2013). By varying the regeneration effectiveness, and assessing spatial correlations between $N_2O$ formation and $NO_x$ storage, and FTIR surface species measurements, it was demonstrated that the primary mechanism for secondary $N_2O$ formation involves reaction between residual $NO_x$ and reductants remaining on the catalyst at the rich-phase end; vs alternate proposed pathways involving nonselective oxidation of ammonia or isocyanates and $NH_4NO_3$ decomposition. In other work, we took the more direct approach of applying SpaciMS to critically assess a global kinetic NSR model (Shwan et al., 2014), which had been originally developed based on effluent measurements (Shwan et al., 2012, 2013), adapted for spatiotemporal analysis using long-cycle (rich\lean: 600 s\900 s) effluent data, and assessed using short-cycle (rich\lean: 5 s\60 s) SpaciMS data (Partridge and Choi, 2009). The model was able to accurately predict several aspects of LNT performance including spatiotemporal lean-phase $NO_x$-storage distribution; spatial $N_2$ formation and temporal evolution along the catalyst axis; spatial $NH_3$ formation and utilization consistent with intermediate-$NH_3$ regeneration, and the temporal evolution of the $NH_3$ profiles; and spatial $H_2$ utilization and temporal nature of the various $H_2$ profiles. Regarding model structure, the spatiotemporal assessment confirmed sensitivity analysis identifying the most important active sites and reaction parameters. SpaciMS provides an ability to critically assess an LNT model structure and function in a detailed way via its ability to accurately predict the spatial and temporal evolution of reactions along the catalyst axis and throughout the rich–lean operation cycle; and which is unmatched by effluent integral assessment. In this way, it enables improved knowledge, and design and development tools.

SpaciMS has also contributed to practical advances in the form of improved operating strategies for reducing $N_2O$ emissions (Mráček et al., 2016).

We found that $N_2O$ is emitted in primary and secondary events corresponding to the start and end of the rich regeneration phase, respectively. Selectivity is associated with the relative PGM redox state, is biased in the order of NO, $N_2O$, $N_2$, and $NH_3$ as the PGM sweeps from oxidized to reduced state at regen onset, and the opposite order at rich-phase termination. Secondary $N_2O$ can account for up to 50% of the total $N_2O$ emissions under conditions of less-efficient regeneration (e.g., due to inefficient reductant, low temperature, etc.) and is generated via reaction between residual $NO_x$ and adsorbed reductants and intermediates. Under normal LNT cycling with c.10% lean-phase $O_2$, PGM rapidly transitions to a more oxidized state at the rich–lean interface resulting in reaction of the intermediate $NO_x$ and adsorbed species with high $N_2O$ selectivity, and producing significant secondary $N_2O$ emissions. By inserting a slightly lean transition region with lower $O_2$ concentration between the rich and lean phases, the rate of PGM oxidation is slowed, correspondingly biasing the secondary reactions to higher $N_2$ and $NH_3$ selectivity. Indeed, we demonstrated that secondary $N_2O$ could be eliminated with a suitable transition region. Like the OBD patent discussed in Section 3.4, this demonstrates a practical benefit of intracatalyst spatiotemporally resolved knowledge of catalyst performance.

## 3.6 Non-SpaciMS Methods for Resolving Distributed Intracatalyst Performance

Besides SpaciMS, alternate experimental methods have been implemented to resolve spatiotemporal reaction distributions. Kočí (Plát et al., 2010; University of Chemistry and Technology, 2017) and Parks (Parks et al., 2005, 2008; Swartz et al., 2006) have made extractive measurements from discrete sections along a modular or continuous catalyst using conventional analyzers as shown in Fig. 14; in both cases extractive measurements were made to conventional FTIR, MS, NDIR, chemiluminescent, etc., analyzers. In Kočí's method, a reactor was formed from three discrete monolith catalyst sections with some separation between each section, which allowed for sampling between each section as well as in the effluent. Parks' method used a continuous monolith catalyst with holes drilled across the diameter at the desired axial sampling locations; the hole size and sampling rate were balanced so that the catalyst space velocity was not impacted by the measurements. An alternate method performs effluent measurements on catalyst cores of different lengths (Bhatia et al., 2009; Choi et al., 2010; Clayton et al., 2008; Li et al., 2009; Prikhodko et al., 2009). In this method, the

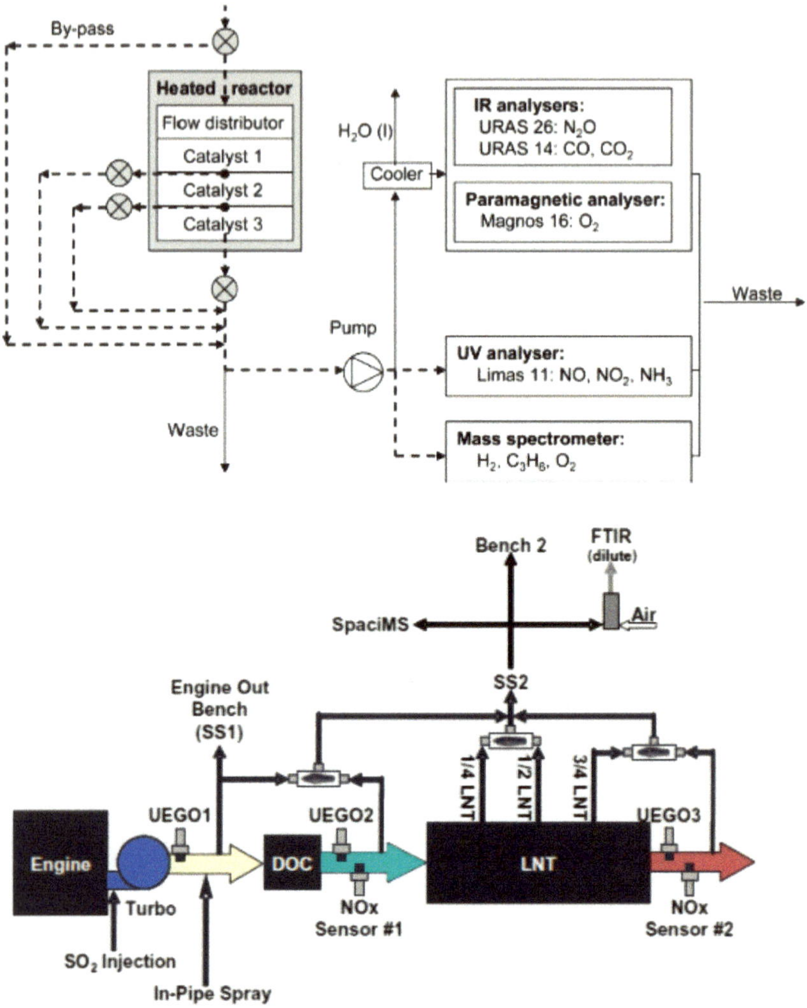


**Fig. 14** Alternate methods for resolving intracatalyst gradients. Extractive sampling between discrete components of a modular catalyst sample reactor (*Top*: T) (Plát et al., 2010; University of Chemistry and Technology, 2017), and within a continuous catalyst reactor (*Bottom*: B) (Parks et al., 2008; Swartz et al., 2006).

spatiotemporal reaction distributions are built up from separate experiments using progressively shorter-length sections of the same catalyst sample and effluent measurements. While this is a very viable method, it can have practical limitations for suitably short sections representative of the catalyst front.

Figure 15 shows spatiotemporal intra-LNT temperature mapping and demonstrates the benefit of combining multiple intracatalyst diagnostics. In this case, temperature was measured via phosphor thermography (PhosphorT) implemented via phosphor-tipped optical fibers (Choi et al., 2006); the optical fibers have the benefit of being small and nonconductive, which enhances their noninvasive nature and accuracy for resolving spatial temperature gradients; vs conductive probes which spatially broaden temperature gradients. In the OSC-cycling case, Fig. 15M, a strong exotherm is observed throughout the regeneration phase and 0.5-in. (1/6*L*) into the catalyst, corresponding to the spatiotemporal extent of the $CO + O_2$ reaction (see Fig. 5M and related discussion). With NSR cycling, Fig. 15B, an additional exotherm occurs at early regeneration times and 1.5-in. (1/2*L*) into the catalyst, corresponding to the spatiotemporal extend of the CO + surface $NO_x$ reaction (see Fig. 5R and related discussion). This demonstrates how correlating the PhosphorT and SpaciMS measurements allows for more detailed decisive information regarding the spatiotemporal extent of the various LNT reactions than from any single diagnostic alone and demonstrates the value in other diagnostics like OFDR (Nguyen et al., 2013; Peng et al., 2017). In separate work, Harold et al. have incorporated spatiotemporal temperature and species analysis to understand the complex dynamical behavior and reaction pathways involved in a combined LNT–SCR catalyst system, further demonstrating the unique and detailed insight enabled by multiple intracatalyst diagnostics (Li et al., 2016).

## 4. SCR APPLICATIONS

Selective catalytic reduction is a common method for selective and continuous $NO_x$ reduction in lean-burn automotive applications, involving injection of a reductant or reductant precursor upstream of the SCR catalyst (Brandenberger et al., 2008; Ciardelli et al., 2005). While many SCR types and reductants have been investigated, the major automotive focus has been on Fe- and Cu-exchanged zeolite catalyst operating with $NH_3$ reductant resulting from decomposition of injected urea. Spatiotemporally resolved intra-SCR measurements have been applied to advance the understanding and technologies associate with these important catalysts. These applications reveal how and where the catalyst is used, and spatial conversion slopes reflect relative rates. Specific examples include that SCR is biased to the

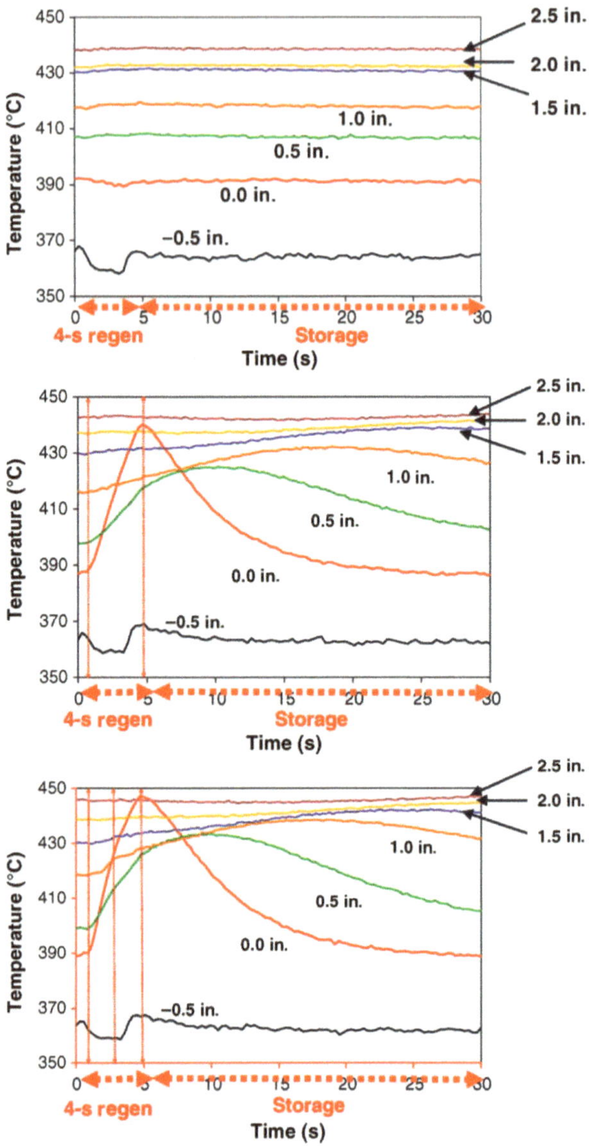


**Fig. 15** Transient temperature distributions during 54-s/4-s lean/rich cycling of a 3-in long Pt/K/$Al_2O_3$-washcoated monolith LNT. Lean-phase gas composition was (*Top*: T) 5% $H_2O$ for neat cycling, (*Middle*: M) 1% $O_2$+5% $H_2O$ for OSC cycling, and (*Bottom*: B) 250 ppm NO+1% $O_2$+5% $H_2O$ for NSR cycling, all with $N_2$ balance. Rich-phase gas composition included 4% CO+5% $H_2O$ in $N_2$ balance; with 0%, 1%, and 8% $O_2$ for neat, OSC, and NSR cycling, respectively (Choi et al., 2006).

catalyst front, $NH_3$ coverage is similarly front biased and practically saturated in the very front regions, and that unused capacity exist downstream and can offset inhibition and less-efficient operation conditions which might lengthen the SCR zone. Spaci-based measurements reveal relative rates of oxidation, adsorption and SCR reactions, the impact of aging on these functions and their interrelation, and have enabled conceptual and actual models of complex SCR and integrated LNT–SCR systems. Recasting spatially resolved measurements allows $NH_3$ adsorption isotherms to be determined under SCR conditions, and further analysis provides insights regarding the number and nature of $NH_3$ adsorption sites, and their variations with aging. And, such analysis has suggested pathways for on-road catalyst state assessment, and related methods for through-life control and optimization. These demonstrate the rich basic and practical information originating from spatio-temporally resolved intracatalyst analysis and create new opportunities for advancing SCR automotive technologies. We discuss these and other specific examples of SpaciMS SCR applications in this section.

The major analytical considerations associated with SpaciMS applications to SCR studies involve avoiding intracapillary ammonium nitrate (AN) formation, and methods for measuring ppm-level $NH_3$ variations in the presence of percent levels of $H_2O$ and/or $N_2$. Sampling $NO_2$ and $NH_3$ can result in AN formation and deposits within sampling capillaries at temperatures below c.195°C; subsequent intracapillary AN reactions can confound intracatalyst SpaciMS studies, in a similar way to the capillary contamination discussed in Section 2. A simple solution is to maintain capillaries above 200°C throughout their length in SCR SpaciMS applications (Ciardelli et al., 2005). Because $H_2O$, $NH_3$, and $N_2$ have parent and fragment ions in the 14–18 $m/z$ range, it can be difficult to measure ppm-level $NH_3$ in the presence of %-levels of $H_2O$ and/or $N_2$. Although we have discussed this elsewhere (Morgan et al., 2016), the literature shows that some need for communication in this area remains. Specifically, $NH_3$ has been measured using chemical ionization MS (Chatterjee et al., 2006; Heeb et al., 2008) and tuned ionization-energy in electron-ionization (EI) MS applications (Epling, 2015; hidenanalytical, n.d.; Li et al., 2016; Olsson, 2015). A simple method for EI-MS applications with %-level $H_2O$ and Ar diluent (only ppm-level $N_2$) is to use 15 $m/z$ for quantifying $NH_3$ (Auvray et al., 2012); with closed ionization sources both NO and $NH_3$ will have signal contributions at 15 $m/z$ ($NO^{2+}$, $NH^+$) and 30 $m/z$ ($NO^+$, $N_2H_2{}^+$), but can be easily differentiated and accounted for in analysis. Indeed, this method allows for experimentally efficient analysis of the major relevant species with a single MS configuration; i.e., without separate experiments with the MS in a dedicated configuration for $NH_3$ measurement and another for the remaining species. A final analytical consideration concerns the tradeoff of improved temporal resolution vs larger capillary size and sampling rate of differential capillary sampling; the relatively slower transients associated with SCR operation may not benefit from the temporal resolution offered by differential sampling, and the less invasive nature of direct capillary sampling may be preferable in many applications.

## 4.1 $NO_x$ and $NH_3$ Reaction Distributions

In the first SpaciMS application to SCR catalysis, we studied variations in spatiotemporal reactions, $NH_3$-capacity utilization, and SCR-onset conversion inflections over a range of temperatures for a 25.4-mm long, model Cu–Beta SCR catalyst under standard SCR conditions. Figure 16T shows distributed NO and $NH_3$ conversion over temperatures ranging from 200 to 400°C and demonstrates how an SCR zone is established at the catalyst front section over which the SCR reaction takes place; the 90% SCR zone is highlighted in Fig. 16T. While the integral conversion is constant and 100% for the range of temperatures investigated, the SCR zone is concentrated to a progressively shorter front section of the catalyst with increasing temperature. The local slope of the conversion distribution reflects the overall SCR rate. Thus, as the SCR rate increases with temperature, the spatial conversion slopes increase, and the SCR zone is correspondingly compressed to the catalyst front. Figure 16T shows relatively uniform SCR throughout an SCR zone in the front c.7/16*L* of the catalyst at 200°C, and much higher initial conversion gradients and spatially nonuniform conversion over a shorter front c.1/4*L* of the catalyst at 400°C. Because SCR cannot proceed beyond the point where $NH_3$ is fully consumed, the corresponding spatial location defines the SCR-zone end. Likewise, as discussed in more detail below, $NH_3$ capacity potential downstream of the SCR zone remains unutilized. Others including Pihl (Pihl and Daw, 2013) and Epling (Hou et al., 2011; Luo et al., 2011a, 2012) have applied SpaciFTIR to make these observations and quantify SCR zone length variations with standard and fast SCR, and aging.

Figure 16T shows stoichiometric NO and $NH_3$ conversion throughout the SCR zone for the range of temperatures investigated despite significant

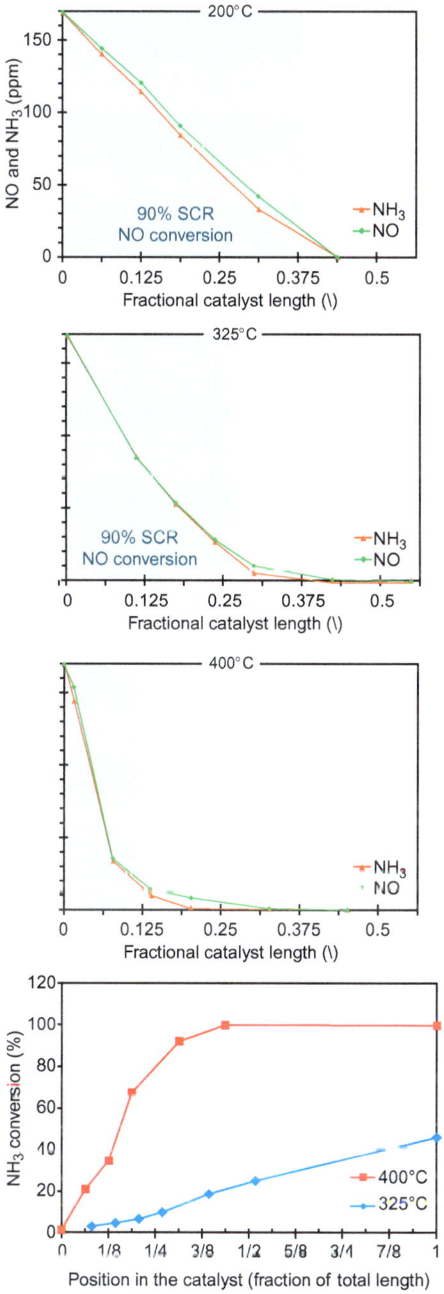


**Fig. 16** (*Top* three plots: T) Variation in spatially distributed NO and $NH_3$ conversion with temperature over a 25.4-mm long, model Cu–Beta catalyst under standard SCR conditions and 30 k $h^{-1}$ space velocity. Results are relative to $NH_3$-limited performance as the actual feed was NO rich. (*Bottom*: B) Neat or $NO_x$-free $NH_3$ oxidation distributions at 325 and 400°C; neat $NH_3$ oxidation was zero at 200°C (Auvray et al., 2012).

NO and $NH_3$ oxidation in the absence of SCR; i.e., neat (i.e., $NO_x$ free) oxidation of $NH_3$ and vice-a-versa. For instance, under neat oxidation conditions at 400°C, $NH_3$ and NO conversion increased uniformly in the catalyst front $1/2L$ reaching c.100% and 22%, respectively, and then approximately plateaued; conversion at 300°C was more uniformly distributed along the catalyst reaching c.45% and 5%, respectively, as shown for neat $NH_3$ oxidation in Fig. 16B. While NO oxidation to $NO_2$ under standard SCR conditions may promote $NO_x$ conversion by enabling the Fast SCR pathway, oxidation of $NH_3$ by oxygen will impede $NO_x$ conversion by competing with the SCR reaction for the limited reductant; for this reason, Kamasamudram et al. have termed this nonselective oxidation of $NH_3$ by oxygen under SCR conditions as parasitic $NH_3$ oxidation ($NH_3$ PO) (Kamasamudram et al., 2010). Despite significant $NO_x$-free $NH_3$ oxidation (cf. Fig. 16B), the stoichiometric $NH_3$ and NO conversion in Fig. 16T indicates that $NH_3$ selectively reacts with NO under SCR conditions for this catalyst; i.e., there is zero parasitic $NH_3$ oxidation. More specifically, this indicates that the SCR reaction is much faster than $NH_3$ oxidation, and thus the high selectivity for the SCR reaction for this model Cu–Beta catalyst (Auvray et al., 2012).

Parasitic $NH_3$ oxidation is not always negligible, and Fig. 17 highlights spatially distributed SCR, neat and parasitic $NH_3$ oxidation for a commercial Cu/SAPO-34 catalyst at 400°C (Partridge et al., 2015). Spatially distributed parasitic $NH_3$ oxidation is apparent for the catalyst in both the degreened

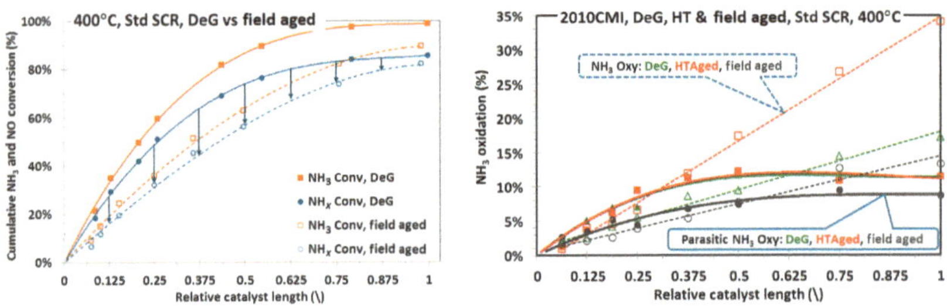


**Fig. 17** SpaciMS evaluation of a monolithic commercial Cu/SAPO-34 catalyst under 200 ppm standard SCR conditions, 60 k $h^{-1}$ space velocity, and different catalyst states. (*Left*: L) Distributed NO and $NH_3$ conversion at 400°C in the degreened (DeG) and field-aged states. Ammonia PO is indicated by the difference in the $NH_3$ and NO distributions at a given catalyst state; *the arrows* highlight aging-induced conversion loss. (*Right*: R) Distributed neat and parasitic $NH_3$ oxidation at 400°C in the DeG, hydrothermally aged (HTAged), and field-aged states (Partridge et al., 2015).

(DeG) and field aged (i.e., sample used in a real vehicle), states in Fig. 17L via the difference in the $NH_3$ and NO conversion; in this case the SCR zone occupies the full catalyst length. In the DeG state, $NH_3$ PO progressively grows in the front half of the SCR zone and then slows in the back half as $NH_3$ is fully consumed. In this case, PO is clearly competing with and limiting the SCR reaction; Epling et al. have observed the same with SpaciFTIR, and termed this SCR limitation by $NH_3$ PO as "quenching" (Luo et al., 2011a). Figure 17R highlights the neat and parasitic $NH_3$ oxidation distributions for the commercial Cu/SAPO-34 catalyst in DeG, hydrothermally aged (HTAged), and field-aged states at 400°C. While the neat oxidation distributions show linear conversion below 100% throughout the catalyst, indicative of spatially uniform washcoat and reaction throughout, the PO distributions grow in the front $1/2L$ and plateau in the back SCR-zone half, characteristic of reaching an $NH_3$ supply limitation; as discussed with Fig. 17L. Although differences in the neat and parasitic $NH_3$ oxidation are great in the effluent, they are comparable in the front c.$3/8L$ where most of the PO-induced SCR limitations arise; in fact, PO appears to be greater than neat $NH_3$ oxidation in the very front catalyst sections where the concentrations are the highest. This is not to say that neat and parasite $NH_3$ oxidation occur through the same reaction pathways and sites, and Kamasamudram et al. have specifically highlighted the differences in these reactions (Kamasamudram et al., 2010). Compared to the DeG state, NO conversion was c.20%–40% lower and $NH_3$ PO c.20% lower in the field-aged state, showing a trend opposite of what might be expected from mitigated $NH_3$ PO competition. We interpreted these results as suggesting that field aging impacts a site common to SCR and $NH_3$ PO, with SCR being more sensitive to the impact, and that $NH_3$ PO is not the major process limiting SCR in the aged state. This demonstrated one example of distributed SCR, neat and parasitic $NH_3$ oxidation measurements giving a more complete description of the competition between the various reactions, and insights regarding selectivity and performance.

In addition to the impacts of PO-induced SCR loss, Fig. 17 highlights some impacts of aging and different aging processes. Aging flattens the conversion distribution and lengthens the SCR zone, in a manner conceptually similar to NSR-zone lengthening by sulfation discussed in Section 3.4. In general, anything that inhibits the SCR reaction will lengthen the SCR zone, Epling et al. observed similar $NH_3$–SCR zone lengthening for hydrocarbon inhibition of Fast and $NO_2$ SCR (Luo et al., 2012), which we discuss in the next section. It is also notable that HTAgeing and field aging have dramatically different impacts on both neat and parasitic $NH_3$ oxidation, indicating that hydrothermal aging does not fully capture the impacts of more complex aging experienced in actual field applications.

## 4.2 Distributed $NH_3$ Capacity Utilization

SpaciMS provides detailed insights regarding the spatiotemporal population and utilization of $NH_3$ capacity during SCR operation and is the focus of this and the next subsection. This section focuses on steady-state capacity utilization and partitioning distributions. Kamasamudram et al. have defined Total (TC) Dynamic (DC) and Unused or Vacant (UC) $NH_3$ capacity components (Kamasamudram et al., 2010); TC is the maximum saturated capacity at a given $NH_{3,g}$ feed level, DC is the fraction of TC used during SCR, UC is the difference between TC and DC, and thus defining the DC + UC = TC balance; these capacity components are quantified via integration of step-transient experiments as described elsewhere (Auvray et al., 2012; Kamasamudram et al., 2010; Partridge et al., 2015). Figure 18 shows the TC and DC distributions at 200 and 325°C for the same model Cu–Beta catalyst shown and discussed in relation to Fig. 16 (Auvray et al., 2012). The TC or saturation capacity is linearly distributed along the catalyst for a uniformly distributed washcoat; deviations from TC linearity in Fig. 18 are likely due to imperfect spatial positioning corrections, and improved methods produce more linear results (cf. Fig. 19). The integral capacity at 325°C is c.30% that at 200°C, and the TC spatial slope is

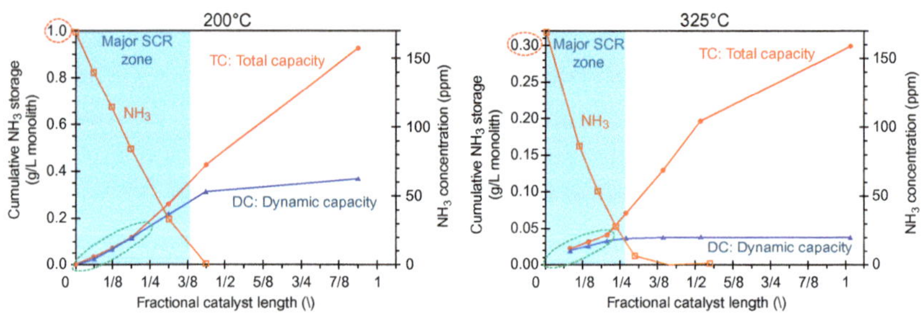


**Fig. 18** Steady-state spatially distributed $NH_3$ conversion, dynamic and total $NH_3$ capacity over a 25.4-mm long, model Cu–Beta catalyst under standard SCR conditions and 30 k $h^{-1}$ space velocity at 200°C (*Left*: L) and 325°C (*Right*: R) (Auvray et al., 2012); same catalyst as in Fig. 16. UC is the difference between TC and DC.

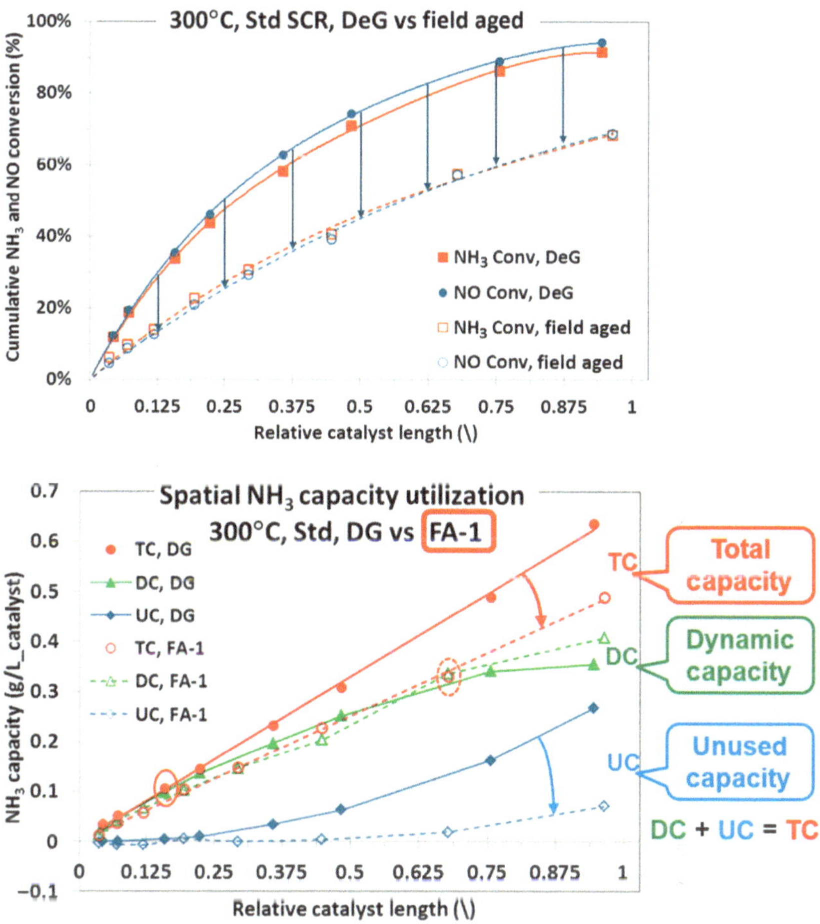


**Fig. 19** SpaciMS evaluation of a monolithic commercial Cu/SAPO-34 catalyst under 200 ppm standard SCR conditions, 60 k $h^{-1}$ space velocity, and different catalyst states; same catalyst as in Fig. 17. (*Top*: T) Distributed NO and $NH_3$ conversion at 300°C in the DeG and field-aged states. (*Bottom*: B) Distributed $NH_3$ capacity utilization and partitioning between TC, DC, and UC, at 300°C in the DeG and field-aged states (Partridge et al., 2015).

correspondingly lower. Interestingly, Fig. 18 shows that DC takes its maximum saturated value equal to TC over much of the SCR-zone front section (cf. the green dash circled region in Fig. 18); this is practically consistent with a low slope region of the $NH_3$ isotherm at higher $NH_3$ partial pressures as exist in the SCR-zone front, and as discussed in more detail in the next subsection. Moreover, this implies that UC is zero in this region which we have independently corroborated (Auvray et al., 2012). The DC separates from TC in the SCR-zone back as the $NH_{3,g}$ is progressively driven to complete conversion at the SCR-zone end; this similarly follows the adsorption isotherm shape where the potential for capacity population is progressively diminished with $NH_{3,g}$ consumption. Beyond the SCR zone where $NH_{3,g}$ is zero, potential $NH_3$ capacity is unutilized. Thus, in this region, DC plateaus and UC grows following TC; although not labeled in Fig. 18, UC can be visualized as the difference between TC and DC. Simply, $NH_3$ capacity potential is only utilized when $NH_{3,g}$ is nonzero, much like NSR capacity can exist downstream of an NSR zone as discussed in Section 3.1. When the environmental conditions or catalyst state changes, in a way that causes $NH_{3,g}$ to exist deeper into the catalyst, this previously unused capacity may become active; an example of this is discussed in Section 4.5 related to hydrocarbon inhibition. This demonstrates how DC exists only in the SCR zone and is biased to the catalyst front, while UC exists in the catalyst back regions.

Catalyst aging can change the distributed capacity utilization, and partitioning between the capacity components. Figure 19 shows the distributed capacity components and conversion for the commercial Cu/SAPO-34 catalyst in the DeG and field-aged states. In addition to the degraded NO and $NH_3$ conversion highlighted earlier, aging results in degraded integral capacity, which changes the TC distribution to a lower-slope state as shown in Fig. 19B. In the DeG state, DC reaches the saturated level in the front c.3/16*L*, and UC progressively increases downstream of the DC–TC separation point. Not surprisingly, the different formulations of the Cu–Beta and Cu/SAPO-34 catalysts produce different adsorption isotherms, and correspondingly different DC–TC separation points; i.e., this point occurs at different $NH_{3,g}$ levels and different points through the SCR zone. Interestingly, for the Cu/SAPO-34 catalyst, the DC–TC separation point occurs relatively early in the SCR zone, where $NH_3$ conversion is only c.45%–50% and increases in a shallow manner downstream. The nature of this DC growth results in little practical DC change with field aging, despite the significant change in conversion. Specifically, the decreasing TC with field aging limits DC in the front regions, but not to a practically significant extent; i.e., the DeG and field-aged DC curves in Fig. 19B are practically identical. This suggests that loss of $NH_3$ capacity is not the limiting factor responsible for the significant SCR conversion loss shown in

Fig. 19T. Of further interest is that although the DC–TC separation point moves significantly downstream with field aging (cf. orange-circled points in Fig. 19B), they occur at a c. common $NH_{3,g}$ concentration. This suggests that the adsorption isotherm is not significantly impacted by field aging; such an impact would change the isotherm shape, as discussed in the next section.

The TC and DC distributed behavior dictates that of the UC via the balance between the capacity components as shown in Fig. 19B. Because DC does not practically change with field aging for the commercial Cu/SAPO-24 catalyst, the TC loss is reflected in a corresponding UC loss. While the catalyst had significant UC in the DeG state, very little UC remained after field aging. Throughout the SpaciMS examples discussed, we have seen many examples for different catalysts and catalyst functions of how the same integral performance can be maintained while the intracatalyst distributed performance changes dramatically. The TC, DC, UC example in Fig. 19B shows a unique example of how a particular catalyst parameter (DC in this case) can remain practically unchanged with aging, while the distribution of other parameters (TC and UC) changes dramatically.

The nature of the UC variation with aging suggests that it could be used as the basis of a practical method for catalyst state monitoring. It is apparent that DC/TC and UC/TC are proportional to the catalyst fraction used for SCR and unused, respectively. We have demonstrated a basic method for predicting catalyst state based on an ANR (ammonia-to-$NO_x$ ratio) pulse, by replicating the UC spatially distributed shape via the ANR-probe signal (Partridge et al., 2016). While the ANR probe was useful in demonstrating the concept, a better manifestation of the concept would use naturally existing drive cycle transients, building on the basic concept with additional spatiotemporal catalyst characterization. Ultimately, such a method for practical on-road catalyst state assessment could be used to predict (e.g., via parameter maps, look-up tables, or more complex models) variations in catalyst parameters (e.g., TC) with aging for use in implementing model-based control strategies for optimizing (e.g., via tuned urea dosing) through-life catalyst performance. A salient point is that in addition to basic catalyst insights, spatiotemporal intracatalyst analysis can provide more general insights suggesting or leading to practical and applied catalyst advances. The general nature and relationship between the capacity components are useful for understanding coverage buildup in pulsed-$NH_3$ applications and the impacts of hydrocarbon inhibition of SCR reactions (Hou et al., 2012; Luo et al., 2012) as discussed in Section 4.5.

## 4.3 Operando $NH_3$ Isotherms

Adsorption isotherms describe how $NH_3$ capacity is populated as a function of local $NH_3$ concentration or partial pressures. Adsorption isotherms are typically measured by exposing the catalyst to different magnitude $NH_3$ steps and integrating to the new poststep steady state, the adsorbed or desorbed $NH_3$ depending on if positive or negative steps are used; the experiment is performed at a range of nonreacting isotherm conditions to map temperature impacts, and other parameters (e.g., [$H_2O$] can be swept too) (Pihl and Daw, 2016). Figure 20T shows isotherms measured in this way for $NH_3$ adsorption at a range of different temperatures on a commercial Cu/SSZ-13 SCR catalyst. An isotherm's shape reflects the adsorption energetics of the adsorption site, and many different isotherm models exist including Langmuir and Temkin (Perry and Green, 2008). The isotherm in Fig. 20T, with the distinct knee at low $NH_3$ concentrations, is consistent with a two-site Langmuir isotherm; the energetics of the two adsorption sites shown in Fig. 20M was determined via additional thermodynamic analysis by Pihl and Daw (2016).

Because SpaciMS measures a range of DC levels and $NH_3$ conversions throughout the SCR zone, the SpaciMS data can be recast to determine the adsorption isotherm under actual SCR operating conditions. While SCR reaction theoretically adds an additional loss term to such an operando isotherm compared to the classical isotherm determined under nonreacting or neat conditions, we have shown that adsorption equilibrates much faster than even Fast SCR, and thus the operando isotherm can be interpreted classically (Partridge et al., 2015; Pihl and Daw, 2013); i.e., DC–TC separation occurs at a common $NH_3$ concentration for both Standard and Fast SCR, and not surprisingly as $NH_3$ adsorption is a nonactivated process. Figure 20L shows such an operando isotherm for the commercial Cu/SAPO-34 catalyst at a range of states and two temperatures. The figure was generated by plotting the local normalized DC (DC/TC) vs the local $NH_3$ partial pressure at the different locations along the catalyst; i.e., combining the data in Fig. 19T and B. The isotherm is only active in the SCR zone where $NH_{3,g}$ is nonzero; thus, the isotherm may be spatially visualized as spread across the SCR zone from high $NH_3$ partial pressures at the SCR-zone front to zero $NH_3$ partial pressure at the back, as indicated by the right-to-left arrow in Fig. 20B. The TC saturation limit can be visualized as a horizontal line at $\theta_{DC}=1$ in Fig. 20B. The DC=TC region at the SCR-zone front discussed in Figs. 18 and 19 occurs

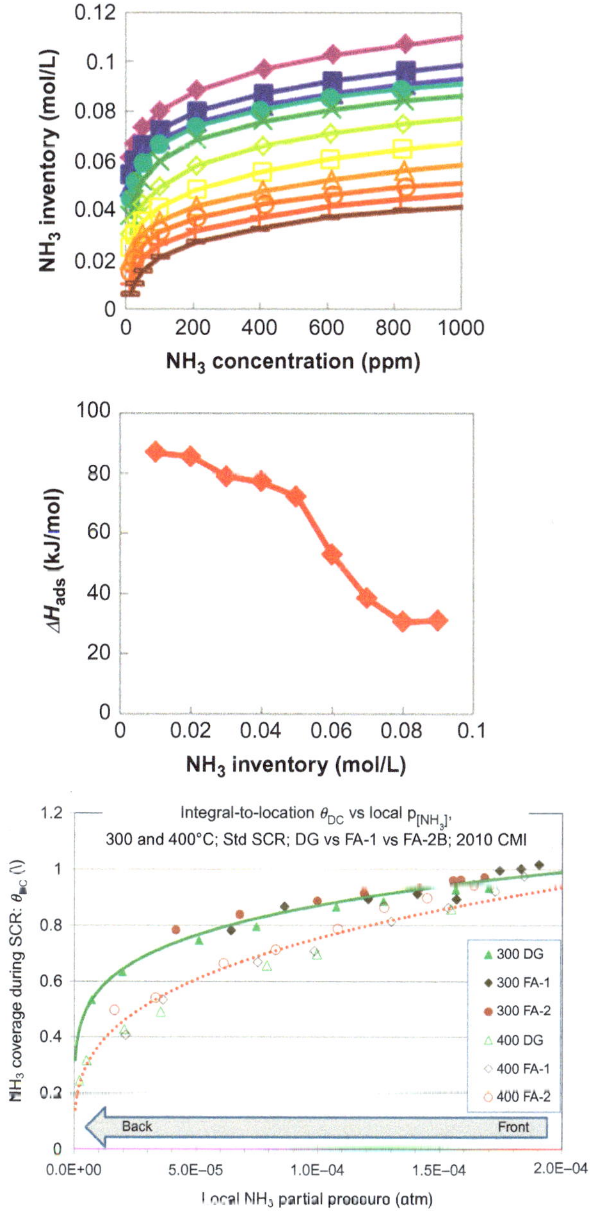


**Fig. 20** (*Top*: T and *Middle*: M) Neat $NH_3$ isotherms of a commercial Cu/SSZ-13 SCR catalyst at a range of temperatures; 150–350°C top-to-bottom, respectively; and corresponding adsorption enthalpy of the two adsorption sites modeled as Langmuir isotherms (Pihl and Daw, 2016). (*Bottom*: B) Operando $NH_3$ adsorption isotherm under SCR conditions determined via SpaciMS measurements of normalized DC and $NH_3$ partial pressure distributions of the commercial Cu/SAPO-23 catalyst at 300 and 400°C and three states: degreened (DG), field aged #1 (FA-1), and field aged #2 (FA-2) (Partridge et al., 2015).

in the low-slope region of the isotherm that is practically parallel and equal to the TC limit in Fig. 20B. The isotherm theoretically and actually continuously falls away from the TC limit from the SCR-zone front and is only practically parallel to TC due to measurement limitations; the DC–TC separation point was measured to occur at a uniform c.85% TC for all temperatures and catalyst states shown in Fig. 20B, confirming the earlier assertion that DC–TC separation at a common $NH_3$ concentration is indicative of an invariant isotherm (Partridge et al., 2015). Beyond the DC–TC separation point, the isotherm slope continuously increases to the point of $NH_{3,g}$ depletion at the SCR-zone back, driving correspondingly progressive UC growth; beyond the SCR zone, the UC spatial distribution parallels that of TC. Imposing a spatial perspective on the adsorption isotherm clarifies the spatially distributed nature of capacity unitization.

The isotherm's shape and its variations give insights regarding the nature of adsorption energetics and their variations with aging. The shape of the operando isotherm is consistent with adsorption via a two-site Langmuir isotherm as measured by Pihl and Daw, cf. Fig. 20T and B (Pihl and Daw, 2016); specifically, the slope variations around a distinct knee at low $NH_3$ partial pressures, and the isotherm flattening with increasing temperature. The isotherm shape is indicative of the nature of the two adsorption sites and their relative population or weighing, and any change in that weighting or nature would be reflected in a corresponding isotherm shape change. For example, selective adsorption-site aging would change the net adsorption energetics by varying the relative population of the two adsorption sites, resulting in a corresponding isotherm shape change. However, the relatively invariant isotherm shape with catalyst state changes (at a given temperature) in Fig. 20B indicates that aging proportionally degrades the two adsorption sites. Thus, it appears that adsorption occurs in the same way throughout the life of the commercial Cu/SAPO-34 catalyst sample, and that a single adsorption model could possibly be used for through-life control; along with a measure of TC loss, possibly via the method for practical on-road catalyst state assessment described in Section 4.2. This demonstrates a new way to use SpaciMS data to generate adsorption isotherms under reaction conditions, which further expands the method utility. Moreover, it provides a pathway for assessing variations in site population and activity, and adsorption energetics with changes in catalyst state and environmental conditions. Integrating the basic insights from the operando isotherm with the catalyst state assessment method, highlights the broad basic and practical insights afforded by spatiotemporally resolved intracatalyst analysis, which are relevant to advancing transportation technology.

## 4.4 SCR Modeling Using Spatiotemporal Data

In addition to the unique modeling-related insights resulting from intra-SCR spatiotemporally resolved characterization described in the other sections, such data have been used explicitly to tune and validate SCR models. In one example, we used SpaciMS 4-Step Protocol data (Auvray et al., 2012) to tune and validate a global kinetic SCR model for a Cu/Beta zeolite catalyst at 200, 325, and 400°C (Auvray et al., 2015). The model included $NH_3$ adsorption and desorption, $NH_3$ and NO oxidation, $N_2O$ formation and SCR reactions; the sorption parameters were determined from microcalorimetry, the various activation energy values from literature, and the steady-state SpaciMS data over a range of conversions was used to tune the preexponential factors for the various reactions. While not included, the spatially resolved results indicated that the model could be improved by incorporating $NO_2$ inhibition of NO oxidation. The difficulty of fitting the $NH_3$ oxidation model at both 325 and 400°C to the steady-state SpaciMS data by tuning the preexponential factor alone, revealed the limitations of using a fixed activation energy; because of greater $NH_3$ oxidation at 400°C corresponding tuning was biased to that data, but resulted in over-predicting conversion at 325°C. Calculated spatial $NH_3$ coverage distributions during SCR were consistent both in terms of shape and spatial extent with corresponding (DC/TC) measurements (cf. Fig. 18). Following tuning of the global kinetic model using the steady-state data, the transient SpaciMS data from the 4-Step Protocol were used for validation. The general NO, $NH_3$, $N_2$, and $N_2O$ transient features were captured by the model at all temperatures, and the results were generally excellent at 200 and 400°C; unsurprisingly due to the fitting and structure compromises, the specific steady-state conversion at 325°C was overpredicted, but the transient shape was generally well predicted. In another example, spatiotemporal SpaciFTIR data were used to guide formulation of and determine kinetic parameters for a multi-$NH_3$-site SCR model (Song et al., 2015). The spatial data were key to determining that a single-$NH_3$-site model could not simultaneously predict both gas-phase and surface species distributions, and that a two-site model was required; furthermore, the spatial data were the basis of a unique method for determining partitioning between the two $NH_3$ surface sites, and activation energy and preexponential kinetic parameters for the many reactions. The resulting model could accurately predict steady state and transient gas and surface distributions during SCR. Beyond the performance details of the specific models, these examples highlight the capability and value of spatiotemporal data for identifying improvement opportunities, guiding formulation, fitting, and validating catalyst models.

## 4.5 Other Intracatalyst Distributed Measurements and Applications

This section discusses several additional uniquely relevant SCR works demonstrating the value of intracatalyst spatiotemporal analysis, but which do not neatly fit in the preceding sections. The first three are by the Epling group and study SCR catalysts reaction pathways, pulsed $NH_3$-capacity loading and redistribution, and hydrocarbon inhibition. Although these works are based on SpaciFTIR measurements, the methods are appropriate to SpaciMS applications, and thus are briefly highlighted. The fourth highlights the Harold group's SpaciMS application to a combined LNT–SCR catalyst system. The fifth work is from Ford and provides a unique SpaciMS application for determining effective washcoat and substrate diffusivity parameters of catalysts in their natural coated-monolith form. Several non-Spaci-based studies resolving distributed intra-SCR and intra-SCR–LNT performance distributions have been published, involve effluent measurements from different catalyst lengths or stacked sections, and are referenced here for completion (Henry et al., 2012; Kim et al., 2014; Raj et al., 2013; Theis et al., 2011; Zheng et al., 2014a, b).

Epling et al.'s first published SpaciFTIR application is uniquely significant in its demonstrating the broad insights available from intracatalyst distributed analysis for understanding reaction details (Luo et al., 2011a); these include relative temperature dependence of SCR vs oxidation reactions, transition between kinetic and thermodynamic limitations, identification of rate-determining steps, and mechanistic routes of inhibition. In this same work, they analyzed spatial concentration and conversion distributions to demonstrate that while Fast and $NO_2$ SCR occur in parallel, Fast and Standard SCR occur in series. Figure 21T shows $NO_x$ conversion distributions for stoichiometric $NO_x$ and $NH_3$ feeds with different $NO_2$ fractions, as indicated in the legend; i.e., the black squares, green triangles, and aqua diamonds represent Standard, Fast, and $NO_2$ SCR, respectively. Based on the spatial slopes, the relative reaction rates rank from the fastest to the slowest, Fast, $NO_2$ and Standard SCR, respectively. Figure 21M shows the spatial concentration distributions for the case where $NO_x = 450$ ppm NO + 150 ppm $NO_2$. The stoichiometric NO and $NO_2$ consumption in

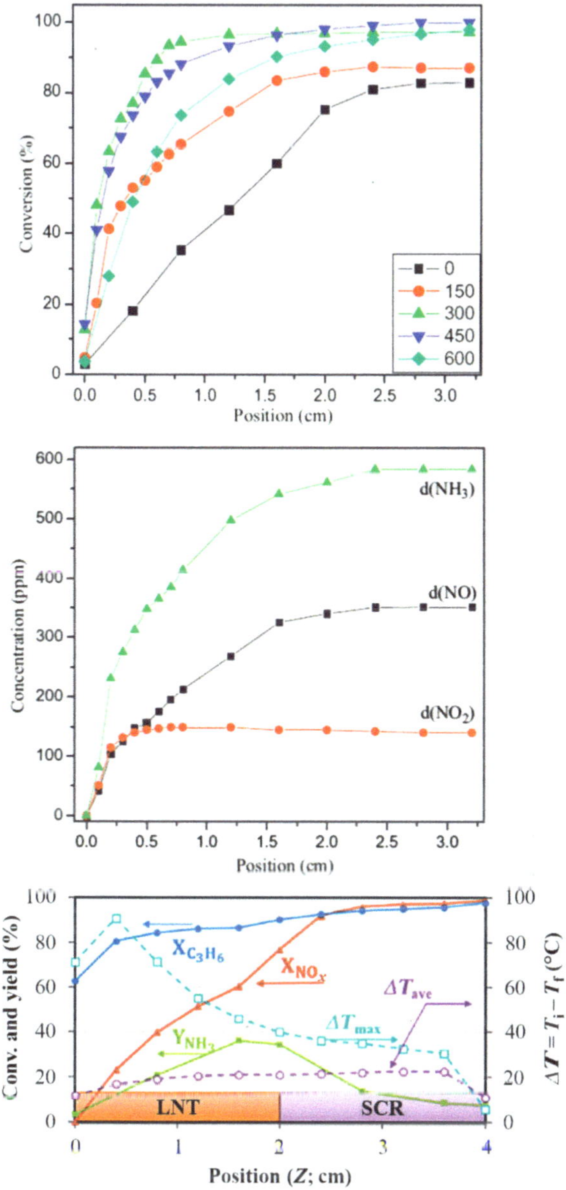


**Fig. 21** (*Top*: T) $NO_x$ conversion distributions for various $NO_2$ feed levels at constant 600 ppm $NO_x$ and 600 ppm $NH_3$ (8% $O_2$, 2.4% $H_2O$, 5% $CO_2$, $N_2$ balance; 300°C). (*Middle*: M) Spatial concentration distributions for the 450 ppm NO + 150 ppm $NO_2$ case of Fig. 21L (Luo et al., 2011a). (*Bottom*: B) Species and temperature distributions through a sequential LNT–SCR catalyst reactor. Lean (60 s): 200 ppm NO, 6% $O_2$, 9% $CO_2$, 7% $H_2O$, balance Ar; Rich (5 s): 6055 ppm $C_3H_6$, 1% $O_2$, 9% $CO_2$, 7% $H_2O$, and balance Ar. GHSV: 90,000 $h^{-1}$ (Li et al., 2016).

the front c.0.5 cm indicates sole Fast SCR reaction to the point where $NO_2$ is depleted. Beyond this point, slower Standard SCR proceeds until $NH_3$ is depleted; the distinctly different rates of the sequential reactions are apparent in Fig. 21T and M, by the different gradients in the different catalyst regions. In this way they prove that Fast and Standard SCR do not occur in parallel, but rather sequentially. Through similar analysis they show that how Fast and $NO_2$ SCR occur in parallel; this is generally apparent in Fig. 21T from the similarity of the Fast SCR and $NO_x = 150$ ppm NO + 450 ppm $NO_2$ distributions.

Epling et al. applied SpaciFTIR to investigate pulsed $NH_3$ capacity buildup in a Cu/chabazite catalyst, and reported spatial distributions and observations consistent with the $NH_3$-capacity discussions above (Hou et al., 2012). Capacity built up from the catalyst front, and progressed down the catalyst length as the front section approached saturation. Investigating $NH_{3,g}$ pulsing characteristics, they found that for equivalent dosing strategies, shorter pulses with higher $NH_{3,g}$ concentrations produced steeper capacity gradients in the catalyst front regions; i.e., greater $NH_3$ coverage in this front region. Moreover, when comparing series capacity buildup followed by an inert hold, and then $NH_{3,g}$ free SCR, they reported more shallow and distributed SCR distributions compared to the initial $NH_3$ capacity distributions. This indicated capacity redistribution during the inert hold period and suggested separate strong and weak adsorption sites which they identified as Lewis-coordinated and Brønsted-coordinated sites, respectively. Notably, this redistribution presumably involved the weakly adsorbed $NH_3$, which was redistributed to downstream previously unoccupied capacity; i.e., occupying previous UC. The reported results are consistent with the adsorption isotherm controlling the capacity distribution and redistribution, and the isotherm shape details being dictated by the nature and relative population of the strong and weak adsorption sites. Their observation that aging impacted the weak sites but not the strong sights, suggests that the adsorption isotherm for that catalyst would have changed with aging; as discussed in Section 4.3.

In a third study by Epling et al., SpaciFTIR was applied to develop a conceptual model for the spatiotemporal impact of $C_3H_6$ inhibition and reaction pathways of a Cu/beta zeolite catalyst (Luo et al., 2012). The work further highlights the spatially distributed SCR concepts discussed earlier, and the general spatiotemporal principles of the conceptual model are similar to the LNT sulfation conceptual model discussed in Section 3.4. In the HC free or uninhibited case, only the catalyst front region is used for SCR; $NH_3$ is fully consumed in this front SCR zone, and thus the downstream capacity remains unused, i.e., UC. Initial $C_3H_6$ exposure poisons some of the SCR sites causing $NO_x$ and $NH_3$ slip beyond the initial uninhibited SCR zone. While $NO_x$ slip is initially observed in the effluent, the slipping $NH_3$ is adsorbed downstream, the SCR zone correspondingly expands downstream, and $NO_x$ conversion is recovered with zero effluent $NH_3$ slip. Furthermore, $C_3H_6$ exposure progressively lengthens the SCR zone, but high $NO_x$ conversion remains as long as the broadened SCR zone is shorter than the catalyst length. Once inhibition progresses to the point that the SCR zone and catalyst exit spatially correspond, $NH_3$ slip is observed. Furthermore, $C_3H_6$ exposure results in increased $NO_x$ and $NH_3$ slip as the SCR zone is pushed out the catalyst back. With this conceptual hydrocarbon inhibition model developed from spatially resolved measurements, the temporal effluent transient's species sequences become obvious. Intra-SCR spatial profiles were also used to conclude that $C_3H_6$ reduction of $NO_2$ to NO proceeds faster than Fast and $NO_2$ SCR rates, and understand the corresponding impacts on reaction pathways. Specifically, causing Fast SCR to become both Fast and Standard SCR via $NO_2$ reduction to NO, and causing $NO_2$ SCR to become both Fast and $NO_2$ SCR for the same reason. Indeed, this paper was rich in examples of general and mechanistic insights enabled by Spaci analysis.

The Harold group applied SpaciMS along with distributed intracatalyst temperature measurements to study a sequential LNT–SCR catalyst (Li et al., 2016). The LNT was $Pt/Rh/BaO/Al_2O_3$, the SCR Cu/SSZ-13, and $C_3H_6$ was used as the reductant. The basic concept involves $NH_3$, formed in and slipping from the upstream LNT, being adsorbed on the SCR, allowing for $NO_x$ conversion across the sequential dual catalyst reactor. They focused on investigating coupling between the LNT and SCR catalysts over a range of operating conditions, identifying optimum component-catalyst lengths, and developing a 10-path reaction network to describe the overall $NO_x$ reduction which included oxidation, water–gas shift, reforming, and $NO_x$ reduction reactions. Figure 21B shows cycle-averaged distributed conversion, yield and temperature at 350°C. The $NH_3$ distribution shows continual buildup throughout the LNT, and then distributed consumption along with additional $NO_x$ reduction in the SCR. A large and distinct exotherm corresponds with major $C_3H_6$ conversion at the LNT face, likely associated with oxygen reduction, and is conducted throughout the remaining dual catalyst system. They identified a non-$NH_3$ $NO_x$ reduction pathway involving $C_3H_6$ in the SCR back, where $NH_3$ is practically depleted. It was anticipated to achieve 50% $NO_x$ conversion in each catalyst element. However, at that 50% $NO_x$ conversion point in the LNT (c.1.2 cm), the $NH_3$ selectivity of $NO_x$ reduction was only 50%–60%, meaning that additional LNT-attributable conversion was required; i.e., there would be insufficient $NH_3$ for converting the remaining $NO_x$ across the SCR with a 1.2-cm long LNT. The length shown in Fig. 21B appears to be well balanced, in that a 1.6–2 cm LNT length would produces an ANR of c. unity at the SCR inlet; i.e., c.35% $NH_3$ yield and equivalent $NO_x$ remaining. In addition to these summary results, they found valuable insights from correlating spatially resolved temperature and species measurements. Via the Spaci species and temperature analysis, they were able to gain a more complete understanding of the reaction pathways in this complex sequential LNT–SCR catalyst system.

Sharma and Dearth from Ford have published a very unique SpaciMS application for determining effective diffusivity of automotive catalyst honeycomb substrates and washcoats (Sharma and Dearth, 2016). Although their application is to SCR washcoats, the method seems generally applicable to any catalyst type, and requires no special sample modification for analytical purposes. Figure 22L describes the method, which involves injecting a tracer near the inlet of a central channel (1), and using SpaciMS to measure axial distribution of the tracer in the adjacent (2), diagonally adjacent (3), and next-adjacent (2-adjacent, 4) channels. The demonstration uses $O_2$ as the tracer, but other tracers could be used. Figure 22R shows the measured $O_2$ axial distribution in the three adjacent channel families. As the adjacent channels share a common wall with the central channel, tracer diffusion is the fastest into that channel, as apparent from the spatial slope

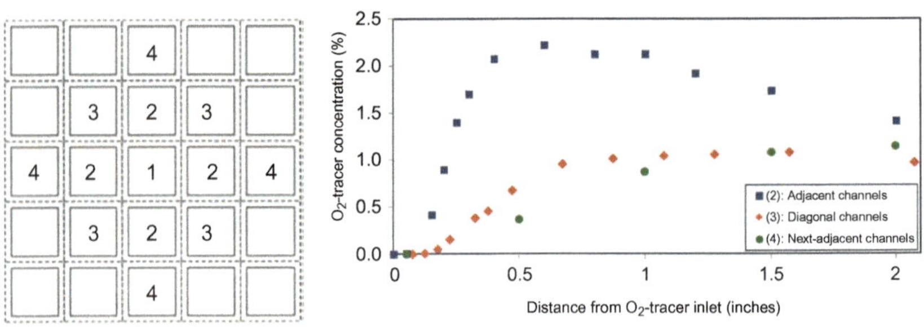


**Fig. 22** (*Left*: L) Family of channels used in the SpaciMS-based method for effective diffusivity measurements. Center (1) where tracer is injected, adjacent (2), diagonal (3) and next-adjacent (4) where SpaciMS measurements are made. (*Right*: R) Tracer concentration (ppm) distributions in the adjacent, diagonal, and next-adjacent channels relative to the tracer injection axial location; for the substrate only at 200°C (Sharma and Dearth, 2016).

of the blue squares in Fig. 22R; tracer concentration in the adjacent channel peaks at c.0.6 in. as tracer outflow into the diagonal and next-adjacent channels via their shared wall surpasses inflow from the central channel. They observed that the spatial location of this peak concentration varied consistently with the expected temperature dependence of diffusion; i.e., shorter at higher temperatures. Using the distributed measurements in the adjacent channel, the known tracer injection concentration and the channel hydraulic diameter, the steady-state tracer distribution in the central channel can be calculated as a function of an overall mass transfer coefficient ($k_{overall}$). The analysis involves performing a mass balance on the adjacent channel, using the tracer distributions in the surrounding channels, determining the best fit $k_{overall}$, and using this value to determine effective diffusivity. By performing the analysis on bare and washcoated substrates, diffusion parameters in the two components could be determined; analysis for two washcoat thicknesses was reported. They determined diffusivity and tortuosity parameters, and based on the diffusivity temperature dependence of the substrate ($T^{1.4}$) and washcoat ($T^{0.5}$) concluded transport followed bulk-phase and Knudsen diffusion, respectively, in these catalyst components. SpaciMS enables this practical technique for determining effective diffusivity parameters of automotive catalyst in their natural state (i.e., actual catalyst samples, rather than analytically specific samples), and improves modeling capabilities by providing accurate values of these critical parameters.

## 5. OXIDATION CATALYST APPLICATIONS

This section highlights the unique insights and advances enabled by SpaciMS application to oxidation catalysts. In addition to their function for hydrocarbon and CO oxidation, oxidation catalysts are an important part of an automotive catalyst systems in that they enhance the performance of other downstream catalysts; for instance, NO oxidation improves $NO_x$ trapping and conversion on LNT catalysts, enables the Fast SCR reaction and thus improves conversion on SCR catalysts, and improves passive and low temperature regeneration of particulate filters. While some of the SpaciMS studies described here are applicable to TWCs, they are included in this section as they focus on oxidation reactions. Because automotive catalyst applications are highly dynamic and cover a broad range of exhaust conditions, the ability to resolve spatiotemporal intracatalyst variations is uniquely enabling and relevant to the associated technology advancement as further demonstrated by the oxidation catalyst studies heighted here.

There are no additional SpaciMS application-specific analytical considerations unique to oxidation catalyst, beyond those discussed in the previous sections. Several non-Spaci methods have been applied to study spatially distributed oxidation catalyst performance including OFDR, thermography, and planar laser-induced fluorescence (PLIF) (Liu et al., 2011; Nguyen et al., 2013, 2015; Sui et al., 2017; Zellner et al., 2015), and postmortem analysis of sectioned catalysts (Beck et al., 1997a, b; Lambert et al., 2009; Winkler et al., 2009; Zotin et al., 2005). While free-space optical diagnostics, such as thermography and PLIF, employ effectively noninvasive light-matter interactions, the required optical access can be invasive, and the modified-reactor conditions profoundly different from those of the native reactor (Morgan et al., 2016); nevertheless, such studies of controlled conditions, albeit different from those of the native reactor, can produce valuable insights as has described for SpaciFTIR (Luo et al., 2011a), and shown in the referenced thermography and PLIF studies. The studies below specifically highlight SpaciMS applications to advancing oxidation catalyst understanding and technology.

In several studies, the Epling group applied SpaciMS and IR-thermal imaging to study various aspects of oxidation catalyst (OxyCat) performance, and gained unique insights by spatiotemporally resolving and correlating intracatalyst distributions and transients. Their early work applied SpaciMS and integral measurements to discover the interrelation and relative sequence of various oxidation catalyst reactions (Irani et al., 2009a, b). As in other applications, reaction gradients were interpreted as reflecting reactivity or rate of reaction, and $NO_x$, $H_2$, CO, and hydrocarbon (HC; propene and dodecane) reactions were studied. As $NO_2$ generation is a significant OxyCat function, NO oxidation performance was assessed in terms of $NO_2$ generation, and found to be "inhibited" by the presence of HCs. Further spatiothermal studies showed that this impact was due to a combination of classic inhibition (e.g., HC and CO species blocking active sites) and selective consumption of the NO-oxidation product via HC oxidation; i.e., HC oxidation selectively occurs through reaction with $NO_2$ vs $O_2$. Once HC lightoff occurred, reducing HC and CO concentrations and the corresponding "inhibition," NO conversion increased. These initial SpaciMS studies identified how ignition followed a back-to-front pattern, which moved progressively toward the front with increasing temperature.

In follow-on studies, the Epling group demonstrated the complimentary nature of SpaciMS and IR thermography for understanding the interrelation between reactions and heat transfer (HT), and how HT is key to the back-to-front ignition patterns (Russel et al., 2010; Shakir et al., 2009). Figure 23T shows the spatiotemporal $C_3H_6$ distributions throughout a 3-in. long, monolith-supported, 1% $Pt/Al_2O_3$ oxidation catalyst; the back-to-front ignition is apparent in each time step, as is corresponding $C_3H_6$ depletion. The measured transient temperature distributions shown in Fig. 23M have contributions from (conduction and convection) HT and exothermic $C_3H_6$ oxidation, which need to be separated to maximize the diagnostic utility. While the transient thermal maps show a similar back-to-front dynamic ignition wave, it's difficult to interpret the reaction nature via the temperature measurements alone; e.g., does the peaking temperature reflect an evolving situation of reversible inhibition and reactant depletion, or are the reaction insights fully hidden in HT effects? Correlating the two intracatalyst measurements shows that the points of peak temperature and $C_3H_6$ depletion coincide in space and time, establishing cause, and effect; i.e., that cooling after the temperature peak is associated with $C_3H_6$ depletion. This understanding allowed development and application of an HT model, which was applied to generate reaction-specific transient temperature maps, as shown in Fig. 23B; i.e., with the HT contributions removed. The true frequencies of the reaction exotherms are apparent in Fig. 23B, as well as how HT had temporally broadened the thermal transients and been responsible for the long postpeak tail. Figure 23B also shows how the transient ignition wave sweeps through the catalyst and stabilizes at the catalyst front as a 0.18-in. thick reaction zone; SpaciMS was further used to confirm this confined catalyst front reaction zone. The improved understanding and analytical techniques were used to study the impacts of operating conditions, and homogeneous and heterogeneous (catalyst front biased) aging on distributed reaction transients and performance. They found that aging resulted in wider reaction zones, more reaction in the catalyst back, and slower movement of the ignition wave through the catalyst. In general, strong coupling was observed between transient reaction and HT distributions, with heat from the exothermic reaction being transferred downstream. Ignition occurred at a given downstream point when the lightoff temperature was reached; the interrelation is apparent in that heat transferred from upstream reactions allows downstream points to reach the lightoff temperature, advancing the reaction front and creating additional heat for further propagation of the ignition wave. Anything that changed the balance between heat generation and loss (reaction vs HT) correspondingly changed the motion and/or speed of the ignition front; e.g., stepped $C_3H_6$ addition, reversible inhibition such as CO desorption and reaction. While the IR

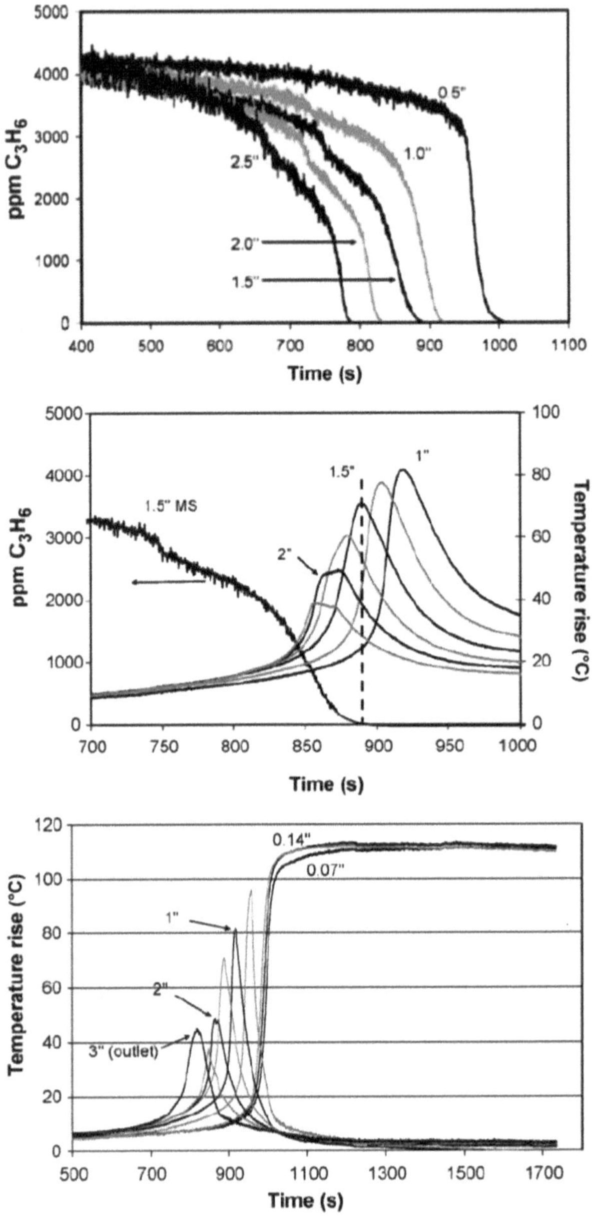


**Fig. 23** Spatiotemporal species and temperature distributions during operation of an oxidation catalyst (Shakir et al., 2009). (*Top*: T) transient propene distributions. (*Middle*: M) Spatiotemporal temperature distributions including reaction and heat-transfer contributions, and propene transient at 1.5-in. into the catalyst. (*Bottom*: B) Spatiotemporal reaction exothermal distributions; i.e., with heat-transfer contributions removed.

thermography is a powerful stand-alone diagnostic, its combination with SpaciMS allowed for significant diagnostic advances, and insights regarding the complex transient distributed multireaction process typical of automotive oxidation catalysis.

The Harold group used SpaciMS to elucidate the origins of some complex dependencies of $CH_4$ oxidation on $O_2$ concentration and temperature over a Pt\Pd\$Al_2O_3$ catalyst, which showed multiple transient and steady-state branches and hysteresis (Bugosha et al., 2015). Figure 24T shows this complex methane-conversion map over a range of $O_2$-$CH_4$ stoichiometries; $CH_4$-rich and $O_2$-rich regions exist for $2p_{O_2}/p_{CH_2}$ values below and above unity, respectively. The map includes transient (B) and metastable (C) branches, a hysteresis loop (A) and classical lean-regime $CH_4$ oxidation over PdO (D), and they interpreted much of these complexities as originating from the state of the precious-group metals. For instance, although an apparent continuum of rich-regime high-conversion transient branches (B) can be accessed depending on the catalyst history or previous environmental conditions, these all slowly decay over several hours to a common single steady-state high-conversion branch; i.e., the solid curve in the B series of Fig. 24T. The SpaciMS measurements in Fig. 24B provide insights into competing reactions, interplay between the various reactions along the catalyst length and their evolution with increasing time on stream, and ultimately evidence of the major factors controlling the complex behavior. Lean-regime $CH_4$ oxidation (branch D in Fig. 24T) involves total oxidation throughout the catalyst length, with the corresponding 2:1 $H_2O$:$CO_2$ product stoichiometry, as shown in Fig. 24Ba. Figure 24Bb and c shows that under rich conditions corresponding to the transient high-conversion branch (branch B in Fig. 24T), multiple reaction zones exist along the catalyst, and multiple reactions can occur including total (R1) and partial (R2) oxidation, steam reforming (R3), and water–gas shift (R4). Even under net-rich conditions, a region exists at the catalyst front dominated by total oxidation. Reactions R2–4 occur in the catalyst back, and a mixed zone exists in the middle where all reactions can occur; the mixed zone spatial start and end locations are defined as corresponding to the onset of $H_2$ production (via R2–4) and depletion of gas-phase $O_2$, respectively. As the rich-regime $CH_4$ oxidation transitions through multiple high-conversion branches toward the steady-state branch, Fig. 24Bb and c shows a change in the spatial extent of the various zones and a general increase in products associated with R2–4. Notably, CO increases throughout the catalyst back where gas-phase $O_2$ is depleted, indicating partial oxidation

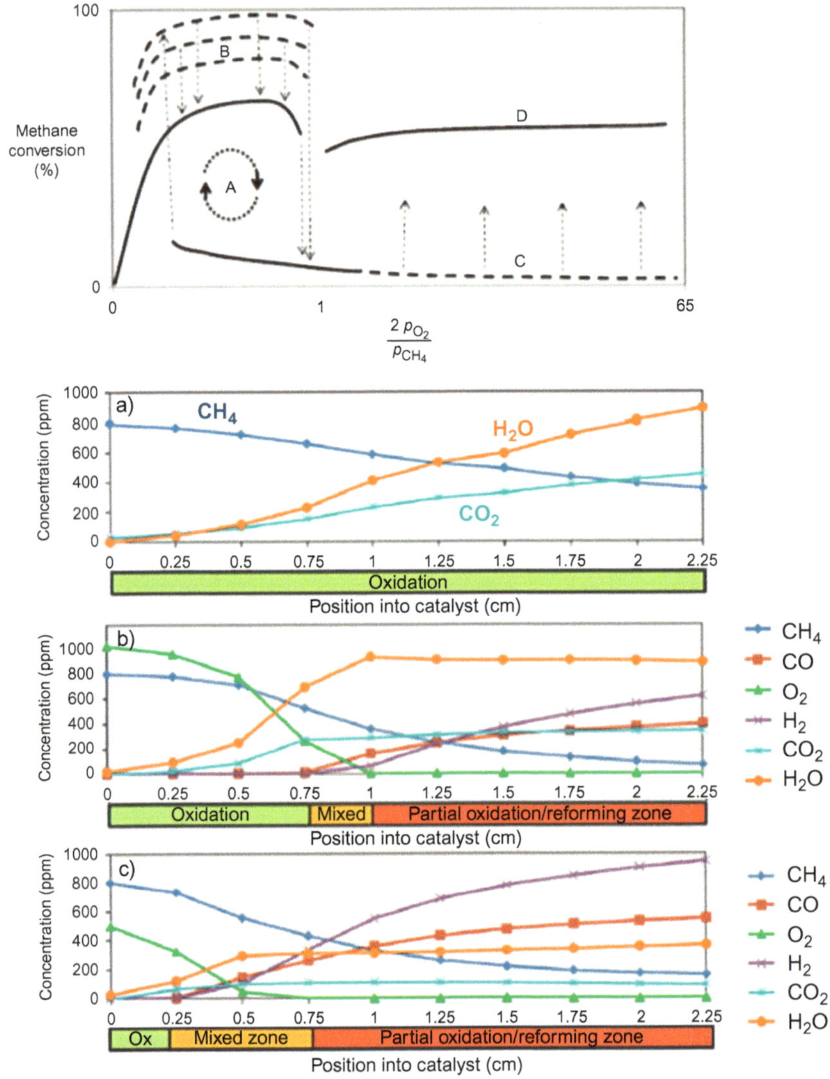


**Fig. 24** Spatiotemporal nature of $CH_4$ oxidation over a $Pt/Pd/Al_2O_3$ catalyst (Bugosha et al., 2015). (*Top*: T) Map of $CH_4$ conversion vs $O_2$ showing transient and steady operation states, and hysteresis. (*Bottom*: B) Spatially resolved species distributions for operation in three different high-conversion branches "B" in Fig. 24L and at a fixed $O_2$ and $CH_4$ concentrations; i.e., fixed $2p_{O_2}/p_{CH_4}$; higher (Ba), medium (Bb), and lower (Bc) conversion branches are shown all within the high-conversion branch "B" of Fig. 24L.

and/or steam reforming. The evolving increase in CO with continued approach to the steady-state branch despite constant $H_2O$ concentration, indicates partial oxidation in this $O_2$-free region; additional SpaciMS measurements not shown here corroborate this conclusion. Enabled by the SpaciMS insights, they conclude that partial oxidation in this back section proceeds via stored bulk oxygen which diffuses to the PdO. With increasing time on stream, bulk oxygen is progressively depleted, active PdO is converted to inactive metallic Pd, and $CH_4$ conversion correspondingly degrades. For similar reasons, numerous transient branches of varying high $CH_4$ conversion can occur after lightoff depending on the previous temperature and $O_2$ exposure, and thus inventory of bulk or lattice oxygen and PdO. Similarly, they attributed the lean-regime transition between the metastable (C) and stable (D) branches to transition of inactive metallic Pd to active PdO with continued $O_2$ exposure, and possibly temperature variations to enhance PdO formation. Furthermore, SpaciMS was used to investigate the role of CO oxidation, its competition with $H_2$ as a function of temperature, and influence on the lean–rich-transition lightoff process and corresponding spatial reaction distributions along the catalyst axis. This work provides an excellent example of SpaciMS providing detailed insights regarding the evolution of reaction pathways, and origins of complex $CH_4$ oxidation over a range of conditions expected in transportation applications.

SpaciMS has been used to resolve the distribution of periodic kinetic oscillations that only exist inside an automotive monolithic catalyst, and which are not apparent in the effluent (Sá et al., 2010). In addition to the intracatalyst oscillating reactivity phenomenon, this work provides a comprehensive SpaciMS historical summary, describes an advanced SpaciMS for species and temperature mapping similar to but preceding the commercial Hiden SpaciMS, and provides experimental and numerical analysis concluding that the capillary and thermocouple probes are effectively noninvasive for the studied application. Although Koci et al. had predicted periodic and distributed oscillations to exist within automotive monolithic catalysts based on modeling (Kočí et al., 2004a, b), measurements of the same in a monolith-supported catalyst had apparently not been demonstrated. SpaciMS was applied to map variations in the kinetic oscillations associated with CO oxidation throughout the 10-cm long Pt-Rh/γ-$Al_2O_3$ catalyst. Figure 25 shows how the oscillations build up in the catalyst front, and slowly decayed beyond the point of CO depletion at

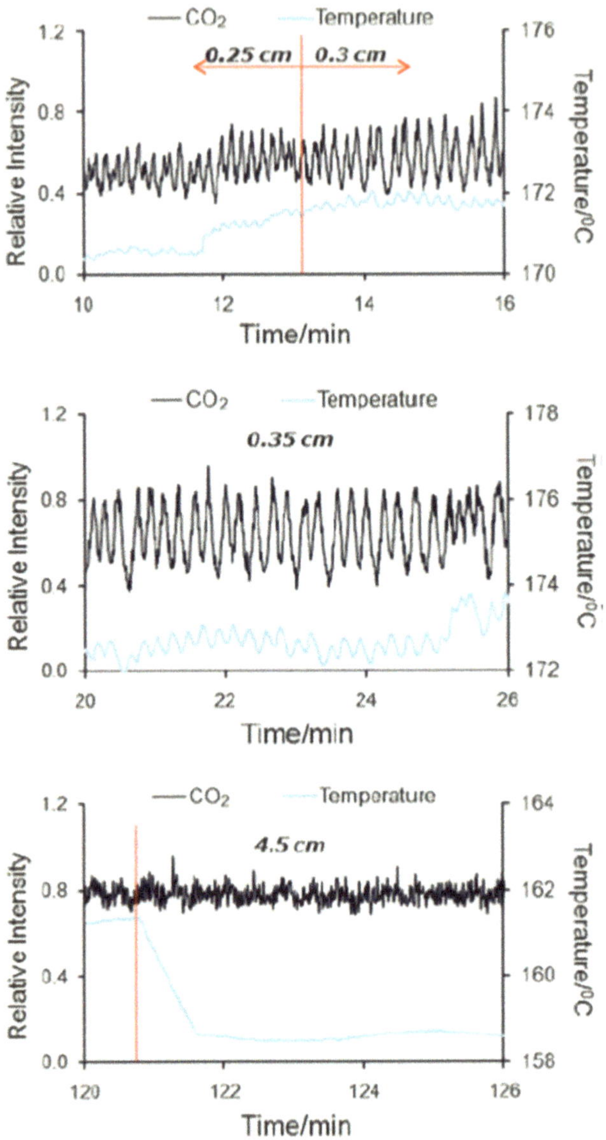


**Fig. 25** Intracatalyst periodic kinetic oscillations associated with CO oxidation over a 10-cm long, monolith-supported Pt-Rh/γ-$Al_2O_3$ catalyst (Sá et al., 2010).

c.0.55 cm, to the point where the concentrations are temporally steady at the c. half-catalyst location; clearly these oscillations are not observed in the catalyst effluent. The oscillations varied with CO:$O_2$ ratio, but no single relationship between oscillation frequency and amplitude and conditions was apparent. However, the phase of the $O_2$, CO, $CO_2$, and temperature dynamics were consistent with the conversion oscillations being driven by corresponding high- and low-CO coverage variations inducing low and high activity, respectively. The behavior suggests a highly dynamic system where the balance between reactive species and active sites can be disrupted by small variations in temperature and/or species concentration; this may be particularly relevant to automotive catalyst applications which are highly dynamic and cover a broad range of exhaust compositions, which might induce such a balance disruption. Presumably such dynamics would lengthen the reaction zone or catalyst length necessary to achieve conversion, compared to a condition where such dynamics did not exist.

In other work, a group at Queen's University Belfast applied SpaciMS to validate a dual-kinetic model under conditions where the transition between mass transfer and kinetic limited reaction is relevant, and which typically exist only inside the catalyst near the front (Stewart et al., 2014a, b). For a given application, their dual-kinetic model allows users to select which reactions are modeled via global kinetic, and which via microkinetics. Although the examples shown here use independent assessment of the global- and microkinetic model sections, their publications provide examples of mixed partitioning of reactions between the two methods. The salient point is that flawed models can give correct effluent results, and that an effluent-limited perspective can obscure model limitations and thus hinder technology advancement. Figure 26 shows such an example, where model and measurement results match deep into the catalyst (Fig. 26M) and in the effluent, but give dramatically different results in the front catalyst regions (Fig. 26T). Although, the models do not follow the measurement transient, they do correctly predict the lightoff temperature and terminal conversion, indicating that the kinetic parameters are correct. In this case, the disparity was due to the reaction being mass transfer limited in the catalyst front, and kinetic limited in the back, with a transition region in between. By incorporating a balance factor to manage the transition from mass transfer to reaction kinetics limited regions, the model accuracy was significantly improved as shown in Fig. 26B. In addition to the dramatic model improvements apparent at the catalyst front, incorporating the balance factor improved model performance deeper into the catalyst, apparently due to integral effects. Without intracatalyst measurements and validation, this model limitation and the opportunity to improve its accuracy and robust nature may not have been recognized. Such an improvement as better managing transitions between different limiting processes is

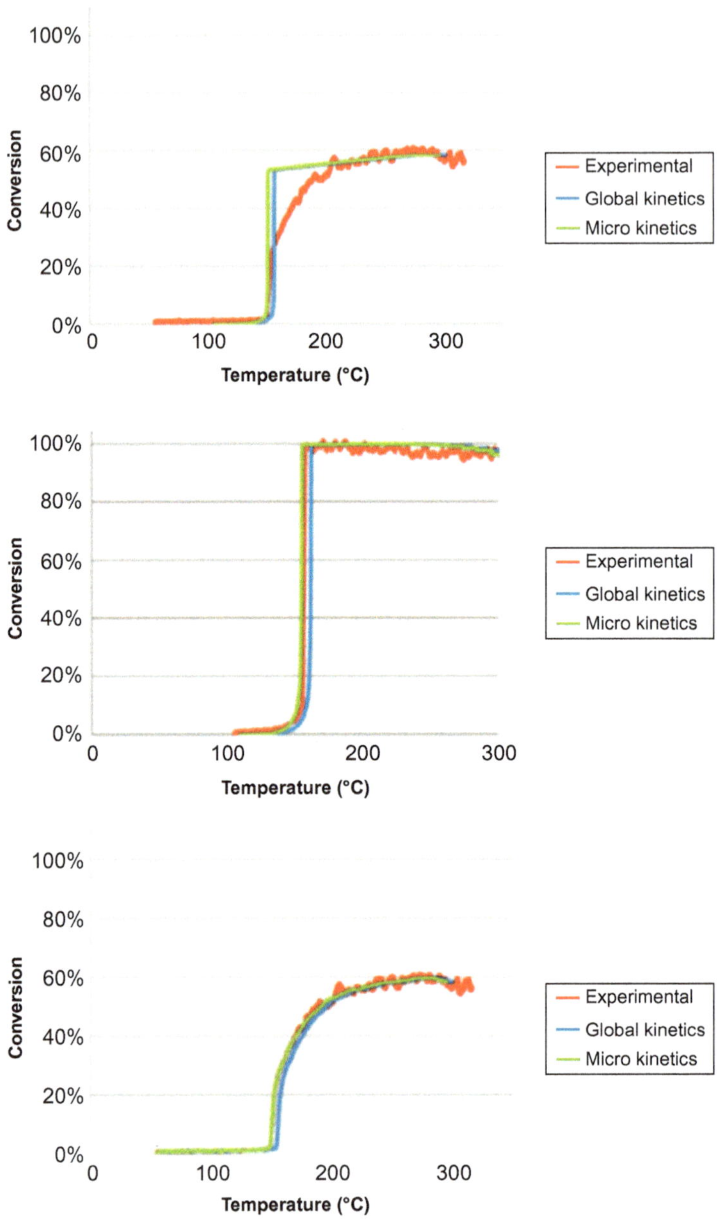


**Fig. 26** Comparison of results from global- and microkinetics models with SpaciMS measurements for CO oxidation over a 30-mm long, Pt-based monolith-supported catalyst (Stewart et al., 2014b). (*Top*: T) 2 mm from inlet face using models without limiting term balance parameter for mass transfer and reaction kinetics. (*Bottom*: B) 26 mm from inlet face using models without balance parameter. (*Bottom*: B) 2 mm from inlet face using models with the balance parameter.

particularly relevant to catalysis in the highly dynamic transportation application. As emphasized in the referenced work, the limiting factor in the cold-start FTP (Federal Test Procedure; a standard drive cycle for emissions compliance assessment) switches from reaction kinetics to mass transfer after c.30 s; thus accurate automotive catalysis models must manage these limiting process transitions in this uniquely dynamic application. SpaciMS provides data and insights for critically assessing automotive catalysis models, and enabling technology advancement.

## 6. PARTICULATE FILTER APPLICATIONS

Although wall-flow particulate filters are increasingly common for both diesel and gasoline applications, there have been very few studies of intrafilter distributed performance under operating conditions. Besides the SpaciMS study described below, there is only one other known study using intrafilter physical probes to resolve performance distributions. That study used c.2.75-mm OD sample probe introduced from the filter exit to extract all flow upstream of the probe tip by controlling the sampling rate via measurement of a tracer injected just downstream of the probe tip (Ganapathi-subbu and Collings, 2003); variations in the sample flow were used to determine axial wall-flow distributions of a clean filter substrate, and it was suggested that the method could be used to determine axial flow distributions of a soot-loaded filter. Radiofrequency is a nonphysical-probe-based sensing technique that has been demonstrated for detecting integral variations in soot loading under full engine-scale operating conditions, and spatial resolution is possible by sweeping RF resonant modes (Feulner et al., 2013; Sappok et al., 2010).

The filter application-specific SpaciMS considerations are related to use of an internal standard and avoiding soot sampling into the capillary. Because the flow rates of direct capillary sampling, as used in the SpaciMS application discussed below, are orders of magnitude less than the typical convective flow rates, we have not observed particulate in the bulk flow being redirected and sampled into the SpaciMS capillaries. The typical method of soot induction results from translating and particularly pushing the capillary tip along a soot-laden catalyst surface; indeed, this was an early lesson learned from the first engine-based SpaciMS application. Thus, while soot induction may be avoided by purging the sampling capillary during translation, a preferred method is to use an array of fixed capillaries and a multiport valve for location-selective sampling. In practice, the capillaries are ideally purged during positioning in a clean filter, and then fixed during soot loading and regeneration. Because varying filter back pressure with soot load can impact the SpaciMS sampling rate, an internal standard is useful for implementing associated corrections; this standard can be an actively injected inert species such as He, or using a naturally abundant species such as Ar in air. The SpaciMS example below uses both an array of multiple fixed capillaries and naturally abundant Ar as an internal standard.

Cummins and ORNL applied SpaciMS and a mass flow model to study distributed $NO_2$ utilization and the evolving soot distributions during passive regeneration of a dual-element partial-flow filter (PFF) in a full-scale engine and catalyst system (Cunningham et al., 2017). Figure 27T shows the catalyst layout and method for SpaciMS capillary introduction; an oxidation catalyst (DOC) generated $NO_2$ to drive passive regeneration of the two downstream PFFs. The PFFs have both through and plugged channels, are in a rear-plugged orientation, and were arrayed with 10 capillaries each for SpaciMS measurements. Figure 27M shows $NO_2$ utilization reflecting the progress and distribution of passive regeneration in the plugged channels. The dashed black line in Fig. 27M is based on measurement of a nonreacting species, and the $NO_2$ distributions converge to this reference distribution as regeneration approaches completion. Regeneration is relatively uniformly distributed throughout PFF-1 at early DeSoot times, proceeds fastest at the PFF-1 front, and is practically complete in the front $1/5L$–$2/5L$ of PFF-1 after 5 h. Regeneration is much more uniformly distributed and proceeds more slowly in PFF-2. Similar distributions were presented for the through channels, but are not shown here. Figure 27B shows the calculated net soot-loading distribution throughout passive regeneration, based on the $NO_2$ mass flow measurements; the net soot is that in both the through and plugged channels. It is clear that the soot loading and gradients are greater in PFF-1. Passive regeneration proceeds much more slowly in the back of each filter, but is faster in the front of PFF-2 than the back of PFF-1. The results indicate that it is very difficult to remove soot from the rear-PFF regions via passive regeneration, but that two series PFFs provide a benefit of enhanced regeneration in the front PFF-2 sections relative to the back of PFF-1. This SpaciMS application provides practical insights for optimizing the design of a PFF system and is applicable to more complex catalyst such as SCR-on-filter.

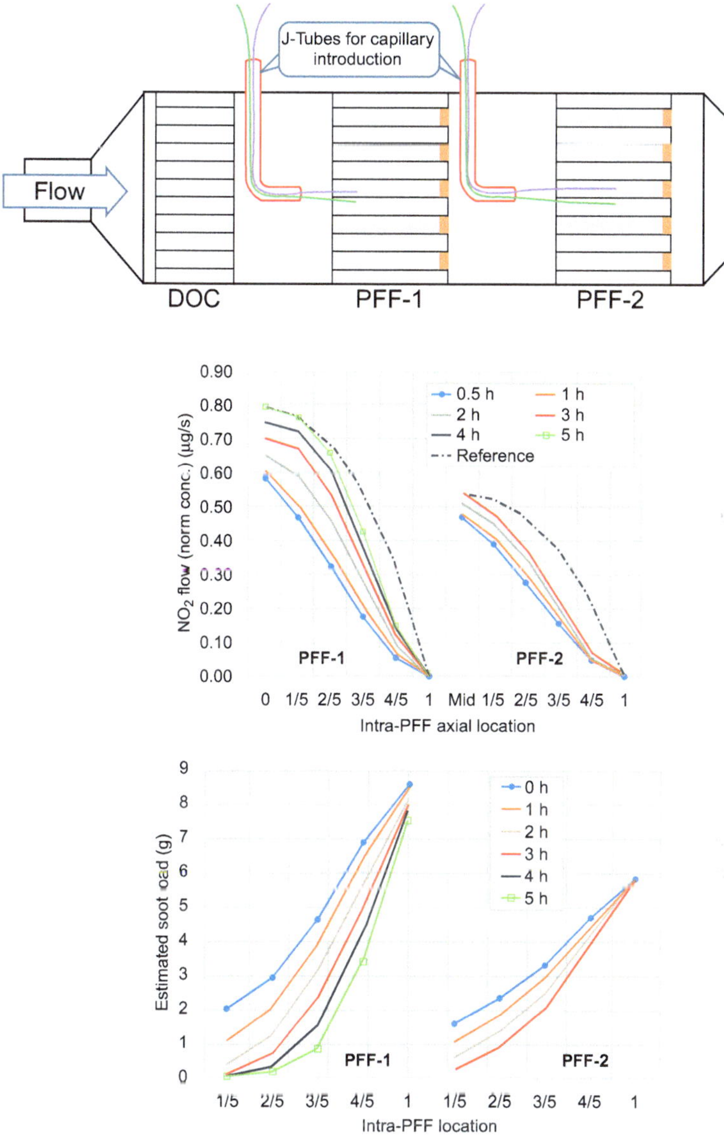


**Fig. 27** Distributed $NO_2$-based passive regeneration of a dual-element PFF (Cunningham et al., 2017). (*Top*: T) Catalyst sequence and schematic, and SpaciMS capillary orientation. (*Middle*: M) $NO_2$ mass flow distribution in the plugged channels of PFF-1 and PFF-2 for various passive regeneration times. (*Bottom*: B) Spatiotemporal variations in soot-loading distribution through the dual-PFF system at various passive regeneration times.

## 7. SUMMARY AND CONCLUSIONS

Although SpaciMS is becoming increasingly mature, areas of potential improvement remain, particularly for automotive catalysis applications in confined monolith channels with transient and complex gas mixtures. This includes methods for designing application-specific sampling configurations with optimized balance between various figures of merit including sensitivity, noninvasive nature, and transient response. Applications with complex gas mixtures and particularly with realistic hydrocarbons can benefit from improved instrument techniques; beyond the strategies discussed herein using specific $m/z$ combinations, diluents (Ar vs $N_2$), and electron-ionization potential tuning, increased application and development of chemical ionization (Sharma and Dearth, 2016) and time-of-flight (Goguet, n.d.) appear promising pathways for addressing these challenges. While internal standards are often implemented to account for varying sampling rate (e.g., due to temperature, viscosity, etc., induced variations), the techniques could be more robust across a broader range of $m/z$ and signal magnitude (concentration) values. Automotive catalysis is inherently transient and many catalyst parameters (e.g., $NH_3$ capacity) are determined via transient analysis, and thus advances that improve SpaciMS transient response are beneficial. Heating to mitigate interactions of polar species with instrument surfaces, which degrades temporal response, has been discussed herein, but further improvement would be beneficial. Strongly oxidizing and reducing species can change the instrument performance causing species-specific impacts; for instance, ionization filament oxidation and reduction can occur over timescales that complicate transient analysis, and while specific filament coatings and formulations are available to mitigate this, in practice the impacts can remain significant in certain applications. Primarily though, SpaciMS and intracatalyst diagnostics more generally will benefit from broader adoption and use of these techniques. As with the relatively small SpaciMS community, where the different users added their unique twists, configurations, and interpretations, so broader adoption will further hasten SpaciMS technology advances, utility and application-related catalyst insights.

The many SpaciMS applications described in this chapter highlight their unique ability to advance a broad range of automotive catalyst technologies. By enabling the spatial and temporal evolution of intracatalyst reactions to be measured under operating conditions, SpaciMS established a new watershed capability and perspective for understanding and advancing automotive catalysis. The different highlighted studies demonstrate specific insights enabled by SpaciMS related to elucidating the network, sequence, and relative rates of reaction during catalyst operation; for instance, the relative rates of CO reaction via $NO_x$ reduction, oxidation, and water–gas shift; species release, redistribution and readsorption; speeds of reductant fronts, and spatial selectivity variations relative to the reductant front location; and reaction-specific sulfation distribution and advancement. In other examples, SpaciMS has allowed the study of phenomena that exist only within the catalyst; these included the intermediate-$NH_3$ pathway for LNT regeneration; oscillations, transients, and hysteresis; and transitions between mass transfer and kinetic limitations. These SpaciMS studies have provided unique basic and practical insights such as the varying sensitivity of different reactions to sulfation; the identity and origin of species involved in $N_2O$ formation and unsteady, transient, or oscillating performance; and ideas for catalyst state assessment for managing LNT desulfation and SCR dosing for enhanced durability and performance. The resulting data and knowledge provide a very advanced ability to critically assess the structure and function of catalyst models in a detailed way, via its ability to accurately predict the spatial and temporal evolution of reactions along the catalyst. One example demonstrated how flawed models can give correct effluent results, and how an effluent-limited perspective can obscure model limitations and thus hinder technology advancement. In that example, SpaciMS made obvious the need for model improvements related to distributed transition between reaction limiting conditions, which were not apparent in the effluent; the resulting model improvement not only enhanced accuracy in the transition and intracatalyst regimes, but also in the effluent. In other modeling examples, SpaciMS was critical to understanding variations in lattice-oxygen inventory as the origin of complex transient catalyst behavior. By enabling more robust, accurate and reliable models, SpaciMS enhances analysis-led-design for faster and lower-cost development of advanced engine catalyst systems. Lower-order versions of these models and associated concepts are applicable to on-road diagnostic and control; similarly, SpaciMS enables virtual and physical sensors, and knowledge of what/when/where to measure specific parameters for implementing advanced control. SpaciMS has also been applied to enhance other diagnostics and has motivated development of other intracatalyst analytical techniques. Indeed, the numerous examples in this and the other chapters herein are a testament to the significance of spatiotemporally resolved intracatalyst diagnostics for advancing automotive catalysis. Through these manifold paths, SpaciMS comprehensively enables automotive systems with advanced performance, efficiency, cost, and durability, and the associated R&D, societal, economical, and environmental benefits.

## ACKNOWLEDGMENTS

The authors thank the U.S. Department of Energy, Vehicle Technologies Office for their long-term support of our research, and related SpaciMS development and applications. We are grateful for our excellent partnership with Cummins Inc., within which the SpaciMS development took place, and particularly Neal Currier, Mike Cunningham, and Alex Yezerets. We gladly acknowledge our partners at Queen's University Belfast, Hiden Analytical, and Y12 National Security Complex for their role in SpaciMS commercialization and the 2008 R&D100 Award, and our university partners in the many SpaciMS applications. We appreciate and recognize the support of our colleagues at ORNL, and notably Pam Olszewski for her administrative help in preparing this chapter.

*Notice*: This manuscript has been authored by UT-Battelle, LLC, under Contract No. DE-AC0500OR22725 with the U.S. Department of Energy. The United States Government retains and the publisher, by accepting the chapter for publication, acknowledges that the United States Government retains a nonexclusive, paid-up, irrevocable, world-wide license to publish or reproduce the published form of this chapter, or allow others to do so, for the United States Government purposes. The Department of Energy will provide public access to these results of federally sponsored research in accordance with the DOE Public Access Plan (http://energy.gov/downloads/doe-public-access-plan).

## REFERENCES

Auvray X, Partridge WP, Choi J-S, et al: Local ammonia storage and ammonia inhibition in a monolithic copper-beta zeolite SCR catalyst, *App Catal B Environ* 126:144–152, 2012. https://doi.org/10.1016/j.apcatb.2012.07.019.

Auvray X, Partridge W, Choi J-S, et al: Kinetic modeling of $NH_3$-SCR over a supported Cu zeolite catalyst using axial species distribution measurements, *Appl Catal Environ* 163:393–403, 2015. https://doi.org/10.1016/j.apcatb.2014.08.003.

Bär M, Zülicke C, Eiswirth M, Ertl G: Theoretical modeling of spatiotemporal self-organization in a surface catalyzed reaction exhibiting bistable kinetics, *J Chem Phys* 96:8595–8604, 1992. https://doi.org/10.1063/1.462312.

Bártová Š, Kočí P, Marek M, et al: New insights on $N_2O$ formation pathways during lean/rich cycling of a commercial lean $NO_x$ trap catalyst, *Catal Today* 231:145–154, 2014. https://doi.org/10.1016/j.cattod.2013.11.05010.1016/j.cattod.2013.11.050.

Bártová Š, Mráček D, Kočí P, Marek M, Choi J-S: Ammonia reactions with the stored oxygen in a commercial lean $NO_x$ trap catalyst, *Chem Eng J* 278:199–206, 2015. https://doi.org/10.1016/j.cej.2014.09.115.

Beck DD, Sommers JW, DiMaggio CL: Axial characterization of catalytic activity in close-coupled lightoff and underfloor catalytic converters, *Appl Catal A Gen* 11:257–272, 1997a. https://doi.org/10.1016/S0926-3373(96)00050-1.

Beck DD, Sommers JW, DiMaggio CL: Axial characterization of oxygen storage capacity in close-coupled lightoff and underfloor catalytic converters and impact of sulfur, *Appl Catal A Gen* 11:273–290, 1997b. https://doi.org/10.1016/S0926-3373(96)00051-3.

Bhatia D, Clayton RD, Harold MP, Balakotaiah V: A global kinetic model for $NO_x$ storage and reduction on $Pt/BaO/Al_2O_3$ monolithic catalysts, *Catal Today* 147S:S250–S256, 2009. https://doi.org/10.1016/j.cattod.2009.07.024.

Brandenberger S, Kröcher O, Tissler A, Althoff R: The state of the art in selective catalytic reduction of $NO_x$ by ammonia using metal-exchanged zeolite catalysts, *Catal Rev* 50(4):492–531, 2008. https://doi.org/10.1080/01614940802480122.

Bugosha GS, Easterling VG, Rusakova IA, Harold MP: Anomalous steady-state and spatio-temporal features of methane oxidation on $Pt/Pd/Al_2O_3$ monolith spanning lean and rich conditions, *Appl Catal Environ* 165:68–78, 2015. https://doi.org/10.1016/j.apcatb.2014.09.058.

Carlsson P-A, Österlunda L, Thormählena P, et al: A transient in situ FTIR and XANES study of CO oxidation over $Pt/Al_2O_3$ catalysts, *J Catal* 226:422–434, 2004. https://doi.org/10.1016/j.jcat.2004.06.009.

Castoldi L, Nova I, Lietti L, Forzatti P: Study of the effect of Ba loading for catalytic activity of $Pt\text{-}Ba/Al_2O_3$ model catalysts, *Catal Today* 96:43–52, 2004. https://doi.org/10.1016/j.cattod.2004.05.006.

Chatterjee D, Burkhardt T, Weibel M, Tronconi E, Nova I, Ciardelli C: *Numerical simulation of $NO/NO_2/NH_3$ reactions on SCR-catalytic converters: model development and applications*, SAE Technical Paper Series, SAE Paper 2006-01-0468, 2006. https://doi.org/10.4271/2006-01-0468.

Choi J-S, Partridge WP, Daw CS: Spatially resolved in situ measurements of transient species breakthrough during cyclic, low temperature regeneration of a monolithic $Pt/K/Al_2O_3$ $NO_x$ storage reduction catalyst, *Appl Catal A Gen* 293:24–40, 2005. https://doi.org/10.1016/j.apcata.2005.06.025.

Choi JS, Partridge WP, Epling WS, Currier NW, Yonushonis TM: Intra-channel evolution of carbon monoxide and its implication for the regeneration of monolithic $Pt/K/Al_2O_3$ $NO_x$ storage-reduction catalyst, *Catal Today* 114:102–111, 2006. https://doi.org/10.1016/j.cattod.2006.02.011.

Choi JS, Partridge WP, Daw CS: Sulfur impact on $NO_x$ storage, oxygen storage and ammonia breakthrough during cyclic lean/rich operation of a commercial lean $NO_x$ trap, *Appl Catal Environ* 77:145–156, 2007. https://doi.org/10.1016/j.apcatb.2007.07.025.

Choi JS, Partridge WP, Pihl JA, Daw CS: Sulfur and temperature effects on the spatial distribution of reactions inside a lean $NO_x$ trap and resulting changes in global performance, *Catal Today* 136:173–182, 2008a. https://doi.org/10.1016/j.cattod.2008.01.008.

Choi J-S, Partridge B, Pihl J, et al: In *Sulfation impact on LNT performance: experimental studies with Umicore CLEERS reference catalyst*, 11th DOE crosscut workshop on lean emissions reduction simulation, University of Michigan, Dearborn, MI, 2008b. http://cleers.org/workshops/workshop11.

Choi J-S, Partridge WP, Lance MJ, et al: Nature and spatial distribution of sulfur species in a sulfated barium-based commercial lean $NO_x$ trap catalyst, *Catal Today* 151:354–361, 2010. https://doi.org/10.1016/j.cattod.2010.01.016.

Choi J-S, Partridge WP, Pihl JA, Kim M-Y, Kočí P, Daw CS: Spatiotemporal distribution of $NO_x$ storage and impact on $NH_3$ and $N_2O$ selectivities during lean/rich cycling of a Ba-based lean $NO_x$ trap catalyst, *Catal Today* 184:20–26, 2012. https://doi.org/10.1016/j.cattod.2011.11.007.

Choi J-S, Pihl JA, Toops TJ, Partridge WP: Reduction of stored $NO_x$ with $CO/H_2$ and hydrocarbons: a spatial resolution analysis. In Lietti L, Castoldi L, editors: *$NO_x$ trap catalysts and technologies: fundamentals and industrial applications*, Cambridge, UK, 2017, The Royal Society of Chemistry, in press.

Ciardelli C, Nova I, Tronconi E, et al: Reactivity of $NO/NO_2\text{-}NH_3$ SCR system for diesel exhaust after treatment: identification of the reaction network as a function of temperature and $NO_2$ feed content, *Appl Catal Environ* 70:80–90, 2005. https://doi.org/10.1016/j.apcatb.2005.10.041.

Clayton RD, Harold MP, Balakotaiah V: $NO_x$ storage and reduction with $H_2$ on $Pt/BaO/Al_2O_3$ monolith: spatio-temporal resolution of product distribution, *Appl Catal Environ* 84:616–630, 2008. https://doi.org/10.1016/j.apcatb.2008.05.018.
Crocker M, Easterling V, Wang J, Dearth M, McCabe RW, Theis J: Laboratory product speciation studies of the LNT + in-situ SCR $NO_x$ emission control concept. In *Directions in engine-efficiency and emissions research (DEER)*, Detroit, MI, 2010, Emissions Control Technologies September 28, 2010. https://energy.gov/sites/prod/files/2014/03/f8/deer10_crocker.pdf.
Cumaranatunge L, Mulla SS, Yezerets A, Currier NW, Delgass WN, Ribeiro FH: Ammonia is a hydrogen carrier in the regeneration of $Pt/BaO/Al_2O_3$ $NO_x$ traps with $H_2$, *J Catal* 246:29–34, 2007. https://doi.org/10.1016/j.jcat.2006.11.008.
Cunningham M, Kim M-Y, Lakkireddy V, Partridge W: In *Axial $NO_2$ utilization measurements within a partial flow filter during passive regeneration*, SAE 2017 world congress, Detroit, MI, 2017. SAE Technical Paper Series, SAE Paper 2017-01-0988. https://doi.org/10.4271/2017-01-0988.
Currier NW, Yezerets A: *Mass spectrometry system and method,* US Patent US7211793 B2; Publication Date, May 1, 2007.
Easterling V, Ji Y, Crocker M, Dearth M, McCabe WR: Application of SpaciMS to the study of ammonia formation in lean $NO_x$ trap catalysts, *Appl Catal Environ* 123–124:339–350, 2012. https://doi.org/10.1016/j.apcatb.2012.05.002.
Epling W. S., University of Houston, personal communication to W.P.P. 2015.
Epling WS, Campbell LE, Yezerets A, Currier NW, Parks JE: Overview of the fundamental reactions and degradation mechanisms of NOx storage/reduction catalysts, *Catal Rev* 46:163–245, 2004. https://doi.org/10.1081/CR-200031932.
Feulner M, Hagen G, Piontkowski A, et al: In-operation monitoring of the soot load of diesel particulate filters: initial tests, *Top Catal* 56:483–488, 2013. https://doi.org/10.1007/s11244-013-0002-9.
Ganapathi-subbu S, Collings N: Techniques for measuring wall flow along the channels of a diesel particulate filter, *Meas Sci Technol* 14:1783–1793, 2003. https://doi.org/10.1088/0957-0233/14/10/307.
A. Goguet n.d. A. Goguet at Queen's University Belfast has developed a TOF-based SpaciMS; https://reactatqub.wordpress.com/xiao-sha/.
Goguet A, Meunier FC, Tibiletti D, Breen JP, Burch R: Spectrokinetic investigation of reverse water-gas-shift reaction intermediates over a $Pt/CeO_2$ catalyst, *J Phys Chem B* 108:20240–20246, 2004. https://doi.org/10.1021/jp047242w.
Goguet A, Stewart C, Touitou J, Morgan K: In-situ spatially resolved techniques for the investigation of packed bed catalytic reactors: current status and future outlook of spaci-FB. In Dixon AG, Deutschmann O, editors: *ACHEM: spatially-resolved operando measurements in heterogeneous catalytic reactors*, 2017, The Royal Society of Chemistry, pp 131–160.
Harold M, University of Houston, personal communication to W.P.P. 2017.
Heeb NV, Saxer CJ, Forss A-M, Bruhlmann S: Trends of NO-, $NO_2$-, and $NH_3$-emissions from gasoline-fueled euro-3-to euro-4-passenger cars, *Atmos Environ* 42:2543–2554, 2008. https://doi.org/10.1016/j.atmosenv.2007.12.008.
Henry C, Kamasamudram K, Currier N, Yezerets A, Castagnola M, Chen HY: *Axially resolved performance of Cu-zeolite SCR catalysts,* SAE Technical Paper Series, SAE Paper 2012-01-1084, 2012. https://doi.org/10.4271/2012-01-1084.
Hettel M, Diehm C, Torkashvand B, Deutschmann O: Critical evaluation of in situ probe techniques for catalytic honeycomb monoliths, *Catal Today* 216:2–10, 2013. https://doi.org/10.1016/j.cattod.2013.05.005.
hidenanalytical, n.d. http://www.hidenanalytical.de/wp-content/uploads/pdf/$NH_3$ Detection in $H_2O$ vapour app note 288.pdf.

Hiden Analytical, ORNL, n.d. The SpaciMS R&D100 award team included Cummins Inc, Queens University Belfast, and Y-12 National Security Complex; (http://www.rd100conference.com/awards/winners-finalists/702/more-complete-catalyst-chemistry/).

Hiden Analytical, SpaciMS: *A system for determination of intra catalyst reactor chemistry,* http://www.hidenanalytical.com/products/for-catalysis-and-thermal-analysis/spacims/.

Horn R, Degenstein NJ, Williams KA, Schmidt LD: Spatial and temporal profiles in millisecond partial oxidation processes, *Catal Lett* 110:169–178, 2006. https://doi.org/10.1007/s10562-006-0117-8.

Hou X, Schmieg SJ, Li W, Epling WS: *Cu-zeolite SCR catalyst thermal deactivation studied with FTIR spatial resolution,* SAE Technical Paper Series, SAE Paper 2011-01-1138, 2011. https://doi.org/10.4271/2011-01-1138.

Hou X, Schmieg SJ, Li W, Epling WS: $NH_3$ pulsing adsorption and SCR reactions over a Cu-CHA SCR catalyst, *Catal Today* 197:9–17, 2012. https://doi.org/10.1016/j.cattod.2012.05.041.

Irani K, Epling WS, Blint R: Spatial resolution of reactant species consumption in diesel oxidation catalysts, *Top Catal* 52:1856–1859, 2009a. https://doi.org/10.1007/s11244-009-9349-3.

Irani K, Epling WS, Blint R: Effect of hydrocarbon species on no oxidation over diesel oxidation catalysts, *Appl Catal Environ* 92:422–428, 2009b. https://doi.org/10.1016/j.apcatb.2009.08.022.

Jacobs G, Davis BH: Low temperature water-gas shift: applications of a modified SSITKA-DRIFTS method under conditions of $H_2$ co-feeding over metal/ceria and related oxides, *Appl Catal A Gen* 333:192–201, 2007. https://doi.org/10.1016/j.apcata.2007.07.029.

Ji Y, Choi J-S, Toops TJ, Crocker M, Naseri M: Influence of ceria on the $NO_x$ storage/reduction behavior of lean $NO_x$ trap catalysts, *Catal Today* 136:146–155, 2008. https://doi.org/10.1016/j.cattod.2007.11.059.

Kamasamudram K, Currier NW, Chen X, Yezerets A: Overview of the practically important behaviors of zeolite-based urea-SCR catalysts, using compact experimental protocol, *Catal Today* 151:212–222, 2010. https://doi.org/10.1016/j.cattod.2010.03.055.

Kim M-Y, Choi J-S, Crocker M: Roles of $C_3H_6$ in $NH_3$ generation and $NO_x$ reduction over a Cu-chabazite SCR catalyst under lean/rich cycling conditions, *Catal Today* 231:90–98, 2014. https://doi.org/10.1016/j.cattod.2013.12.021.

Kočí P., University of Chemistry and Technology, Prague, personal communication to W.P.P. 2017.

Kočí P, Nevoral V, Záhrubský M, Kubíček M, Marek M: Nonlinear dynamics of automobile exhaust gas converters: the role of nonstationary kinetics, *Chem Eng Sci* 59:5597–5605, 2004a. https://doi.org/10.1016/j.ces.2004.07.115.

Kočí P, Kubíček M, Marek M: Modeling of three-way-catalyst monolith converters with microkinetics and diffusion in the washcoat, *Ind Eng Chem Res* 43:4503–4510, 2004b. https://doi.org/10.1021/ie034137k.

Kočí P, Bártová Š, Mráček D, et al: Effective model for prediction of $N_2O$ and $NH_3$ formation during the regeneration of $NO_x$ storage catalyst, *Top Catal* 56:118–124, 2013. https://doi.org/10.1007/s11244-013-9939-y.

Lambert CK, Cheng Y, Dobson D, et al: *Post mortem of an aged tier 2 light-duty diesel truck aftertreatment system,* SAE Technical Paper Series, SAE Paper 2009-01-2711, 2009. https://doi.org/10.4271/2009-01-2711.

Li J, Currier NW, Yezerets A, Chen HY, Hess HS: *Evaluation of spatially resolved performance of $NO_x$ adsorber catalysts,* SAE Technical Paper Series, SAE Paper 2009-01-0275, 2009. https://doi.org/10.4271/2009-01-0275.

Li M, Easterling VG, Harold MP: Spatio-temporal features of the sequential $NO_x$ storage and reduction and selective catalytic reduction reactor system, *Catal Today* 267:177–191, 2016. https://doi.org/10.1016/j.cattod.2016.01.020.

Liu Y, Harold MP, Luss D: Spatio-temporal features of periodic oxidation of $H_2$ and CO on Pt/$CeO_2$/$Al_2O_3$, *Appl Catal A Gen* 397:35–45, 2011. https://doi.org/10.1016/j.apcata.2011.02.008.

Luo T, Gorte RJ: A mechanistic study of sulfur poisoning of the water-gas-shift reaction over Pd/ceria, *Catal Lett* 85:139, 2003. https://doi.org/10.1023/A:1022125209602.

Luo JY, Hou X, Wijayakoon P, Schmieg SJ, Li W, Epling WS: Spatially resolving SCR reactions over a Fe/zeolite catalyst, *Appl Catal Environ* 102:110–119, 2011a. https://doi.org/10.1016/j.apcatb.2010.11.031.

Luo JY, Al-Harbi M, Pang M, Epling WS: Spatially resolving LNT desulfation: re-adsorption induced by oxygen storage materials, *Appl Catal Environ* 106:664–671, 2011b. https://doi.org/10.1016/j.apcatb.2011.06.033.

Luo JY, Oh H, Henry C, Epling W: Effect of $C_3H_6$ on selective catalytic reduction of $NO_x$ by $NH_3$ over a Cu/zeolite catalyst: a mechanistic study, *Appl Catal Environ* 123–124:296–305, 2012. https://doi.org/10.1016/j.apcatb.2012.04.038.

Morgan K, Touitou J, Choi J-S, et al: Evolution and enabling capabilities of spatially resolved techniques for the characterization of heterogeneously catalyzed reactions, *ACS Catal* 6:1356–1381, 2016. https://doi.org/10.1021/acscatal.5b02602.

Mráček D, Kočí P, Marek M, Choi J-S, Pihl JA, Partridge WP: Dynamics of $N_2$ and $N_2O$ peaks during and after the regeneration of lean $NO_x$ trap, *Appl Catal Environ* 166–167:509–517, 2015. https://doi.org/10.1016/j.apcatb.2014.12.002.

Mráček D, Kočí P, Choi J-S, Partridge WP: New strategy for driving the selectivity of $NO_x$ reduction to $N_2$, $NH_3$ or $N_2O$ during lean/rich cycling of a lean $NO_x$ trap catalyst, *Appl Catal Environ* 182:109–114, 2016. https://doi.org/10.1016/j.apcatb.2015.09.002.

Nguyen H, Harold MP, Luss D: Optical frequency domain reflectometry measurements of spatio-temporal temperature inside catalytic reactors: applied to study wrong-way behavior, *Chem Eng J* 234:312–317, 2013. https://doi.org/10.1016/j.cej.2013.08.074.

Nguyen H, Harold MP, Luss D: Spatiotemporal behavior of Pt/Rh/$CeO_2$/BaO catalyst during lean–rich cycling, *Chem Eng J* 262:464–477, 2015. https://doi.org/10.1016/j.cej.2014.09.103.

Nguyen H, Peng PY, Luss D, Harold MP: Assessing intrusion by the capillary during spatially resolved mass spectrometry measurement, *Chem Eng J* 307:845–859, 2017. https://doi.org/10.1016/j.cej.2016.08.101.

Nova I, Castoldi L, Lietti L, Tronconi E, Forzatti P: How to control the selectivity in the reduction of $NO_x$ with $H_2$ over Pt-Ba/$Al_2O_3$ lean $NO_x$ trap catalysts, *Top Catal* 42–43:21–25, 2007. https://doi.org/10.1007/s11244-007-0144-8.

Olsson L., Chalmers University of Technology, personal communication to W.P.P. 2015.

Parks J, Huff S, Pihl J, Choi JS, West B: Nitrogen selectivity in lean $NO_x$ trap catalysis with diesel engine in-cylinder regeneration, *SAE Transact J Fuels and Lubr* 114(4):1753–1765, 2005. SAE Paper 2005-01-3876. https://doi.org/10.4271/2005-01-3876.

Parks J, West B, Swartz M, Huff S: *Characterization of lean $NO_x$ trap catalysts with in-cylinder regeneration strategies,* SAE Technical Paper Series, SAE Paper 2008-01-0448, 2008. https://doi.org/10.4271/2008-01-0448.

Partridge B, Choi J-S: In *Ammonia formation and utilization in lean NOx trap catalysts: experimental determination of reaction pathways*, 11th DOE crosscut workshop on lean emissions reduction simulation, Dearborn, MI, 2008, University of Michigan. http://cleers.org/workshops/workshop11.

Partridge WP, Choi J-S: $NH_3$ formation and utilization in regeneration of Pt/Ba/$Al_2O_3$ $NO_x$ storage-reduction catalyst with $H_2$, *Appl Catal Environ* 91:144–151, 2009. https://doi.org/10.1016/j.apcatb.2009.05.017.

Partridge WP, Storey JME, Lewis SA, et al: Time-resolved measurements of emission transients by mass spectrometry, *SAE Transact J Fuels Lubr* 109:2992–2999, 2000. SAE Paper 2000-01-2952. https://doi.org/10.4271/2000-01-2952.

Partridge WP, Lewis SA, Ruth MJ, Muntean GG, Smith RC, Stang JH: Resolving EGR distribution and mixing, *SAE Transact J Engines* 111:2699–2704, 2002. Transactions. https://doi.org/10.4271/2002-01-2882.

Partridge WP, Toops TJ, Green JB, Armstrong TR: Intra fuel cell stack measurements of transient concentration distributions, *J Power Sources* 160:454–461, 2006. https://doi.org/10.1016/j.jpowsour.2006.01.016.

Partridge WP, Choi JS, Parks J, et al: *Cummins/ORNL FEERC CRADA: NOx control & measurement technology for heavy duty diesel engines,* Bethesda, MD, 2008, DOE Vehicle Technologies Program Annual Merit Review. http://info.ornl.gov/sites/publications/Files/Pub10341.pdf.

Partridge B, Choi J-S, Pihl J, et al: In *Understanding the distributed intra-catalyst impact of sulfation on water gas shift in a lean $NO_x$ trap catalyst*, 12th DOE crosscut workshop on lean emissions reduction simulation, Dearborn, MI, 2009, University of Michigan. http://cleers.org/workshops/workshop12.

Partridge WP, Kim M-Y, Pihl JA, et al: In *Field-ageing impacts on a commercial automotive Cu/SAPO-34 SCR catalyst: focusing on $NH_3$ capacity utilization*, 2015 DOE crosscut workshop on lean emissions reduction simulation, Dearborn, MI, 2015, University of Michigan. http://cleers.org/workshops/workshop2015.

Partridge WP, Salazar M, Pihl JA, et al: In *Cummins-ORNL\FEERC emissions CRADA: $NO_x$ control & measurement technology for heavy-duty diesel engines, self-diagnosing smart catalyst systems*, DOE vehicle technologies program annual merit review, Washington, DC, June 9, 2016, 2016. http://energy.gov/sites/prod/files/2016/06/f32/ace032_partridge_2016_o.pdf.

Peng P-Y, Nguyen H, Harold MP, Luss D: Spatio-temporal phenomena in monolithic reactors measured by combined spatially-resolved mass spectrometry and optical frequency domain reflectometry. In Dixon AG, Deutschmann O, editors: *ACHEM: spatially-resolved operando measurements in heterogeneous catalytic reactors*, 2017, The Royal Society of Chemistry, pp 83–130.

Perry RH, Green DW: *Perry's chemical engineers' handbook,* New York, 2008, McGraw-Hill.

Pihl J, Daw S: In *Spatially resolved $NH_3$ storage distributions on a commercial small pore Cu zeolite SCR catalyst*, 2013 DOE crosscut workshop on lean emissions reduction simulation, Dearborn, MI, 2013, University of Michigan. http://cleers.org/workshops/workshop2013.

Pihl J, Daw S: In *Energetics of $NH_3$ storage on zeolite SCR catalysts*, DOE crosscut workshop on lean emissions reduction simulation, Dearborn, MI, 2016, University of Michigan. http://cleers.org/workshops/workshop2016.

Pihl JA, Parks JE, Daw CS, Root TW: *Product selectivity during regeneration of lean $NO_x$ trap catalysts*, 2006. SAE Technical Paper Series, SAE Paper 2006-01-3441.

Plát F, Bártová Š, Štěpánek J, Kočí P, Marek M: Dynamics of a combined DOC-NSRC-SCR exhaust gas after treatment system with periodic regenerations, *Ind Eng Chem Res* 49:10348–10357, 2010. https://doi.org/10.1021/ie100590s.

Prikhodko VY, Nguyen K, Choi J-S, Daw CS: Axial length effects on Lean $NO_x$ Trap performance, *Appl Catal Environ* 92:9–16, 2009. https://doi.org/10.1016/j.apcatb.2009.07.030.

Raj R, Harold MP, Balakotaiah V: NO inhibition effects during oxidation of propylene on Cu-chabazite catalyst: a kinetic and mechanistic study, *Ind Eng Chem Res* 52(44):15455–15465, 2013. https://doi.org/10.1021/ie4009343.

Roy S, Baiker A: $NO_x$ storage-reduction catalysis: from mechanism and materials properties to storage-reduction performance, *Chem Rev* 109:4054–4091, 2009. https://doi.org/10.1021/cr800496f.

Russel A, Epling WS, Hess H, et al: Resolved temperature and gas species changes in a lean-burn engine emissions control catalyst, *Ind Eng Chem Res* 49:10311–10322, 2010. https://doi.org/10.1021/ie1005299.

Sá J, Fernandes DLA, Aiouache F, et al: SpaciMS: spatial and temporal operando resolution of reactions within catalytic monoliths, *Analyst* 135:2260–2272, 2010. https://doi.org/10.1039/C0AN00303D.

Sappok A, Bromberg L, Parks J, Prikhodko V: *Loading and regeneration analysis of a diesel particulate filter with a radio frequency-based sensor,* SAE Technical Paper Series, SAE Paper 2010-01-2126, 2010. https://doi.org/10.4271/2010-01-2126.

J. W. Schwank and A. R. Tadd, "Catalytic reforming of liquid hydrocarbons for on-board solid oxide fuel cell auxiliary power units," Editors, James J. Spivey and Kerry M. Dooley; Published by The Royal Society of Chemistry, London, UK, 2010, Catalysis vol. 22. Chapter 2, pp 56–93; https://doi.org/10.1039/9781847559630-00056 (ORNL SpaciMS work referenced on pp 65–67, and References 103–109).

Shakir O, Yezerets A, Currier NW, Epling WS: Spatially resolving concentration and temperature gradients during the oxidation of propylene on $Pt/Al_2O_3$, *Appl Catal A Gen* 365:301–308, 2009. https://doi.org/10.1016/j.apcata.2009.06.033.

Sharma M, Dearth M: Effective diffusivity measurement on a monolithic reactor using SPACI-MS technique, *Appl Catal Environ* 188:177–188, 2016. https://doi.org/10.1016/j.apcatb.2016.01.048.

Shwan S, Jansson J, Korsgren J, Olsson L, Skoglundh M: Kinetic modeling of H-BEA and Fe-BEA as $NH_3$-SCR catalysts-effect of hydrothermal treatment, *Catal Today* 197:24–37, 2012. https://doi.org/10.1016/j.cattod.2012.06.014.

Shwan S, Nedyalkova R, Jansson J, Korsgren J, Olsson L, Skoglundh M: Influence of hydrothermal ageing on $NH_3$-SCR over Fe-BEA-inhibition of $NH_3$-SCR by ammonia, *Top Catal* 56:80–88, 2013. https://doi.org/10.1007/s11244-013-9933-4.

Shwan S, Partridge W, Choi J-S, Olsson L: Kinetic modeling of $NO_x$ storage and reduction using spatially resolved MS measurements, *Appl Catal Environ* 147:1028–1041, 2014. https://doi.org/10.1016/j.apcatb.2013.10.023.

Song X, Parker GG, Johnson JH, Naber JD, Pihl JA: A SCR model calibration approach with spatially resolved measurements and $NH_3$ storage distributions, *Emiss Control Sci Technol* 1:98–107, 2015. https://doi.org/10.1007/s40825-014-0003-6.

Stewart J, Douglas R, Goguet A, Stere C: Detailed validation of an automotive catalysis model using spatially resolved measurements within the catalyst substrate, *Can J Chem Eng* 92:1535–1541, 2014a. https://doi.org/10.1002/cjce.21993.

Stewart J, Douglas R, Goguet A, Stere CE, Blades L: *A mathematical approach to the balancing of mass transfer and reaction kinetics in dual kinetic model for automotive catalysis,* SAE Technical Paper Series, SAE Paper 2014-01-2821, 2014b. https://doi.org/10.4271/2014-01-2821.

Sui R, Es-Sebbar ET, Mantzaras J, Bombach R: Homogeneous ignition during fuel-rich $H_2/O_2/N_2$ combustion in platinum-coated channels at elevated pressures, *Combust Flame* 180:184–195, 2017. https://doi.org/10.1016/j.combustflame.2017.02.033.

Swartz M, Huff S, Parks J, West B: *Intra-catalyst reductant chemistry and $NO_x$ conversion of diesel lean $NO_x$ traps at various stages of sulfur loading,* SAE Technical Paper Series, SAE Paper 2006-01-3423, 2006. https://doi.org/10.4271/2006-01-3423.

Theis JR, Dearth M, McCabe R: *LNT+SCR catalyst systems optimized for $NO_x$ conversion on diesel applications,* SAE Technical Paper Series, SAE Paper 2011-01-0305, 2011. https://doi.org/10.4271/2011–01-0305.

Touitou J, Morgan K, Burch R, Hardacre C, Goguet A: An in situ spatially resolved method to probe gas phase reactions through a fixed bed catalyst, *Cat Sci Technol* 2:1811–1813, 2012. https://doi.org/10.1039/c2cy20141k.

Touitou J, Burch R, Hardacre C, et al: An in situ spatially resolved analytical technique to simultaneously probe gas phase reactions and temperature within the packed bed of a plug flow reactor, *Analyst* 138:2858–2862, 2013. https://doi.org/10.1039/c3an00250k.

United States Environmental Protection Agency: Greenhouse gas emissions, understanding global warming potentials, US EPA, Washington, D.C. https://www.epa.gov/ghgemissions/understanding-global-warming-potentials.

University of Chemistry and Technology, Prague; Monolith Research Group, http://www.vscht.cz/monolith (2017).

West B, Huff S, Parks J, et al: Assessing reductant chemistry during in-cylinder regeneration of diesel lean $NO_x$ traps, *SAE Transact J Fuels Lubr* 4:1975–1985, 2004. SAE Paper 2004-01-3023. https://doi.org/10.4271/2004-01-3023.

Winkler A, Ferri D, Aguirre M: The influence of chemical and thermal aging on the catalytic activity of a monolithic diesel oxidation catalyst, *Appl Catal Environ* 93:177–184, 2009. https://doi.org/10.1016/j.apcatb.2009.09.027.

Yezerets A., Popuri S.S., Currier N.W., Epling W.S., Millington P.J., Lafyatis D.S., n.d. "Method for on-board diagnostics of diesel engine $NO_x$ adsorber catalysts," Cummins Inc., US Patent No. US 20080168824 A1.

Zellner A, Suntz R, Deutschmann O: Two-dimensional spatial resolution of concentration profiles in catalytic reactors by planar laser-induced fluorescence: NO reduction over diesel oxidation catalysts, *Angew Chem Int Ed* 54:2653–2655, 2015. https://doi.org/10.1002/anie.201410324.

Zheng Y., Liu Y., Harold M.P., Luss D., "LNT–SCR dual-layer catalysts optimized for lean $NO_x$ reduction by $H_2$ and CO," Appl. Catal. Environ. 148–149, 311–321, 2014a. https://doi.org/10.1016/j.apcatb.2013.11.007.

Zheng Y, Luss D, Harold MP: Optimization of LNT–SCR dual-layer catalysts for diesel $NO_x$ emission control, *SAE Int J Engines* 7(3):1280–1289, 2014b. SAE Paper 2014-01-1544. https://doi.org/10.4271/2014-01-1544.

Zotin FMZ, Gomes ODM, de Oliveira CH, Alcover AA, Cardoso MJB: Automotive catalyst deactivation: case studies, *Catal Today* 107–108:157–167, 2005. https://doi.org/10.1016/j.cattod.2005.07.111.

CHAPTER TWO

# Spatio-Temporal Phenomena in Monolithic Reactors Measured by Combined Spatially-Resolved Mass Spectrometry and Optical Frequency Domain Reflectometry

**Po-Yu Peng, Hoang Nguyen, Michael P. Harold[1], Dan Luss[1]**
University of Houston, Houston, TX, United States
[1]Corresponding authors: e-mail address: mharold@uh.edu; dluss@uh.edu

## Contents



## Abstract

Measuring the spatiotemporal temperature and reacting species concentrations of a reacting system under realistic conditions is essential to optimize the reactor design and operation. Conventional measurements of the inlet and outlet values provide only a partial picture of the reactor behavior. We describe the novel method coherent optical frequency domain reflectometry (c-OFDR) which provides unparalleled information about the spatiotemporal temperature in a catalytic reactor. c-OFDR launches a laser beam through an optical fiber positioned inside a reactor. The spatial dependence of temperature is determined through the measured spectral shift of reflected scattering light frequency. The advantages of c-OFDR include high spatial and temporal resolution (3 mm and 0.9 Hz), broad operating range (up to 800°C), noninvasive measurement (~125 μm OD), stationary, and chemically resistant (gold-coated) sensor.

*Advances in Chemical Engineering*, Volume 50
ISSN 0065-2377
https://doi.org/10.1016/bs.ache.2017.07.001

In addition to describing the theory and standard operation of c-OFDR, we demonstrate c-OFDR for several applications involving reaction in a catalytic monolith during both steady-state and transient operation. The combination of c-OFDR with spatially resolved capillary inlet mass spectrometry (SpaciMS) provides detailed characterization of a catalytic reactor under periodic operation. Combined SpaciMS and c-OFDR successfully follow the cyclic lean/rich behavior during temperature programmed oxidation experiments. SpaciMS invasiveness is assessed through systematic measurement of the temperature by c-OFDR and concentration by SpaciMS. In an ideal adiabatic reactor, the temperature peak and complete reactant conversion should occur at the same position. Our experiments enabled generation of a map of operating conditions under which this did not occur which helps identify when the SpaciMS method is invasive.

## 1. INTRODUCTION

Alan Turing predicted that the interaction between a slowly diffusing "local" activator and a rapidly diffusing "long range" inhibitor, may create a spatial concentration pattern (Turing, 1990). He pointed out that reaction–diffusion interaction could be the cause of numerous patterns observed in biological and chemical systems.

Measurements of spatiotemporal patterns in catalytic systems are usually carried out in either the *microscopic regime*, in which the reaction on a well-defined single crystal surface is studied under vacuum, or the *macroscopic regime* in which a reaction on a polycrystalline catalyst is studied under high pressure. Spatiotemporal phenomena have been studied in the microscopic regime using surface sensitive techniques such as electron energy loss spectroscopy (EELS), Auger electron spectroscopy (AES), and low energy electron diffraction (LEED). Among these, a major tool is photoemission electron microscopy (PEEM), in which a low energy photoelectron emitted to the surface is used to infer the work function (Bethge and Klaua, 1983). Because the work function strongly depends on the adsorption or reaction layers, it can reveal the distribution of chemical species. Measurement of the work function was successfully used to study the oscillatory patterns of CO oxidation on a platinum catalyst surface. In those experiments, the work function of chemisorbed oxygen is higher than that of CO, and oxygen covered areas are darker than CO-covered areas (Rotermund et al., 1990). In the macroscopic regime, measurements of spatiotemporal patterns elucidate how chemical species are distributed on a catalyst surface. Measurements of changes of either temperature and/or concentration follow how the chemical species interact either with each other or the catalyst. Furthermore, localized patterns may be distributed along the length of a reactor.

Often the chemical reactions are accompanied by the generation or consumption of heat. Therefore, the change of temperature and concentration is interrelated. In fact, the extent of a single overall chemical reaction can be followed by measuring either the concentration or the temperature change. Temperature is an effective indicator of the reaction progress. For example, in an oxidation reaction, the reactant conversion usually undergoes a dramatic increase at the light-off or ignition temperature. Following temporal temperature and concentration variations is one of the most common tools to observe how a chemical reaction evolves in a batch system, while the spatial variations are of interest in a distributed flow system. Spatiotemporal measurements provide information on how the reaction proceeds and on the rate of the various transformations. Mathematical models based on these measurements may be used to predict the behavior under other conditions. These predictions can save expense, effort, and time. The development of "accurate" spatiotemporal measurements is essential to follow, understand, and predict the progress of a chemical reaction. On a catalyst surface many advanced spatiotemporal measurement techniques have been developed, which we highlight below.

## 2. SPATIALLY RESOLVED TEMPERATURE MEASUREMENTS

### 2.1 Review of Spatiotemporal Temperature Measurements

Several techniques exist for measuring spatiotemporal patterns. Thermocouples inserted at various locations inside a fixed-bed or monolith reactor can measure the temperature at several positions (see Fig. 1) (Jaree et al., 2001; Kapička and Marek, 1989). The number of needed thermocouples depends on the desired spatial resolution and the impact of the thermocouples on the system behavior. For example, the invasive insertion of thermocouples into a reactor may generate leaks at the drilled reactor wall (van Sint Annaland and Nijssen, 2002). To minimize the impact of inserted thermocouples, especially in a small reactor, many researchers have used a moving thermocouple in a thermowell to measure the spatial temperature under steady-state conditions (see Fig. 2) (Hartig and Wicke, 1963; Puszyński and Hlavacek, 1984; Zohour et al., 2013). As it takes time to move the thermocouple, this measurement cannot accurately determine rapid changes in spatial temperature profiles, such as might be encountered during ignition or a moving temperature front. Moreover, inaccurate temperature readings can occur

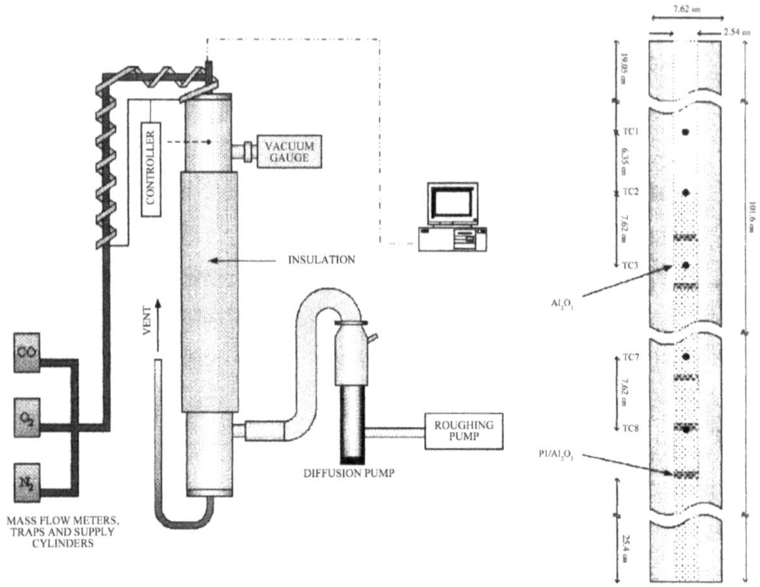


**Fig. 1** Overview of the experimental unit showing feed provision, method of inducing the temperature disturbance, the reactor bed with insulation and vacuum system (A); and structure of the segregated bed showing the location of the thermocouples (TC1, TC2, etc.) and Pt catalyst layers separated by inert (alumina) layers (B) ( Jaree et al., 2001).

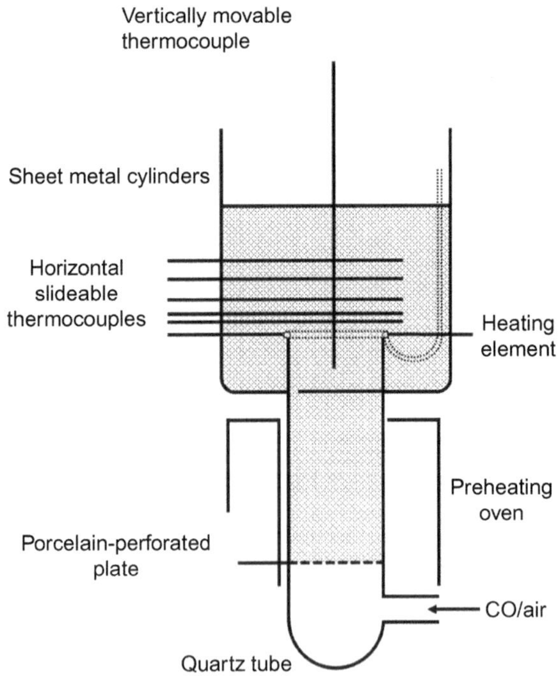


**Fig. 2** The schematic illustration of the application of the moving thermocouple. *Adapted from Hartig H, Wicke E: Brennzonen in Kontaktschichten bei der katalytischen CO-oxydation,* Z Phys Chem *38:265–284, 1963.*

when thermal equilibrium is not achieved between the probe and the catalyst. Another drawback is that heat conduction along the thermowell can affect the measured local temperature so that the measured value is affected mainly by that of the fluid.

All surfaces emit infrared radiation at an intensity that increases with the temperature. *Infrared thermography* (IRT) uses a camera to measure the infrared radiation intensity in order to determine the surface temperature. The advantages of this technique include noninvasive measurements, fast sampling speed, and lack of an upper temperature limit. IRT has successfully measured spatiotemporal temperature patterns on catalytic surfaces on which exothermic reactions occurred. The first report of IR imaging of catalytic surface temperatures was by the group of Schmitz (Brown et al., 1985). Their imaging system was sensitive semiconductor (InSb) detector and had a spatial resolution of ~0.1 mm. IR thermography has been applied to measure temperature patterns on various catalytic surfaces, such as wafers, pellets (Kellow and Wolf, 1990, 1991), ribbons, and disks (Lobban and Luss, 1989; Lobban et al., 1989). For example, Fig. 3 shows infrared images of the

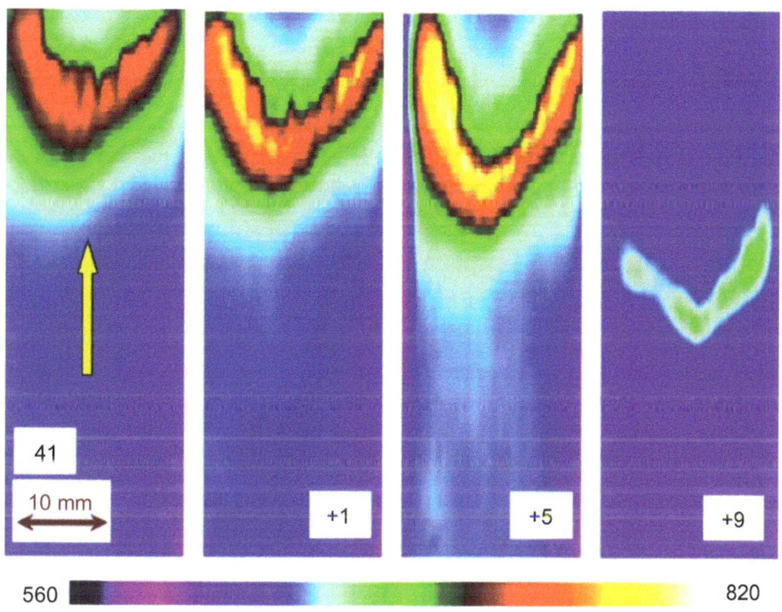


**Fig. 3** Thermal images of motion of hot region during DPF regeneration (Chen et al., 2009).

motion of a hot zone during the regeneration of a diesel particulate filter (Chen et al., 2009).

IRT measurements of high surface temperature usually require placing an infrared transparent sapphire window between the object and the camera (Russell et al., 2010). The sapphire window is set to separate the experimental environment (e.g., high temperature or high pressure) from the outside environment and should allow IR to be transmitted to a camera for detection. A good window selection can rule out the unwanted wavelength noise. Local infrared temperature measurements were conducted by sliding an optical probe over different locations and using a pyrometer to determine the temperature (Bitsch-Larsen et al., 2008).

*Phosphor thermography* is another temperature measurement technique relying on radiation detection. It exploits the fact that the fluorescent excitation decay rate of an inorganic phosphor is temperature dependent and circumvents the time delay between measurements at different positions and the impact of conductive probe spatial smearing (Allison and Gillies, 1997). Processing the fluorescent decay profiles using a Boltzmann distribution analysis yields spatiotemporal temperature profiles (see Fig. 4A). Fig. 4B shows how phosphor thermography measures the temperature on the tip of an optical fiber on which phosphor is deposited. A pulsed laser excites the optical fiber and the fluorescent decay is monitored to determine the temperature. An advanced system can measure the temporal temperature with a time resolution of picoseconds over a wide temperature range from cryogenic up to 2000 K. Fig. 4C shows how by coating a fluorescent material (YAG:Dy) on a 2D ceramic plate the phosphor thermography can measure spatiotemporal temperature profiles during the heating—of the plate (Heyes et al., 2006). This technique relies on measuring the fluorescent excitement by the laser. A potential problem is the eventual decay of the phosphor fluorescent excitability due to the material's degradation. Currently, this technique is applied in the temperature determination of engines, hypersonic wind tunnel experiments, pyrolysis studies, and droplet/spray/gas. We are not aware of its application in studies of catalytic reactions, but it has the potential to be used in the future.

The distributed temperature sensor (DTS) is a recently developed technique to measure the spatial temperature along the length of an optical fiber. This method uses the inherent backscatter or birefringence in the optical fiber to monitor the local temperature. The advantages of this technique include fast measurement (order of a second), and an exact record of the

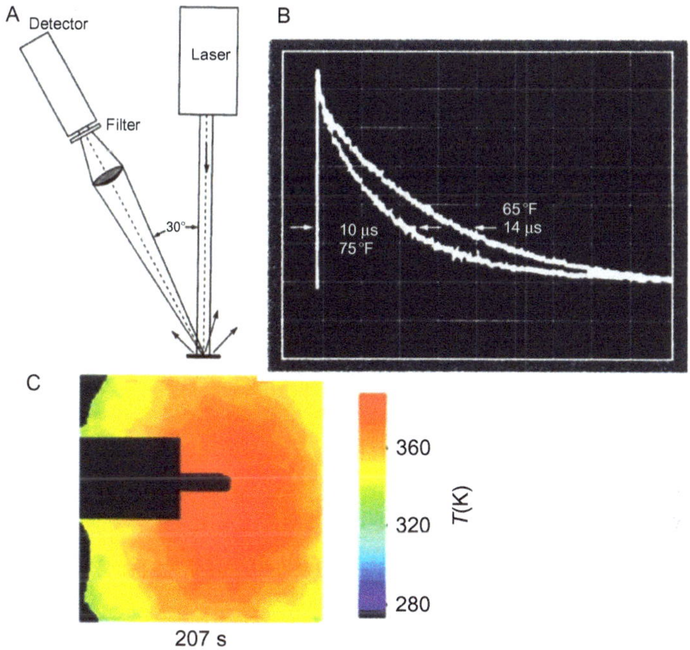


**Fig. 4** (A) The experimental arrangement for determining the fluorescence signal strength (Allison and Gillies, 1997). (B) Phosphor decay characteristics at 65 and 75 °F ($La_2O_2S$:Eu at 514 nm) (Allison and Gillies, 1997). (C) The thermal images of a heated 110 mm × 240 mm ceramic plate by Phosphor thermography (Heyes et al., 2006).

location and magnitude of any thermal events. An early DTS is the quasi-DTS, made by a series of grating sensors along a fiber. For example, extrinsic Fabry–Perot interferometers utilize small air gaps within a fiber (Murphy et al., 1991). Intrinsic Fabry–Perot interferometers require the construction of two reflecting surfaces in the optical fiber generally by use of a fusion splice (Yoshino et al., 1982). Bragg gratings are produced by UV exposure causing periodic changes in the index of refraction of the fiber core (Meltz et al., 1989). The high manufacturing cost of a series of grating sensors has prevented its application to catalytic reaction systems (to the best of our knowledge). We describe next a novel, robust DTS technique called c-OFDR (coherent optical frequency domain reflectometry), developed by Luna Innovation of Technology, Roanoke, Virginia (Childers et al., 2001). We were the first to use it to measure the thermal behaviors inside a honeycomb monolith reactor in which exothermic reactions were carried out.

## 2.2 Optical Backscatter Reflectometer

### 2.2.1 Measurement of Spatiotemporal Temperature by c-OFDR

An optical backscatter reflectometer (OBR) measures the spatiotemporal temperature along an optical fiber by launching a laser beam into the optical fiber and measuring the local reflective spectral scatter. The OBR technique determines the spatial temperature profile and/or strain based on the change in the Raman, Rayleigh, or Brillouin reflective spectral scatter in various wavelength ranges. OBR techniques include: OTDR (optical time domain reflectometry), typically used for long range (kilometer scale) low resolution measurements, OLCR (optical low coherence frequency domain reflectometry) used for submillimeter range, high sensitivity, and short inspection time. c-OFDR is used for measurements in the range between those of the OTDR and OLCR techniques.

The first developed and commercialized OBR method was OTDR in the late 1980s. It measures the spatiotemporal temperature with a temperature resolution of 1 K, spatial resolution of one to a few meters, and at frequency of about 1 Hz by using stimulated Raman scatter (SRS) distributed probe. There are several successful industrial applications of this method including the monitoring of cable temperature (Nakamura et al., 1992), monitoring of temperature in power generating plants (Hartog, 1995), detecting leaks in dams and dikes (Khan et al., 2008), monitoring temperature in drilled oil and gas wells (Smolen and Van der Spek, 2003), automatic fire detection (Glombitza and Hoff, 2004), among other uses. Although it is useful to use OTDR in long range temperature measurements, it is unable to measure the spatiotemporal temperature at a fine spatial and temporal resolution. In contrast, OFDR has attracted great attention as a high resolution tool. The first implementation of an OFDR was by Froggatt and Moore (1998) at NASA's Langley Research Center and was later developed and commercialized by Luna Innovation of Roanoke, Virginia (Childers et al., 2001).

Our research was the first to measure spatiotemporal temperature in a honeycomb-monolith using an c-OFDR 4600 system developed by Luna Innovation (Soller et al., 2005). A fine optical fiber (outer diameter ~100–400 μm) is inserted into a monolith channel of a reactor. One end of the optical fiber is supported by a holder while the other end is loosely laid within the monolith channel. The small fiber diameter minimizes the space it occupies while the reflective spectral scatter avoids the need to move the sensor. The c-OFDR enables temperature measurements at a spatial resolution of up to 3 mm at a frequency of 0.9 Hz with temperatures as high as 800 °C. This spatial and spectral resolution is an order of magnitude higher than other distributed fiber sensor measurements. We describe these and other results later, but first highlight the theoretical and operational features of OFDR.

### 2.2.2 Theory

#### 2.2.2.1 Optical Fiber Structure and Light Transmission Features

Optical fibers are usually made of glass or polymer. While a polymer fiber generally has a lower material and manufacturing cost, it has a higher signal loss and narrower light bandwidth than a glass fiber. Hence it is not used for sensing applications. A glass fiber is composed of a core, a cladding material, and a coated polymer or metallic jacket. The cladding has a smaller reflective index than the core material to confine the light in the fiber. There are single- and multiple-mode glass fibers. The core diameter (~50 μm) and numerical aperture (the capability of an optical fiber to capture light) of the multiple mode fiber are larger than that of the single-mode fiber (SMF). This allows the fiber to accept a broader light spectrum, but has a lower transmission speed and higher signal distortion over a long distance due to modal dispersion. The Luna instrument can produce a fiber with a gradual change in the refractive index between the core and the cladding (graded index fiber) rather than a sharp change (stepped index fiber). Graded index fibers have a lower modal dispersion by gradually bending rays toward the center. This allows the rays to travel at the same group velocity. On the other hand, the multiple mode fiber has limited sensing applications due to the complicated analysis of the reflected spectrum.

The SMF is designed to minimize the dispersion for a narrow range of transmission wavelengths by reducing the core diameter (~9 μm) and the NA of the fiber. In the SMF, light waves are distributed in space in the same way (single mode), resulting in a single parallel ray of light along the fiber length. This feature is desirable for distributed sensing application. However, the difficulty of launching light into the fiber makes its application costly. Hence a good laser source with a narrow wavelength, i.e., an injection laser diodes (ILDs), is needed (Crisp, 2005).

The c-OFDR 4600 system uses a single mode optical fiber as a sensing probe. To protect the fiber from chemical exposure, it is coated with a thin layer of either polyimide or gold. The polyimide coating allows the fiber to operate at temperatures below 350 °C. The more expensive gold coating enables the fiber to operate up to 800 °C without losing its integrity.

#### 2.2.2.2 Reflected Spectral Scatter in the Optical Fiber

The manufacturing process produces a reflective optical fiber glass with an amorphous core structure, in which a silicon atom tetrahedron unit cell is surrounded by four oxygen atoms. The fiber may exhibit density variations along its length during the manufacturing process. A high energetic structure exists between two adjacent tetrahedral unit cells due to thermal equilibrium in the "melt" stage of glass. Quenching the silica fiber to room temperature causes these structures to be trapped during the final stage, and the crystalline structure remains in its high energy structure. Microscopic structural irregularities arise because of these high energy structures. Structural and index fluctuations can also be generated by dopant molecules (i.e., Ge) introduced into the $SiO_2$ lattice structure (Kersey, 1996).

Scattering occurs when the incident light encounters a geometrical change such as the grain boundary between the microscopic crystallites or a different local refractive index of the fiber core. The scattered light intensity varies according to the local physical properties such as fiber density, which is a function of its temperature and strain. This scattering is exploited to spatially resolve the temperature measurements.

When light is introduced into a reflective optical fiber, it can scatter the light in multiple directions. The backward reflective scattering is of primary interest. Three types of backward scattered light exist in an optical fiber; specifically, Rayleigh, Raman, and Brillouin types (see Fig. 5A, Güemes et al., 2010). Raman scattering is caused by a quantum effect that is related to how the silica bonds absorb and emit the incoming photon energy (see Fig. 5B). Via an intermediate quantum state, a term labeled Stoke scattering occurs if the material absorbs energy and the emitted photon has a lower energy than the incoming photon. Another term called Anti-Stokes Raman occurs if the material loses energy and the emitted photon gains energy. These two energy difference outcomes are formed due to the material's two resonant states. The energy gap between the Stoke and Anti-Stoke ($\triangle E$) follows a simple $\exp(-\triangle E/kT)$ relationship. Measuring this ratio enables estimation of the temperature. Furthermore, this ratio is independent of other external stimuli such as strain, humid, electromagnetic field, etc. In contrast to Raman scattering, Brillouin scattering is generated by an interaction between the propagating light photon and the acoustic waves in the silica

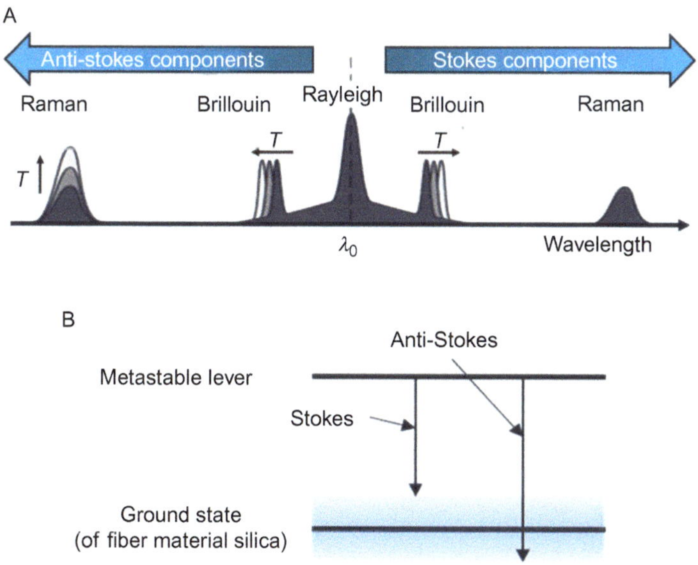


**Fig. 5** (A) Wavelengths of the backscattered light. $\lambda_0$ is a wavelength of incoming light. (B) Raman scattering process in optical fiber. *Reproduced with permission from Di Sante R: Fibre optic sensors for structural health monitoring of aircraft composite structures: recent advances and applications,* Sensors *15:18666–18713, 2015 published by MDPI.*

fiber, generating frequency-shifted components. The frequency shift is directly related to the local fiber density and depends on both the temperature and the strain. Rayleigh scattering is the strongest signal among the three and is one of the main intrinsic losses of energy mechanisms in pure silica. Consider two electrons forming a dipole along the propagation path of a wave. The Rayleigh scattering process consists of an incident wave exciting the dipole into oscillation, resulting in some absorption of the incident power. The dipole then reradiates the power. As the frequency of oscillating dipole increases, a larger fraction of the incident power is scattered by the reradiation. Because the dipole oscillator length is shorter than the incident wavelength. Index fluctuations arising from concentration or density fluctuations in the optical fiber are much shorter than an excited wavelength. Therefore, Rayleigh scattering always occurs in an optical fiber, and these scatter profiles depend on environmental stimuli such as temperature and strain.

The fiber backscattering Rayleigh profile enables the c-OFDR 4600 to measure a distributed temperature profile. When the fiber is exposed to a temperature rise, the local density (or refractive index) changes and the reflected wavelength is shifted toward a longer wavelength. The typical order of magnitude of temperature sensitivity of wavelength of 1550 nm is of 10 pm/°C without a spectral shape modification and without hysteresis (Hill and Meltz, 1997). Thus, monitoring this shift enables determination of the temperature.

### *2.2.3 Coherent c-OFDR*

The c-OFDR technique utilizes a Mach–Zehnder interferometer to measure the Rayleigh backscattering profile of the fiber (see Fig. 6). The interferometer consists of a test arm connected to the optical fiber, and a reference arm, which has a mirror at its end to reflect the light. A narrowband and tunable laser source (TLS) transmits coherent optical radiation into both arms over a predetermined range of frequencies. The Rayleigh backscatter of each reflector in the test arm interferes with the backscattered signal from the reference arm and produces interference fringes. A spectrum analyzer measures the intensities of these interference fringes. The c-OFDR 4600 interferometer has two improvements over the Mach–Zehnder instrument.

The first improvement is how it deals with the polarization dependence problem. In some cases, small amounts of birefringence along the fiber length may cause the state of polarization (SOP) of the reflected signal to be undetectable on the spectrum analyzer. This occurs as the intensity of the measured signal is calculated based on the dot product of the SOP of the probe light with respect to the reference light. If the SOP of the measured and reference fields are orthogonal to the detector, the interference is zero and cannot be detected. To overcome this problem, the c-OFDR 4600 uses a polarization diversity detection technique (Kobayashi et al., 1991). It uses a polarizing beam splitter (PBS) followed by a recombining coupler to achieve a polarization independent of the fiber. The PBS allows the interference field to be split evenly between states "s" and "p," ensuring that an

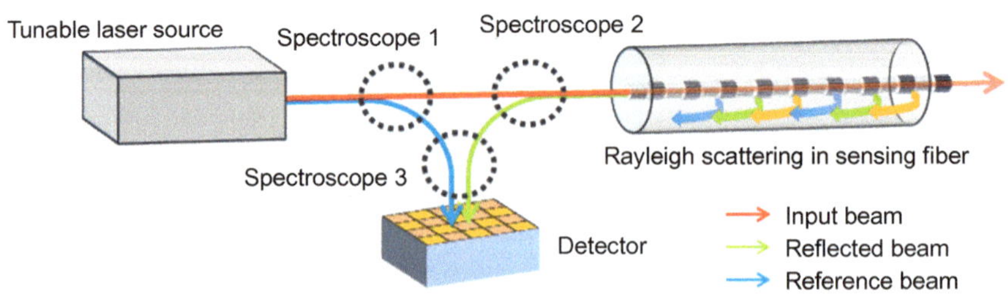


**Fig. 6** Operation principle of c-OFDR.

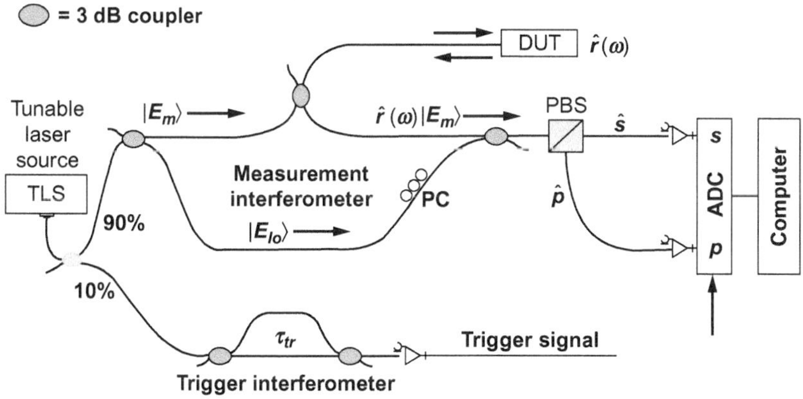


**Fig. 7** Optical network of the OBR 4600 system (Soller et al., 2005).

interference signal is detected by at least by one of the s and p detectors for any reflected fields from the optical fiber. The second improvement deals with nonlinearity in the optical frequency modulation. In the c-OFDR setup, a TLS is the key component. The TLS may fluctuate during frequency tuning and result in nonlinear sampling. This problem can be avoided by a feedback control. The c-OFDR 4600 uses a frequency sampling method in which the time-varying optical frequency is simultaneously measured by an auxiliary interferometer (Soller et al., 2005). The fringe signals at the detector output of the auxiliary interferometer trigger data acquisition on detectors "s" and "p" to sample the incoming data at equal frequency increments (Childers et al., 2001). This approach restricts the maximum length of the optical fiber to about 30 m. This maximum length is determined by the physical time delay difference of the auxiliary interferometer used for data triggering. Fig. 7 shows the schematic optical network used in the c-OFDR 4600 system (Soller et al., 2005).

The advantage provided by c-OFDR is its spatially resolving capability. Therefore, before showing the maximum spatial-resolution capability, we first explain the generation of the spatial spectrum data. Spatially resolved profiles are generated depending on the time delay of the reflective scatter among segments along with the optical fiber. Fig. 8 is a schematic illustration of the generation of the discrete spatial spectrum data. After the laser is launched to the optical fiber, the laser beam passes through the optical body and is partially scattered by the intrinsic structure of the optical fiber. The scattered light is reflected back to the machine and captured by the optical detector. The further is the scattering point from the detector the longer is the time delay between the launching point and the detection. The travel

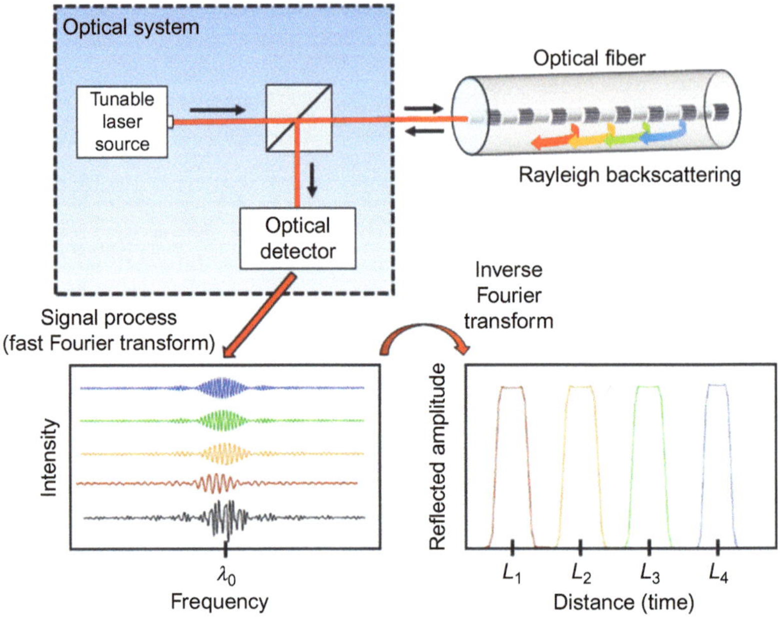


**Fig. 8** The schematic illustration of the process to generate the discreet spatial spectrum data.

distance is twice that of the physical distance from the launching point to the particular point along the fiber. A Fast Fourier Transform algorithm transforms the wavelength signals into frequency-based packets that are differentiated by time. Then a spectral shift analysis is done of the intensity–frequency packets to determine the temperature change. Those details are highlighted below.

The spectral shift data are converted to temperature rise (from the reference) as follows. When the temperature of the *n*th reflector site is increased, its density decreases. The Rayleigh profile is elongated in this spatial domain and shifted toward a longer wavelength in the frequency domain. This shift is proportional to the applied temperature and/or strain. From the Rayleigh spectral shifts, one can use Eq. (1) which relates the spectral shift changes to temperature fluctuations and mechanical stress applied to the fiber. When the external stress caused by fiber fluctuations is negligible and the induced strain ($\varepsilon$) is just a function of temperature, the spectral shift ($\Delta\nu$) satisfies the relation

$$-\frac{\Delta\nu}{\nu} = K_T \Delta T + K_\varepsilon \varepsilon \tag{1}$$

where $\Delta\nu$ is the spectral shift, $\nu$ is the mean optical frequency. $K_T$ and $K_\varepsilon$ are the temperature and strain calibration constants, respectively.

The following procedure is used to determine the spatial temperature profile (Froggatt and Moore, 1998; Zhou et al., 2012). The dependence of the temperature on the spectral shift is determined by a cross-correlation algorithm consisting of the following steps. First, the fiber spatial Rayleigh profile at a known ambient temperature is measured as the "reference state." The actual measurement is carried out for a given operating condition (measurement state) to record the fiber's spatial Rayleigh profile. Then, the reference and measurement states are transformed to the wavelength domain by an inverse Fourier transformation algorithm (IFT). A cross-correlation analysis is then used to measure the similarity between the two states (reference and measurement state) and determine the magnitude of the cross-correlation peak shifts as a function of position. A positive/negative shift of the cross-correlation peak indicates an increase or decrease of the temperature.

These results are then recorded and plotted as a function of the distance along the fiber according to the aforementioned time delay. The spatial resolution of the distance increment $\Delta z$ is determined by the time delay increment $\Delta\tau$, which follows the linear relation $\Delta z = c \cdot \Delta\tau/2n$ where $c$ is the speed of light and $n$ is the group index (the ratio of the vacuum velocity of light to the group velocity in a medium), while the factor of 2 accounts for the double-pass nature of the light reflection. Since the signal detected during the time increment $\Delta\tau$ should also be the minimum time that the laser monitor interferometer is able to process, $\Delta\tau$ is determined by the scanning rate and the number of sample points of the laser monitor interferometer. The expression $\Delta\tau = \frac{\tau_{lm}}{S} = \frac{1}{S\delta\nu_{lm}}$ relates the time delay to $\tau_{lm}$, which is the optical time delay between paths in the laser monitor interferometer and is the inverse of the laser monitor frequency $\nu_{lm}$, and $S$, which is the number of sample points of the laser monitor interferometer. For example, for a 2.5 MHz laser monitor interferometer frequency, $2 \times 10^6$ points data record, and $n = 1.48$ of average group index in silica, the $\Delta z$ is about 20 μm. There are some other environmental factors which could cause the spatial resolution to be shorter, such as noise, laser tuning deficiency, and unbalanced dispersion in the measurement arm. Nevertheless, the high spatial resolution of c-OFDR enables good measurement of the spatiotemporal temperature (Kersey, 1996; Soller et al., 2005).

### 2.2.4 Calibration and Testing Procedure for c-OFDR 4600 System

#### 2.2.4.1 Calibration of the c-OFDR 4600 System

There are three calibration procedures of the c-OFDR 4600.

**i.** *Reflectivity calibration*

The reflectivity is calibrated by measuring the reflectivity magnitude of a very strong reflector. For the calibration, we used a spectrally flat, gold-plated polished fiber sensor provided by Luna Innovation. All subsequent reflectivity values are measured with respect to this reference value.

**ii.** *Calibration of polarization controller*

The polarization controller must be calibrated before the measurements. In an ideal situation, the signal intensity from the measure arm should be redistributed evenly between the s and p detectors of the PBS. This is accomplished by blocking the signal in the measurement path and monitoring the DC power level of the s and p detectors. The polarization controller is adjusted such that an equal amount of power is incident on the s and p detectors.

**iii.** *Effective temperature response coefficient calibration*

Each optical fiber responds differently to an external stimuli. Thus, $K_T$ is determined as the ratio between the Rayleigh spectral shift and the actual temperature rise during a set of preliminary experiments. In these experiments, the actual temperature rise as measured by a K-type stainless steel sheathed thermocouple (OD = 0.5 mm). To create a uniform environment between the optical fiber and the thermocouple, an optical fiber and an thermocouple were held by a 431.8 mm long quartz tube (ID = 3.52 mm, OD = 6.04 mm) along an electrically heated plate to control the temperature. A contact between the optical fiber and thermocouple tip was avoided by inserting the thermocouple at the opposite end of the quartz tube to prevent thermal or stress interactions. A linear relation between the thermocouple-measured temperature and the corresponding spectral shift was determined over the temperature range of 25–338.6 °C as the one shown in Fig. 9 that satisfies the relation:

$$T - 25\left[^{\circ}\mathrm{C}\right] = -0.6017\Delta\nu[\mathrm{GHz}] - 1.6841 \quad R^2 = 0.9997 \tag{2}$$

The high correlation factor $R^2$ indicates that $K_T$ is constant under these conditions and can be determined by the above relation.

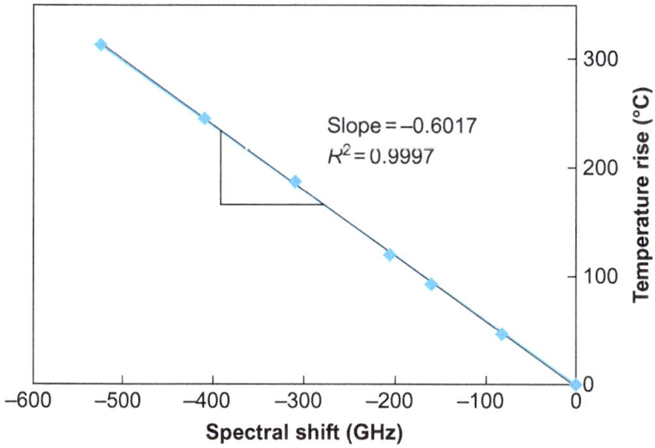


**Fig. 9** Relationship between spectral shift and temperature difference (Nguyen et al., 2013).

iv. *Determination of the object position along the optical fiber*

In order to calibrate the optical fiber held inside the monolith reactor, we used a laser pointer to heat one point along the fiber. This created a hot spot at that position. This enabled us to determine the location of all other points relative to that heated by the laser.

### 2.2.4.2 Spatial Temperature Measurement Testing

Preliminary experiments were conducted to determine the accuracy of the temperature measurements by the c-OFDR technique (Nguyen et al., 2013). The optical fiber was inserted into the center of monolith channel from the reactor upstream. A type-K thermocouple (OD = 250 μm) was inserted from the downstream in a channel adjacent to that housing the optical fiber. No gas was fed to the monolith in a static mode experiment. The 7.5 cm long monolith was heated to 100°C along the first 5 cm by an electrical heating tape. After the system stabilized the spatial temperature was measured by c-OFDR along the optical fiber. At the same time, a fine thermocouple was slowly moved toward the upstream at a 1 cm steps to obtain the reference spatial temperature profile.

The time-averaged temperature profiles of both the fiber and the thermocouple under no-flow condition from six temperature measurements are shown in Fig. 10A. The error bars were determined by the standard

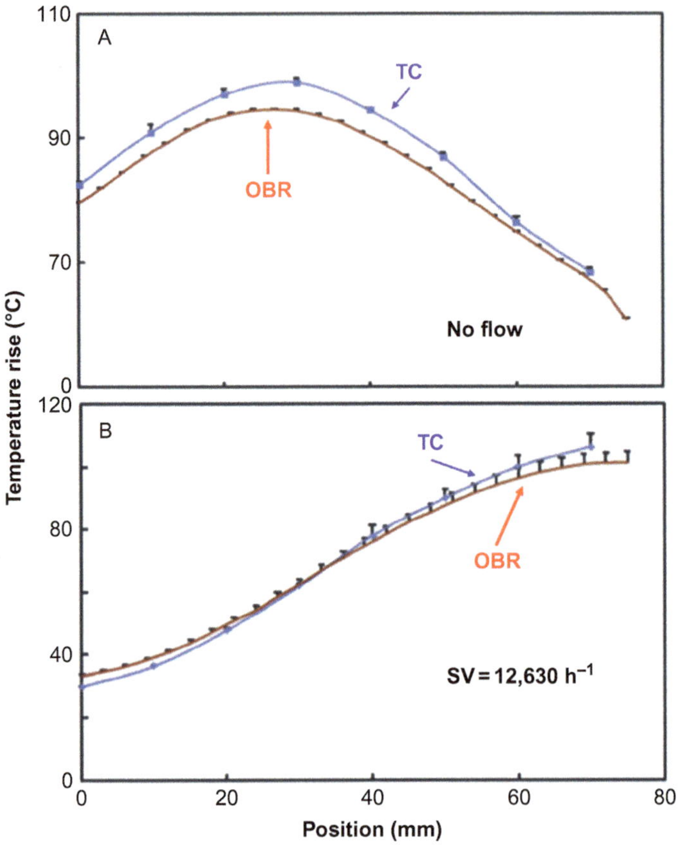


**Fig. 10** Temperature measurement inside a monolith channel by thermocouple and the OBR (A) under no flow condition (the heating tape = 100°C) and (B) under flow condition (the heating tape = 200 °C, SV = 12,630 $h^{-1}$) (Nguyen et al., 2013).

deviation and are only 0.25°C by the optical fiber measurement. Both time-averaged temperature profiles show a good agreement between the two measurement methods. Under flow at a 12,630 $h^{-1}$ space velocity, the same experiment was conducted by raising the heating tape temperature to 200 °C. In Fig. 10B, the time-averaged spatial temperatures show a good match between two methods. The error bars of the optical fiber's measurements became larger under the flow condition when closing the optical fiber's free end, its maximum standard deviation of 3.5 °C is as 10 times larger than that under the no-flow condition. Although this larger standard deviation is satisfactory, it points out the potential error caused by the flow-induced vibration on the free end of the fiber.

#### 2.2.4.3 Practical Considerations of c-OFDR Operation

Occasionally a strong reflection exists in the optical fiber (i.e., a very steep spatial temperature gradient) due to mixing between different reflectors along the fiber. This may generate artificial (ghost) peaks in the Rayleigh spectrum. A simple way to identify these ghost peaks is to reduce the local oscillation reflection by reducing the gain of the interferometer. The intensity of the real peaks monotonically decreases with the gain, while the ghost peaks remain constant. Comparing the two traces enables distinguishing between the real and ghost peaks (Oberson et al., 2000).

Another concern is the inaccurate temperature measurement caused by the flow-induced vibration. As mentioned, the fiber used in our study was held at one point but its other end was unsupported in the channel. Thus, the flow can vibrate the fiber, especially at high flow rates. This may lead to inaccurate local temperature determination, due to the impact of vibrations on the optical fiber strain. According to Eq. (1), the local optical fiber strain directly affects the Rayleigh spectral shift. This may lead to erroneous prediction of the temperature based on the reflected signal. Therefore, at high flow rate condition, the optical fiber's vibration may lead an erroneous interpretation of the reflected signal. One potential solution is fixing the loosely free end by hanging with a physical holder to dampen the vibrations.

## 2.3 Applications of c-OFDR

Our interest in using spatiotemporal measurement by the c-OFDR method was motivated by its high spatial, temporal, and temperature resolution. The wide operating range and easy-to-install optical fiber broadens its applicability to various catalytic reaction conditions. One potential application is continuous, nondestructive spatiotemporal temperature measurements within the flow channel of a structured reactor such as a monolith or foam substrate. Specifically, we applied the c-OFDR technique first to measure the spatial temperature profiles of an exothermic reaction in an adiabatic monolith catalytic reactor. Later we studied the transient behavior in heterogeneous catalytic reactors. We highlight below the capability of the spatiotemporal temperature measurement by the c-OFDR technique.

### *2.3.1 Steady-State Spatial Temperature Profile of a Catalytic Exothermic Reaction*

Steady-state experiments were conducted using a 3 in. long monolith which had a 60 g/ft$^3$ PGM loading (Pt:Pd = 2:1 (g) $\asymp$ 1:1 (mol)), 0.5 g/in.$^3$ beta zeolite loading and 0.7 g/in.$^3$ alumina loading. The feed composition was

0, 970, 1400, and 1730 ppm propylene and 10% $O_2$ with the balance being inert carrier Argon. The feed temperature was 100 °C and the total gas space velocity was 12,630 $h^{-1}$. Fig. 11 shows for several propylene feed concentrations the spatial temperature profiles spanning the distance from the monolith front upstream to the downstream. For a feed devoid of propylene, a uniform temperature of 100°C existed along the entire length. No temperature rise was observed, as no net heat was generated or lost under adiabatic operation. For a feed containing propylene, a temperature rise occurred in the monolith reactor. For instance, for a feed containing 1400 ppm propylene, the temperature gradually increased from the 100 °C feed value, to a peak of 189 °C about 45 mm from the upstream. This temperature profile exhibits a typical behavior of an exothermic reaction in an adiabatic monolith reactor. No heat loss appeared from the position of the peak temperature to the exit. We can use the temperature rise ($\Delta T_3$ as labeled in Fig. 11) to calculate the conversion, knowing system parameters such as the heat of reaction and heat capacity. We calculated the theoretical temperature rise by using the relation of:

$$\Delta T = \frac{(-\Delta H)\langle C_f \rangle X}{\sum \theta_i C_{pi}} \tag{3}$$

where $-\Delta H$ is the heat of reaction (1930 kJ/mol for propylene complete conversion into $CO_2$) and $\langle C_f \rangle$ is the reactant (propylene) feed

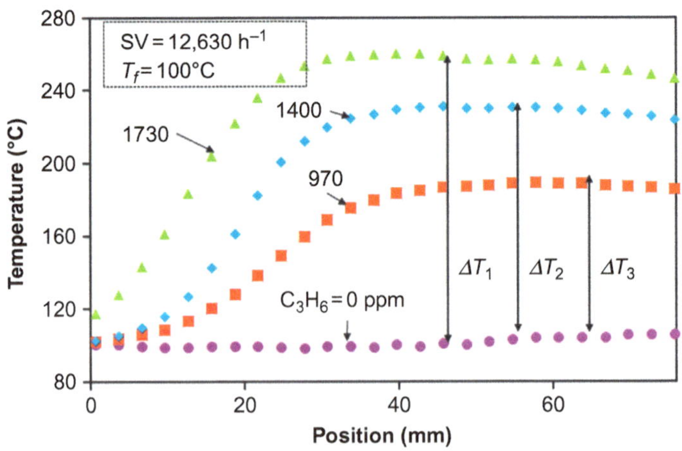


**Fig. 11** The spatial temperature profiles of the propylene oxidation in an adiabatic steady-state condition. ($T_{inlet}$ = 100 °C, 0, 970, 1400, and 1730 ppm propylene, 10% $O_2$ balanced with Argon, space velocity of 12,630 $h^{-1}$.)

concentration. $C_{pi}$ is the heat capacity of gas species $i$. Using a heat capacity of 20.79J/mol/K (that of Argon), the theoretical $\Delta T(970)$ of 90 °C is very close to the measured temperature rise of 89 °C ($\Delta T_3$). The theoretical temperature rise is in good agreement with experimental measurements at the other propylene feed concentrations. For example, $\Delta T(1400)$ and $\Delta T(1730)$ are 130 and 160 °C, respectively, in good agreement with the measured temperature values ($\Delta T_2 = 130$ and $\Delta T_1 = 159$ °C). The data also show that the positions of the peak temperature moved upstream as the propylene feed concentration increased. This trend indicates that the oxidation is of positive order with respect to propylene. Modeling of the system (results not shown here) indicates that the trend is the result of nonisothermal behavior. Specifically, axial conduction from the reaction zone upstream enhanced the light-off at the higher propylene concentrations. This trend is not observed in a shorter monolith which has a more uniform temperature. These experiments demonstrate the ability to measure the steady-state spatial temperature at various propylene concentrations.

### 2.3.2 Spatiotemporal Temperature Measurements of Wrong-Way Behavior

The wrong-way behavior is a counterintuitive transient reactor temperature response to a sudden change in the feed conditions, such as a sudden decrease of the feed temperature or increase of the feed velocity (Mehta et al., 1981). We demonstrate here the capability of c-OFDR to measure these transient temperature profiles. Initially we conducted steady-state ammonia oxidation in a 400 cpsi, 3 in. long monolith reactor. (Details of the experimental system are reported in Nguyen et al., 2013.)

The feed contained 1.2 vol.% ammonia, 12.5% of oxygen, and the balance Argon. The feed temperature was 143 °C and the space velocity was 12,630 $h^{-1}$. The initial steady-state temperature profile is shown in Fig. 12 by the curve labeled as at $t = 0$ s. A sudden decrease of the feed temperature from 143 to 100 °C at $t = 0$ s led to the evolution of transient temperature profiles shown by the set of curves in Fig. 12. Owing to the shift to cooler feed, the upstream temperature is cooled below the initial steady state values. However, the temperature profiles at $t = 40, 80, 160, 343$ s reveal formation of a peak temperature higher than the initial steady state. This measured inverse response was predicted by earlier simulations. It is caused by the different speed of propagation of the temperature and concentration disturbances (Il'in and Luss, 1992). These experiments show that the c-OFDR

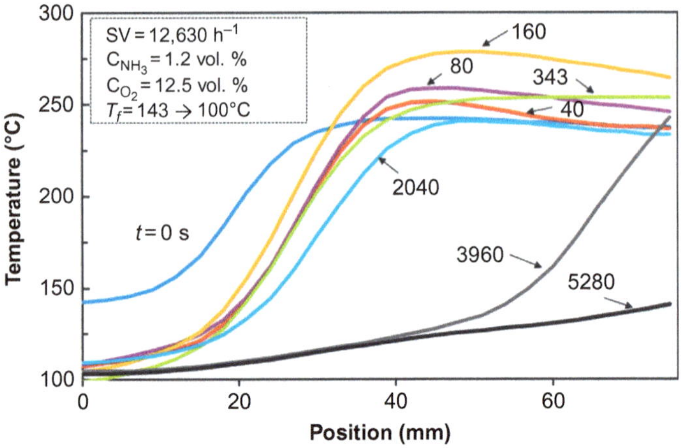


**Fig. 12** Wrong-way transient temperature rise caused by a sudden decrease of the feed temperature by 35 °C. Initial and final steady-state temperature are denoted as $t=0$ and 5280 s, respectively (Nguyen et al., 2013).

method enables a detailed mapping of transient temperatures inside a monolith.

# 3. SPATIOTEMPORAL TEMPERATURE AND CONCENTRATION MEASUREMENT BY c-OFDR AND SpaciMS

## 3.1 Spatial Concentration Measurement by SpaciMS

Measurement of spatiotemporal concentration profiles is essential to elucidate complex catalytic reactions, particularly those that involve multiple pathways and products. Numerous techniques have been developed to measure these. For example, temporal analysis of products (TAP) method provides detailed insight into catalytic reaction pathways and kinetics (Kumar et al., 2010). However, its operation under vacuum conditions, limits the applicability of its kinetics findings to catalysts operation under practical, high pressure recently deconditions. Moreover, TAP is not amenable to resolving spatial profiles.

The spatially resolved capillary inlet mass spectrometer (SpaciMS) captures the spatial concentration under practical operating conditions at various locations inside a reactor. SpaciMS was developed in 2005 by Choi, Partridge and et al. at Oak Ridge National Laboratory (ORNL) (Partridge et al., 2000). The method involves the insertion of a fused-silica capillary tube into a monolith channel to suck the reaction mixture for its composition analysis by a mass spectrometer at the capillary end. SpaciMS has been used to study a variety of catalytic reaction systems. For example, it enabled an in-situ measurements of $NO_x$ storage and reduction inside a monolith under atmospheric pressure. The technique found that sulfation increased ammonia production due to the shrinkage of the downstream oxygen storage zone (Choi et al., 2008). Concentration measurements by SpaciMS were used, among others, to study the steady-state features for an exothermic reaction (Bugosh et al., 2015), resolve species concentration in a sequential catalytic reaction system (Li et al., 2016), and to validate a global kinetic model (Shwan et al., 2014). These insightful findings demonstrate the usefulness of SpaciMS measurements. A remaining challenge is to determine if and to what extent can the probe blockage and suction impact the measured concentration.

## 3.2 Combined Spatiotemporal Measurements by SpaciMS and c-OFDR

Simultaneous in situ spatiotemporal concentration and temperature measurements in complex catalytic reaction systems provide insight that cannot be attained by use of only one of the two measurements (Bugosh et al., 2015; Choi et al., 2005, 2008; Li et al., 2016; Morgan et al., 2016; Shwan et al., 2014). While c-OFDR can measure the spatiotemporal temperature features, the SpaciMS can measure either the temporal concentration at one location or that at several locations but at different times. Thus, spatiotemporal concentration measurements by the SpaciMS of a periodic behavior (repeated experiments) require repeated measurements at many locations. This approach relies on the consistency of the repeated experiments. The combined use of a c-OFDR with SpaciMS can provide simultaneous spatiotemporal temperature and concentration profiles under steady-state or periodic conditions. The spatiotemporal temperature profiles by c-OFDR are, in general, more accurate than those of concentration profiles by the SpaciMS. We were the first to demonstrate the advantages of the combined use of SpaciMS and c-OFDR to measure spatiotemporal features of periodic behavior in monolith reactors (Nguyen et al., 2015, 2017).

The spatiotemporal concentration and temperature measurements by the SpaciMS and c-OFDR need to be conducted in a way that minimizes their impact on the system behavior. The SpaciMS uses a fused silica capillary probe and the c-OFDR an optical fiber. Both need to be inserted either in the same channel or adjacent channels in the monolith. Fig. 13 shows a case in which a 125 μm optical fiber and 170–363 μm capillary probe were

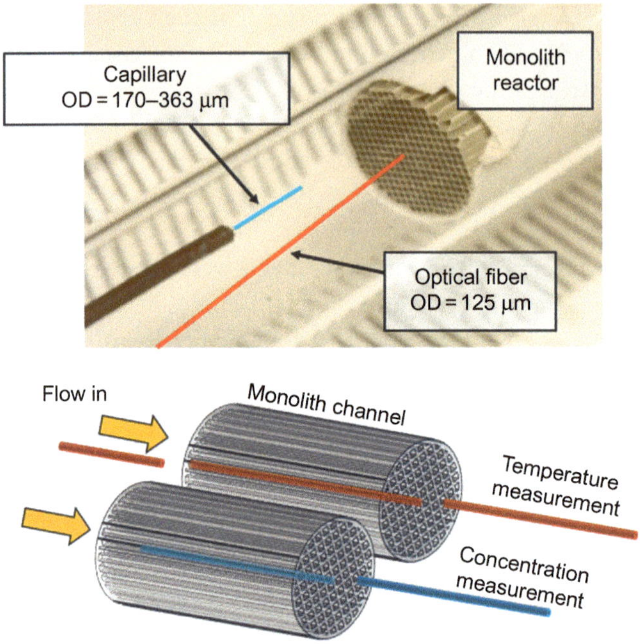


**Fig. 13** Concentration profile is measured with the SpaciMS. In the adjacent channel, temperature profile is measured with the optical fiber (Nguyen et al., 2017).

inserted into a monolith. The alignment direction of the optical fiber is independent of the flow direction. However, the capillary probe is usually inserted opposite to the flow direction for more convenient sampling. The impact of the probe diameter on the flow will be discussed later.

### *3.2.1 Combined Steady-State Concentration and Temperature Measurements*

A mixture of a 0.2 vol.% $C_3H_6$, in 10 vol.% $O_2$ and balance Argon was oxidized in both 100 and 600 cpsi monolith with a (2.3 g/in$^3$) Pt/$Al_2O_3$ washcoat. The space velocity was 12,630 h$^{-1}$ and the feed temperature was 210 °C. Fig. 14A shows the results for the 100 cpsi monolith in which both the optical fiber and capillary were inserted into the same channel, while Fig. 14B shows the results for the 600 cpsi monolith in which the two probes were inserted in adjacent channels. The propylene concentration monotonically decreased in the flow direction until it was depleted at 45 mm. The corresponding temperature monotonically increased and reached its peak at 45 mm, at which position the propylene was completely converted. Both the concentration and temperature stayed constant from

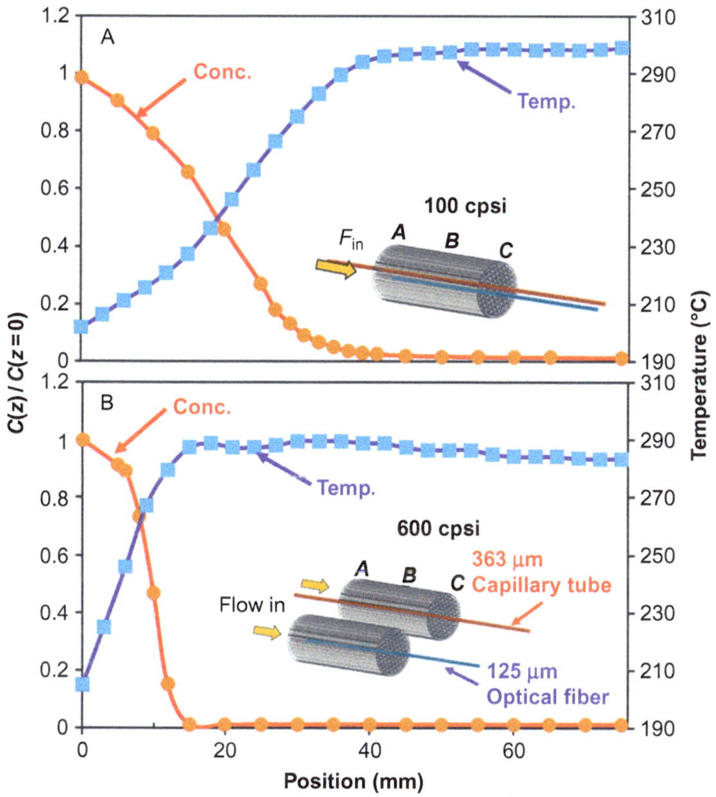


**Fig. 14** Spatial temperature and concentration profiles of the propylene combustion reaction (0.2 vol.% $C_3H_6$, 10 vol.% $O_2$, $SV=12{,}630\,h^{-1}$) on (A) 100 cpsi monolith reactor where optical fiber and capillary tube were in the same channel and (B) 600 cpsi monolith reactor where capillary tube was in adjacent channel to the optical fiber (Nguyen et al., 2017).

that point until the reactor exit. The temperature profiles reflect the extent of reaction (propylene conversion); for instance, more heat is generated as more propylene is consumed. For a single reaction, the temperature rise is proportional to the conversion and feed concentration (Eq. 2).

When using small monolith channels (600 cpsi), the optical fiber and capillary probe had to be placed inside adjacent channels to avoid interaction between the two sensors. Fig. 14B shows a good match between the spatial concentration and temperature profiles in the 600 cpsi monolith. Both concentration and temperature reached their extremum at the same position (at 15 mm). This shows that inserting the optical fiber and the capillary probe inside adjacent channels of a small channels monolith enables adequate matched concentration and temperature profiles. The conversion in the 600 cpsi exceeds that in the 100 cpsi due to the enhanced mass transport in the thinner washcoat (i.e., constant washcoat loading).

## 3.3 Applications

### *3.3.1 Invasiveness of SpaciMS Measurements*

Some investigators have raised questions about the accuracy of the concentration measurements by the SpaciMS due to the intrusive nature of the capillary; specifically, flow blockage, suction rate, and boundary layer transport limitations. When a capillary probe is inserted inside a channel it blocks part of the cross section of the channel. This decreases the flow rate in the channel and increases the residence time in the probe-occupied channel from those in the others. Sá et al. used computational fluid dynamics (CFD) to study the impact of blockage, assuming that the flow was equally distributed among all monolith channels (Sá et al., 2010). They concluded that the probe blockage impact was negligible. On the other hand, Hettel et al. argued that the assumption of equal flow distribution among all the monolith channels may not be adequate (Hettel et al., 2013). Using CFD to study the catalytic partial oxidation of methane, they demonstrated that the probe blockage impact was significant even if the probe occupied just 3.5% of the channel area. Moreover, the conversion in the channel with the probe exceeded that in the channels without a probe. The position of the probe within the channel cross section affects the blockage impact. A probe in the channel center has the largest blocking impact (Hettel et al., 2013).

An increase in the probe suction rate increases the total flow rate into that channel. The impact of this decrease in the channel residence time may compensate for the blockage impact. That is, the shift of the concentration profile by the blockage may be compensated for by the suction. This may help to explain why some researchers found that the impact of the SpaciMS probe intrusion was negligible, while others disagree. The ability of the suction to compensate for the blockage depends on the magnitude of the blockage. Hence, it is important to predict and quantify both the blockage and the suction effects. Gossler et al. (2016) evaluated through computational modeling the impact of a cylinder within a monolith channel of different shapes. Their results show the expected result that the flow rate through a channel decreases as the probe occupies an increasing fraction of the channel. For the case of a square cross-section channel, the flow blockage is the largest when the probe in the geometric center of the channel while the least blockage occurs if the probe is in the corner. It would be interesting to investigate the impact on the solution if suction is included.

A calculation method reported in Nguyen et al. (2017) predicts that a higher total volumetric flow rate and larger capillary diameter increases the blockage impact.

It is essential to determine when a SpaciMS accurately measures the spatiotemporal concentration, and what is the impact of the probe axial position inside the monolith channel and the blockage and suction effects. To investigate this we conducted propylene oxidation in four Pt–$Al_2O_3$ washcoated monoliths of different channel densities (100, 200, 400, and 600 cpsi) over a range of space velocity (12,438–18,657 $h^{-1}$). The spatial temperature was measured by c-OFDR using a 125 μm outer diameter optical fiber (Nguyen et al., 2013). The corresponding spatial concentration was measured by a SpaciMS using two capillary probes (170 and 363 μm). We conducted 37 experiments to gain insight and provide some guidelines about the accuracy of the measured concentrations. We studied the impact of the axial distance, defined as $\Delta Z$, between the position at which of the reactant was completely depleted and the temperature attained its peak. Note that under adiabatic operation of a single overall reaction that distance should be zero ($\Delta Z = 0$). Thus, a very small $\Delta Z$ implies negligible probe intrusion, while a large $\Delta Z$ implies a strong intrusion effect. We tried to correlate the measured $\Delta Z$ as a function of $Z'$, defined as the dimensionless position at which the propylene conversion is complete, and $\log(\phi)$, where $\phi$ is the dimensionless ratio between the transverse diffusion time ($\tau_d$) and the reaction time ($\tau_r$):

$$\phi = \frac{R_{\Omega 1}^2 \times k_1}{D_f \times C_T} \qquad (4)$$

$$\phi = \tau_d / \tau_r$$

where $R_{\Omega 1}$ is the effective flow transverse diffusion length, $D_f$ is the diffusivity of the limiting reactant species, $C_T$ is the total gas phase concentration, and $k_1$ is the temperature-dependent reaction rate constant. The equation derivation and fitting results are reported in Nguyen et al. (2017).

Fig. 15 is a color map of $\Delta Z$ as function of $Z'$ and $\log(\phi)$. The main features of the $\Delta Z$ map are:

- The smallest values of $\Delta Z$ are mostly located in the downstream region ($Z' > 0.6$).
- The region of no apparent intrusion (black region in Fig. 14) is distributed in the range of $Z' > 0.6$ and $\log(\phi) > \sim 3$.

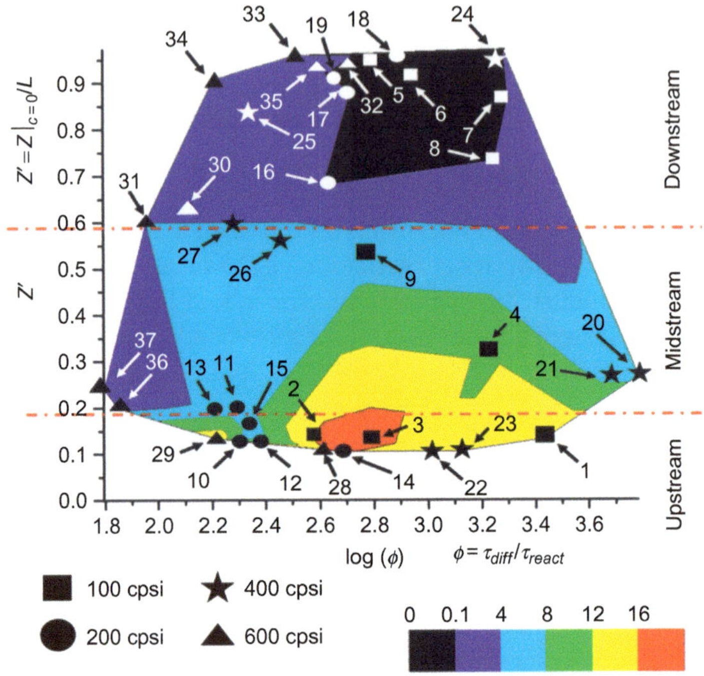


**Fig. 15** 3D map of log($\phi$) vs $Z'$ vs $\Delta Z$ for 37 experiments which were conducted on 100, 200, 400, and 600 cpsi monolith reactors (Nguyen et al., 2017).

- The $\Delta Z$ values are in the range of 4–20 mm in the midstream region, except one blue region when log($\phi$) $< \sim 2.1$.
- The largest $\Delta Z$ values are located in the upstream region ($Z' < 0.2$).

The $\Delta Z$ map indicates that a large mismatch between the temperature and concentration profiles occurred in the upstream region. This mismatch is due to an inaccuracy of the SpaciMS measurements. The probe blockage effect is affected by the intrachannel flow rate. Thus, it is expected that the blockage effect will be more dominant at higher flow rates and larger capillary diameters. However, the map indicates that relatively small $\Delta Z$ values are observed at many high flow rates and large capillary diameters. In contrast, larger $\Delta Z$ were observed at low flow rates and small capillary diameters. These unexpected results suggest that $\Delta Z$ is affected not only by the blockage and suction but also by other effects.

Another potential problem is the difficulty of interpreting data for which the concentration and temperature profiles are not in agreement. Such a mismatch may result when the temperature measured by the fiber is at a location different from that measured by the SpaciMS and there are sharp temperature and concentration transverse gradients in that cross section.

The propylene oxidation reaction is very fast in the upstream so that the solid temperature is likely to exceed that of the fluid by more than 10°C. This will cause a negative temperature gradient in the gas next to the surface as the c-OFDR is likely to measure the fluid temperature due to its larger contact area with the fluid. However, the SpaciMS probe measures the fluid concentration next to solid surface. Therefore, when the difference between the solid and fluid temperature is significant, the $\Delta Z$ value will also be high. We define the $\Delta Z$ caused by the gradient between solid and fluid as the transverse gradient effect.

Fig. 16 demonstrates the impact of the transverse gradient effect in a 100 cpsi monolith reactor. Fig. 16A shows that a small $\Delta Z$ was obtained

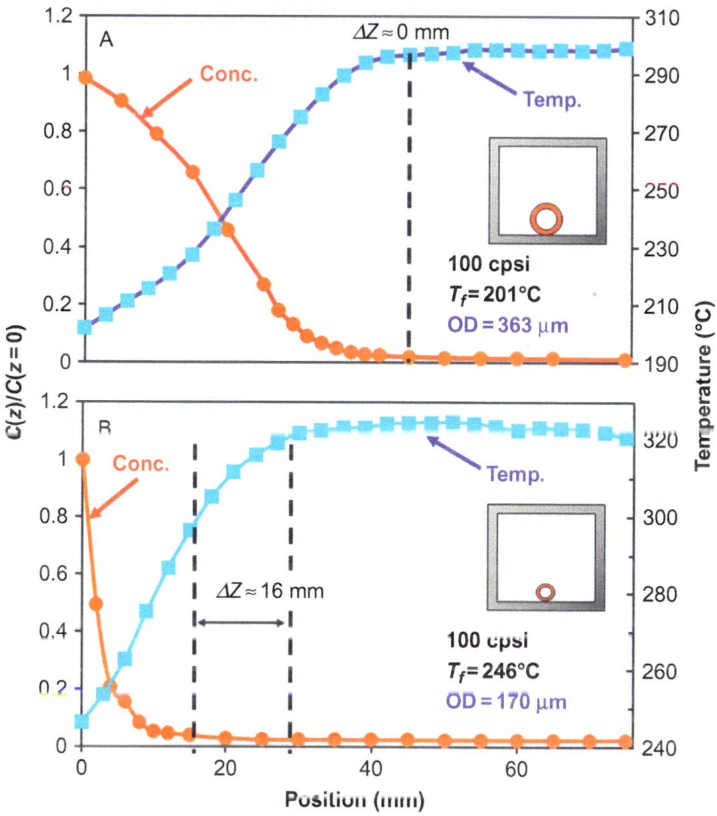


**Fig. 16** The propylene concentration and temperature profiles were measured in a 100 cpsi reactor with the condition of $C_3H_6 = 0.2$ vol.%: (A) SV = 13,993 h$^{-1}$, $T_f = 201$ °C with a capillary tube (OD = 363 μm). (B) SV = 12,438 h$^{-1}$, with a capillary tube (OD = 170 μm). Both the optical fiber and the capillary probe are in the same channel (Mukai et al., 2004).

using a large diameter capillary probe (363 μm) which occupied 3% of the channel area at a space velocity of 13,993 $h^{-1}$. Fig. 16B describes a case in which a smaller capillary probe (170 μm OD), led to a high $\Delta Z$ value of 16 mm at the lower space velocity of 12,438 $h^{-1}$. A thinner capillary probe and a lower space velocity are expected to decrease the blockage effect. By comparing the two cases in Fig. 16, we conclude that the conditions expected to decrease the blockage effect did not reduce the $\Delta Z$ values. The high $\Delta Z$ observed in Fig. 16B should be considered as being caused by the transverse gradient effect. The occurrence of a transverse gradient does not mean that the concentration measurement is inaccurate but that a disguised result can be caused by a fast surface reaction.

We can conclude the determining the conditions under which the SpaciMS measurements are accurate cannot be predicted just by the location in the two-dimensional map of $Z'$ vs log($\phi$). Mathematical modeling may be used to determine the conditions (space velocity, mass, heat transfer, etc.) which are most likely to generate a transverse gradient. While the $\Delta Z$ map does not provide a perfect guideline about the measurements accuracy, it points out some operating conditions that lead to inaccurate results.

### 3.3.2 Spatiotemporal Features of Lean/Rich Cycling

Lean/rich cycling is used to improve catalyst performance by promoting certain reactions and inhibiting others, particularly when there are some competing reactions. For example, cerium oxide (Ceria) stores $O_2$ in an $O_2$-rich environment and releases it in an $O_2$-deficient environment. The three-way catalyst (TWC) is used in the emission after treatment of stoichiometric gasoline engines. The TWC conventionally contains precious metal(s) and ceria which promotes the conversion of CO, nonmethane hydrocarbons (HC), and $NO_x$. The combination of oxygen storage and cycling enables this performance. During the rich part of the cycle stored oxygen is released, promoting the conversion of $NO_x$ while during the lean part of the cycle oxygen is stored and CO and HCs are converted. Understanding the details of the reaction pathways and product distribution spanning the lean + rich cycle enables optimization of the catalyst and its operation.

Fig. 17 is a schematic illustration of the HC (propylene) oxidation on a $Pt/Rh/CeO_2/BaO$ catalyst during simulated lean/rich cycling. In this experiment, we focus on the storage and release of oxygen, together with the propylene oxidation. During the lean phase, $O_2$ is fed and stored by

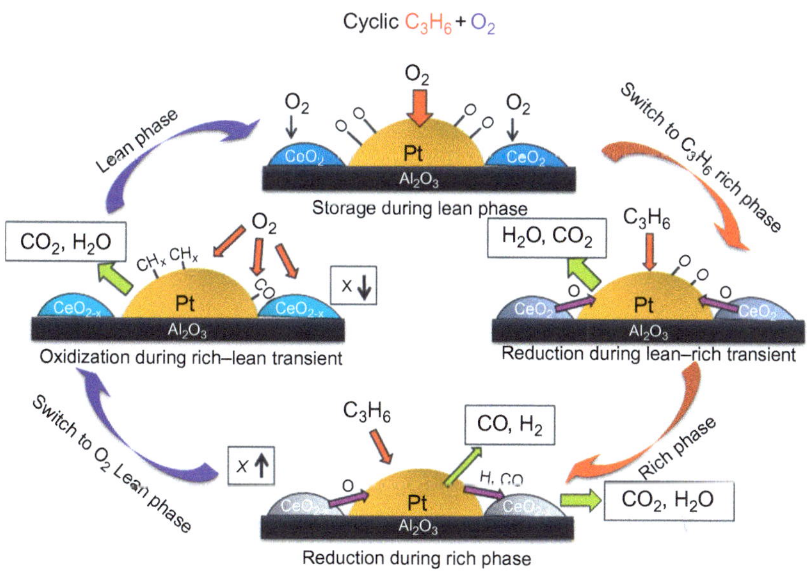


**Fig. 17** Schematic representation of the cyclic mechanism for the propylene oxidation reaction (Nguyen et al., 2015).

the initially reduced ceria and partially by the Pt surface. Upon switching to a rich phase, propylene is gradually fed and reacts with the oxygen on the Pt surface. Since the Pt surface is fully covered by oxygen atoms, propylene is easily converted to $CO_2$ and $H_2O$. Meanwhile, ceria continuously provides oxygen atoms to the Pt surface, preventing an oxygen deficiency. During the rich phase, when $C_3H_6$ is continuously fed into the catalyst, it becomes difficult for ceria to maintain a continuous supply of oxygen to the Pt. As a result, while a fraction of the propylene is fully oxidized, the remaining fraction is either unconverted or converted to partial oxidation products CO and $H_2$. During the switch back from the rich to lean phase propylene that accumulates on the catalyst surface reacts with freshly supplied $O_2$, leading to deep oxidation products $CO_2$ and $H_2O$.

The continuous cycling improves the $CO_2$ yield, the desired product in this application. Although the use of cycling is relatively simple, additional factors also determine the catalyst performance. For instance, the formed CO is strongly adsorbed and inhibits its own and propylene oxidation, the extent of which decreases with increasing temperature. The exothermic water gas shift (WGS) reaction occurs at a temperature as low as 200 °C while the reverse WGS reaction occurs at high temperatures. The steam reforming (SR) reaction occurs at temperatures in the 350–400 °C range.

This temperature-related coupling suggests that a complicated reaction behavior may occur during an exothermic HC oxidation reaction, especially when the temperature profiles in the reactor are not uniform. On the other hand, the concentration distribution for each chemical species further determines which reaction dominates. Clearly, knowledge of the spatiotemporal features during the cyclic HC oxidation reactions is essential to explain how the chemical species interact, when the oxygen is released, and where the reactions occur.

An example illustrating the advantages of simultaneous measurement of the spatiotemporal temperature and concentration of an exothermic reaction under lean–rich cycling is the oxidation of propylene on a Pt/Rh/$CeO_2$/BaO catalyst. The initial feed temperature was 300 °C and the space velocity 12,630 $h^{-1}$. A rich feed containing 1.2 vol.% propylene was fed for a short time (5 s) followed by a lean feed containing 10 vol.% $O_2$ fed for a longer time (30 s). The cycle was repeated until a stable cyclic state was achieved. Fig. 18 displays the spatiotemporal temperature profiles and the $H_2$, $CO_2$, and $C_3H_6$ concentrations at the cyclic state. The transient behaviors were the same in each cycle. The profile labeled as 0 s is the start of the R/L switch and the 30 s the start of the L/R switch. The R/L switch is shown on the left-hand side and L/R switch on the right-hand side in Fig. 18. In Fig. 18H, as the rich phase was first fed, the propylene concentration profile transformed from a uniform 0 vol.% profile to a monotonically decreasing profile. Accumulation of propylene began in the upstream and it was completely consumed around the 6 cm position. Meanwhile, $H_2$ gradually was formed. The $H_2$ profiles at first monotonically decreased between 31 and 32 s but attained a local maximum at 33 and 33.7 s. Fig. 19H shows that for a ceria-free catalyst, the $H_2$ monotonically grows along the catalyst and has a higher total $H_2$ concentration. Comparing both catalysts shows that the addition of ceria indirectly inhibited the WGS reaction: $H_2$ and CO were consumed at the catalyst downstream by either conversion to $H_2O$ and $CO_2$ or storage by the ceria. During the R/L switch as $O_2$ is just fed (see Fig. 18C), the $CO_2$ highest concentration is at around the 60 mm position. A large amount of $CO_2$ formation affects the temperature rise in Fig. 18A. The temperature profiles exhibited a local maximum at the 60 mm position at the 3 s time. In contrast, in ceria-free catalyst the reaction had monotonically increasing $CO_2$ and $H_2$ profiles in both the $CO_2$ and $H_2$. More detailed experimental conditions are discussed in Nguyen et al. (2015).

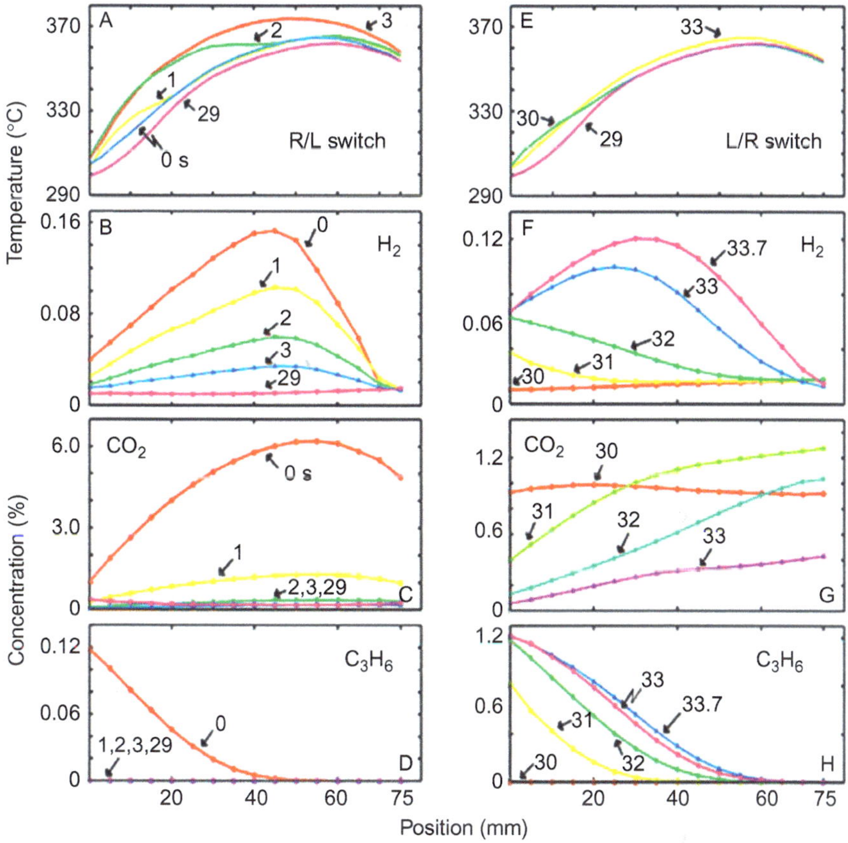


**Fig. 18** Spatiotemporal profiles for ceria-containing catalyst (PRBC) in a periodic lean–rich steady state. $t=0$ is the starting time for feeding $O_2$ into the reactor (R/L switch). $t=29$ indicates the time at which the L/R switch begins. We did not report $CO_2$ profile at $t=33.7$ s because the measurement was unreliable (Nguyen et al., 2015).

In the past, catalyst performance was studies by measuring the feed and effluent chemical species concentration and temperature. It was very difficult to assess the impact of the behavior inside the catalyst due to the lack of spatiotemporal information. We were aware of the propylene reduction by the ceria, but details about the reaction extent including primary and secondary products were unavailable. This introduced uncertainty into the design and optimization of the catalyst formulation and reactor. Information about and understating of the spatiotemporal behavior is critical for a reliable design of a better operating strategy and reactor.

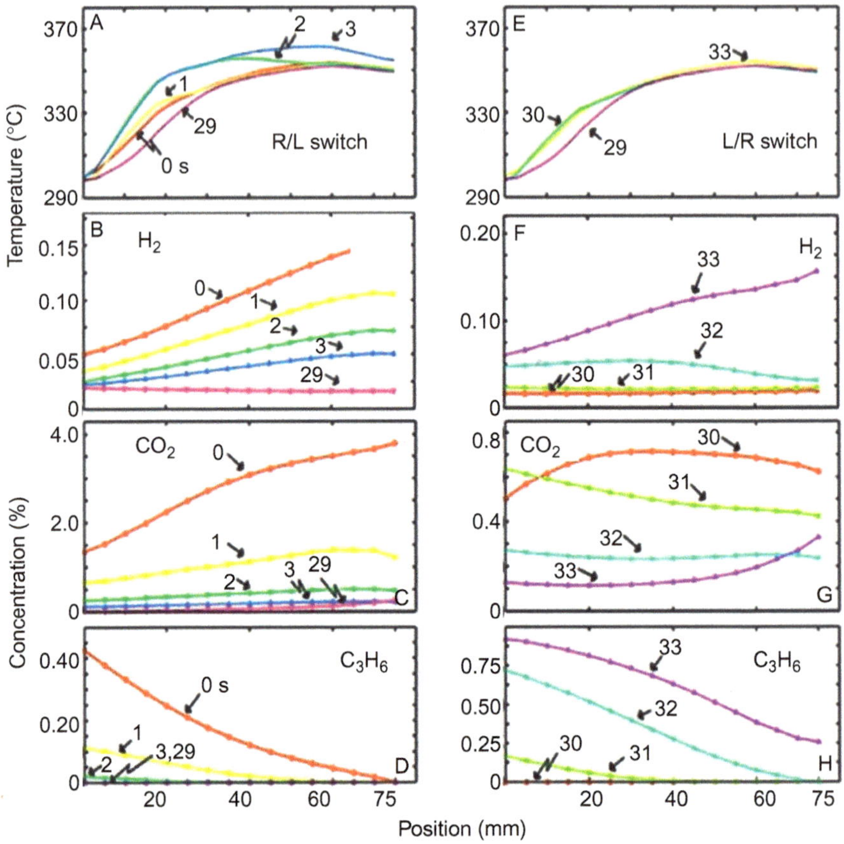


**Fig. 19** Spatiotemporal profiles for nonceria containing catalyst (PR) in a periodic lean–rich steady state. $t=0$ is the starting time for feeding $O_2$ into the reactor (R/L switch). $t=29$ indicates the time at which the L/R switch begins (Nguyen et al., 2015).

### *3.3.3 Spatiotemporal Features of the Catalytic Hydrocarbon Trap*

Simultaneous SpaciMS and c-OFDR measurements are a powerful tool to determine the spatiotemporal features of a reaction system. Thus far we have illustrated this tool for an adiabatic exothermic reaction conducted at steady-state and transient cyclic storage and reaction. To broaden the applicability of this method, it should be generally applied to follow the spatiotemporal behavior of a reaction system that is exposed to continuously changing conditions. Moreover, there is a need for a strategy that interprets the spatiotemporal data.

A reactor system that is subject to transient conditions and for which the spatiotemporal features are not well known or understood is the catalytic HC trap used to treat diesel vehicle exhaust. The role of the HC trap is to compensate for insufficient HC oxidation capability of the diesel oxidation catalyst (DOC) during engine cold-start ($T < 150°C$). This device has been described mostly in the patent literature and to a lesser extent in the open literature (Ballinger et al., 1997; Engler et al., 1993; Heimrich et al., 1992; Yamamoto et al., 2000; Zones et al., 2013). The DOC generally performs well for HC or CO oxidation at temperatures above the light-off temperature ($T > 200$–$300°C$). However, during the cold-start a large fraction of HCs are not oxidized and emitted from the DOC to the environment (Mukai et al., 2004). In developing a HC trap-capable DOC, a zeolite is added to store HC during the cold-start and to release them when the catalyst is hot. A trap is effective if it stores more than 90% of the HC during the cold-start (Chang et al., 2014; Elangovan et al., 2005, 2006; Li et al., 2005; Mukai et al., 2004; Murakami et al., 2004). Many types of zeolites, having different pore sizes and structures have been tested as HC trap components. The emitted diesel HCs are typically linear, long chain alkanes (C16–C24) while gasoline emitted HCs include aromatics and medium molecular weight including branched alkanes (C6–C10) (Kaiser et al., 1991). Many studies have shown that HC trapping capacity can be improved by using different types of zeolites (Elangovan et al., 2006; Iliyas et al., 2007; Migliardini et al., 2015; Navlani-García et al., 2013a, b; Ogura et al., 2007; Puertolas et al., 2014; Serra et al., 2012), changing the zeolite structure (Navlani-García et al., 2013a, b), tuning the surface chemistry (Westermann et al., 2014), and coating methods (Li et al., 2005). Bugosh and Harold recently studied the impact of zeolite loading on the light-off of CO in a mixture with dodecanal (Bugosh and Harold, 2017). Knowledge is still lacking about the interaction of stored-HC with other emission species, and how it may affect the light-off value during the cold-start. An enhanced understanding of the trapping, desorption, and oxidation processes may help to develop more efficient traps.

We used simultaneous SpaciMS and c-OFDR measurements to study the transient behavior of HC trapping and oxidation during the cold-start. First, a feed containing dodecane was used to determine the trapping capacity of the catalyst. Inert gas flow was then used to remove weakly bonded HC. Measuring the spatially resolved carbon species concentration during a temperature programmed oxidation (TPO) revealed multiple of HC light-off events. To enhance our understanding of the zeolite beta (BEA) impact on the HC oxidation, we studied the reactions under steady-state conditions at various space velocities. The interpretation of the conversion benefitted from the detailed spatiotemporal data.

#### 3.3.3.1 Temperature Programmed Oxidation (TPO)

A TPO was used to verify the system ability to trap and oxidize the unwanted HC from the ambient temperature to the reaction temperature. We expected that adding BEA to trap large HC molecules will decrease the hydrocarbon release during the temperature ramp. To examine this, before the TPO start, we stored the HC in a 400 cpsi, 3 in. long catalyst containing precious metal (Pt) and HC adsorbent BEA at the initial temperature of 60°C. (Details of the catalyst are reported in Bugosh and Harold, 2017.) Fig. 20 exhibits a typical two-point measurement result, starting with the adsorption then desorption and oxidation. A feed contained 150 ppm dodecane ($C_{12}$) and 10 vol.% $O_2$ with the balance Argon was fed at a space velocity of 12,438 $h^{-1}$. The feed $C_{12}$ concentration is denoted by a yellow line, the effluent $C_{12}$ concentration is labeled by a blue line, the effluent $CO_2$ concentration is denoted by a red line, and the spatially averaged catalyst temperature by a green line.

At the start of the experiment, $C_{12}$ concentration was not detected in the effluent due to its trapping by the catalyst. As the catalyst adsorption sites gradually filled by the $C_{12}$, the effluent $C_{12}$ concentration increased and

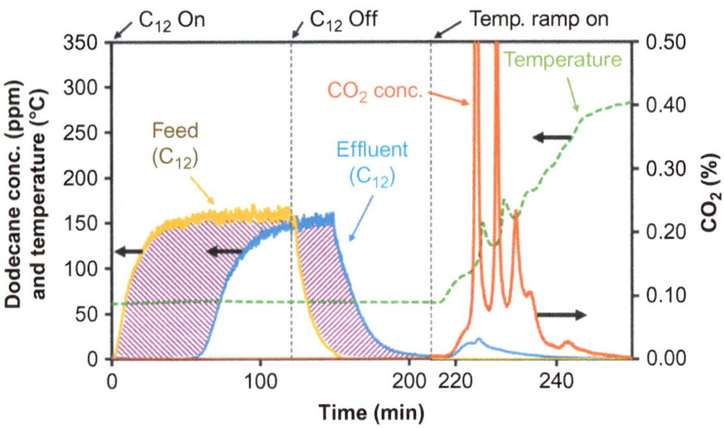


**Fig. 20** $C_{12}$ and $CO_2$ concentration measured during adsorption and desorption followed by a TPO in a monolith catalyst with Pt/Pd/$Al_2O_3$ washcoat and added β-zeolite. Feed during saturation: 150 ppm dodecane and 10% $O_2$, the balance Ar. Feed during desorption and TPO: 10% $O_2$, balance Ar.

eventually reached the feed concentration of 150 ppm. By integration of the depletion area between the feed and effluent $C_{12}$ lines, the $C_{12}$ uptake of the catalyst was determined as 118 mg $C_{12}$/g (WC). This uptake is either by physical or by chemical adsorption. When the catalyst was saturated, the feed was switched to an inert feed of 10% $O_2$ and 90% Ar to remove the physically bonded $C_{12}$. The decrease of the effluent $C_{12}$ concentration is a result of the release of this weakly bonded $C_{12}$. The depletion time of the physically bonded $C_{12}$ is that at which effluent $C_{12}$ concentration vanishes. The amount of weakly bound $C_{12}$ desorbed by purging (of 45.4 mg $C_{12}$/g (WC)) was less than the amount of $C_{12}$ adsorbed. This indicates that some $C_{12}$ has been strongly trapped by the catalyst and could be removed at a higher temperature. No $CO_2$ formed during the purge.

Upon commencement of the temperature program procedure, an unexpected transient behavior occurred. During the 5 °C/min ramp the effluent $CO_2$ profile showed three peaks and a few minor peaks. The multiple $CO_2$ peaks indicate multiple $C_{12}$ light-offs. This behavior differs from the conventional single HC oxidation light-off during TPO. The temperature profiles confirm the occurrence of multiple $CO_2$ peaks, as the exothermic $CO_2$ formation leads to an instantaneous temperature rise. We conjecture that the cause of this may be intermittent $C_{12}$ release at different positions along the catalyst, but the data do not explain the mechanism causing this behavior. We describe below how the cause for the occurrence of multiple $CO_2$ formation peaks was resolved by use of simultaneous SpaciMS and c-OFDR measurements.

#### 3.3.3.2 Spatially Resolved Hydrocarbon Trapping

We measured the $C_{12}$ breakthrough curves at various positions inside the monolith channel by SpaciMS to check if the multiple $C_{12}$ light-off were due to a nonuniform $C_{12}$ trapping. Fig. 21 shows the temporal $C_{12}$ concentration at five axial positions: 0, 18, 36, 54, and 76 mm. The five breakthrough curves are evenly distributed, indicating that the $C_{12}$ is uniformly accumulated on the catalyst. Thus, the multiple light-off behavior was not due to inhomogeneous $C_{12}$ trapping. As no zeolite was loaded upstream of the catalyst, the delay at 0 mm curve indicates a measurement error by the MS. We conjecture that it was caused by the dead volume within the suction line. The delayed measurement affected only the $C_{12}$ reading but not the other light molecular weight species, such as $CO_2$. This error region was subtracted during the $C_{12}$ uptake calculations. To determine the $C_{12}$ uptake, we integrated the area between the feed and measured

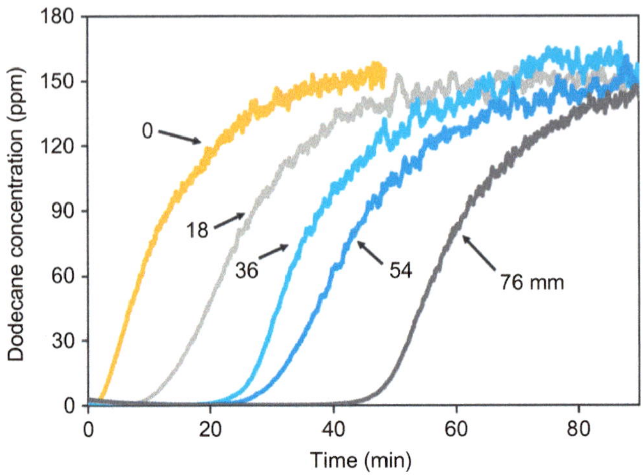


**Fig. 21** Temporal effluent dodecane concentration during hydrocarbon storage experiments ($T=60$ °C, 150 ppm $C_{12}$, 10% $O_2$, balance Ar, SV = 12,438 $h^{-1}$).

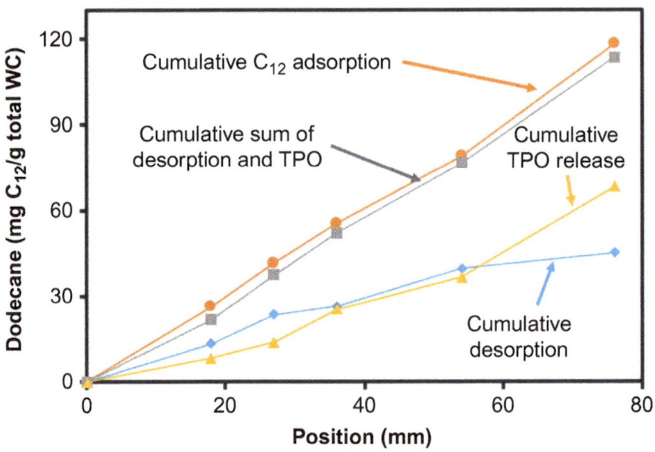


**Fig. 22** Cumulative amount of Dodecane (or equal molar $CO_2$) per total washcoat weight during the adsorption, desorption, and TPO.

$C_{12}$ concentration at various positions. Fig. 22 describes the calculated cumulative $C_{12}$ uptake at as a function of position. The linear $C_{12}$ uptake indicates that the $C_{12}$ trapping ability is spatially independent.

### 3.3.3.3 Spatially Resolved TPO

SpaciMS and c-OFDR measurements were conducted during TPO to check whether the multiple $C_{12}$ light-off behavior is spatially dependent.

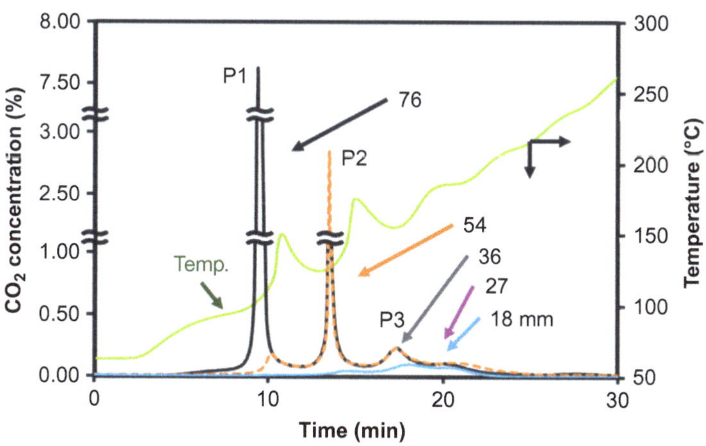


**Fig. 23** $CO_2$ concentration at different locations in the $C_{12}$ prestored catalyst during TPO experiment with a feed of 10% $O_2$, balance Ar, SV=12,438 h$^{-1}$, and heating-up rate of 5.0 °C/min.

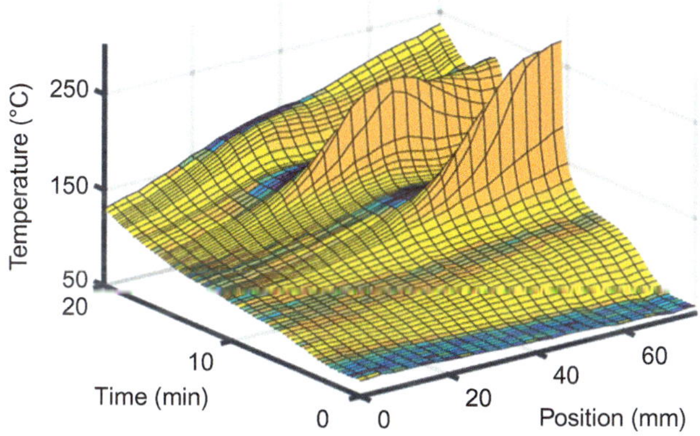


**Fig. 24** Spatially resolved temporal temperature profile with $C_{12}$ prestored catalyst with the feed of 10% $O_2$ balanced with Ar.

The measured results are shown in Figs. 23 and 24. We use the $CO_2$ concentration as a measure of the dodecane oxidation. Fig. 23 shows the temporal $CO_2$ concentration at various positions along with the spatially averaged temperature. The number of temporal $CO_2$ peaks in the upstream section is smaller than those in the downstream section. For example, at position 18 mm three peaks occur between 14 and 21 min. These are much lower in intensity (~0.1% $CO_2$) than the four peaks observed for position

54 mm between 10 and 21 min. Thus, the intensity and the number of peaks depend on the axial position. The axial position affects the peak number and intensity but not the timing, indicating $C_{12}$ light-off occurred at different positions. According to the data, the first peak occurred at the downstream (76 mm) and generated the highest $CO_2$ peak intensity (~7.5% $CO_2$). The second and third peaks were much smaller than the first, indicating that most of the $C_{12}$ was consumed downstream.

As shown in Fig. 24, which is a 3D plot of temperature vs time and position shows the position and timing of the temperature rise. The first temperature rise peak occurred in the downstream, the second peak in the midstream, and the third peak in the upstream. The time intervals between the peaks are the same for the $CO_2$ light-off. The height of the temperature rise is in agreement with the decreasing $CO_2$ peaks intensity. In addition, the c-OFDR temperature measurements display many details about the spatial exothermic reaction during the TPO. For example, the temperature rose faster in the downstream in the beginning, suggesting that the time sequencing of $C_{12}$ light-off might be caused by the temperature gradient. The higher downstream temperature triggers the first $C_{12}$ light-off while the cooler upstream sections were not ready to light-off the $C_{12}$. Instead of a continuous oxidation of all the stored $C_{12}$, the $C_{12}$ release was intermittent. This behavior suggests a competition between the strong BEA uptake of the $C_{12}$ and the $C_{12}$ oxidation. Although the catalyst downstream was active, the $C_{12}$ uptake onto vacant adsorption sites in the cooler upstream section prevented the occurrence of a continuous reaction. The stored $C_{12}$ in the midstream and upstream could ignite only after these locations were sufficiently hot to cause a light-off.

The amounts of $C_{12}$ desorption during the purge and $C_{12}$ consumption during the TPO were calculated by integrating the area beneath the effluent $C_{12}$ and $CO_2$ concentrations. Based on carbon species conservation, we assumed that a $CO_2$ concentration multiplied by 12 is equivalent to the $C_{12}$ concentration. Fig. 22 shows that the total released amount by summation of the desorbed and the TPO released amounts are linear with position, indicating a uniform $C_{12}$ trap. The total hydrocarbon released amount was close to the $C_{12}$ uptake, indicating $C_{12}$ was completely released from the catalyst. Thus, we conclude that no other side products formed. Also, the linear-spatial dependence of the TPO release proves that the multiple peaks occurrence was not caused by a nonuniform $C_{12}$ trap, but was due to the spatially interactive $C_{12}$ readsorption/light-off along the catalyst.

### 3.3.3.4 Spatial Features of Oscillating Trapping and Oxidation

As reported above, the competition between the HC trapping and oxidation led to an intermittent light-off behavior during TPO in a HC prestored catalyst. This intermittent light-off behavior occurred even under steady-state conditions of constant feed temperature and space velocity. We refer to this intermittent behavior as periodic oscillations.

We followed this behavior by SpaciMS and c-OFDR measurements during experiments conducted with a feed temperature of 135°C at various space velocities (9329, 12,438, and 18,657 $h^{-1}$) using the same as the one used in the previous section. The feed contained 150 ppm $C_{12}$ and 10 vol.% $O_2$ in Argon. The transient concentration and temperature profiles are shown in Fig. 25A–C

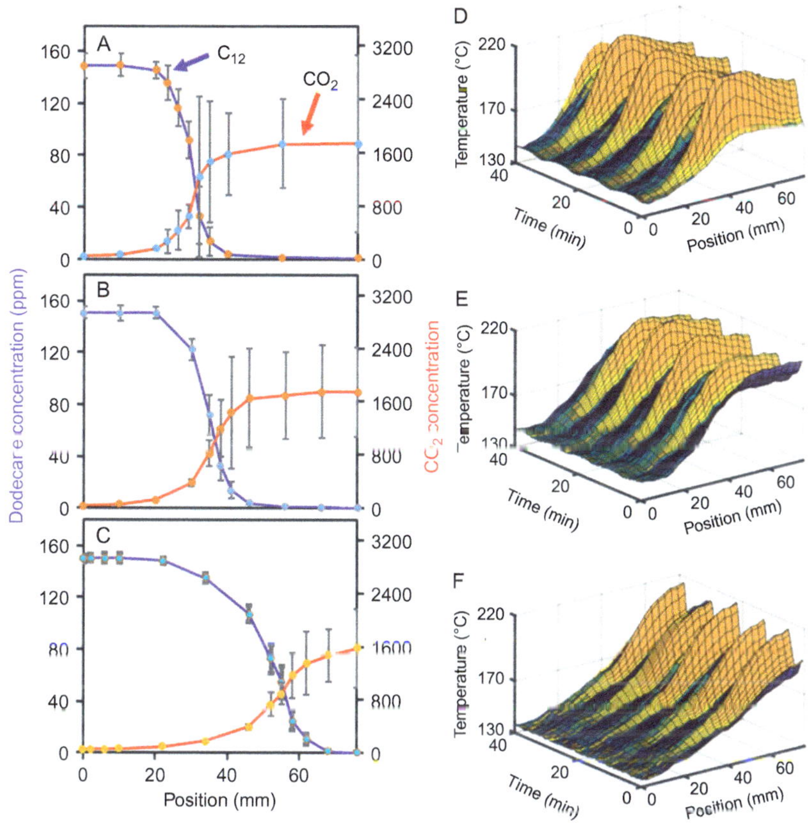


**Fig. 25** Spatially resolved $C_{12}$ and $CO_2$ concentration and spatiotemporal temperature profiles at steady state at 135°C feed temperature at SV of (A) and (D) for 9329, (B) and (E) for 12,438, (C) and (F) for 18,657 $h^{-1}$ with the feed of 150 ppm $C_{12}$ and 10% $O_2$ balanced with Ar.

and D–G, respectively. Fig. 25A–C displays the $C_{12}$ and $CO_2$ concentration as a function of the position in the reactor at various space velocities. The amplitude of the periodic oscillating concentration is denoted as the vertical segment intersecting each data point. The spatiotemporal temperature profiles are presented in a 3D plot to convey the time-dependent periodic oscillating behavior. Fig. 25A shows that at a space velocity of 9329 $h^{-1}$, the $C_{12}$ started to decrease at the position of 20 mm and was depleted at 40 mm. The $CO_2$ concentration started to grow at 20 mm and had a peak of 1800 ppm at 40 mm (complete consumption of the 150 ppm $C_{12}$ feed). These results indicate that $C_{12}$ was completely converted into $CO_2$ and that the reaction occurred at a relatively small region between 20 and 40 mm. At higher space velocity, the breakthrough curves were shifted to the downstream and the reaction regions expanded. Fig. 25B shows that for a space velocity of 12,438 $h^{-1}$, the $C_{12}$ decreased at 20 mm and was completely oxidized at 45 mm. The $CO_2$ formed at 20 mm and saturated at 45 mm. Fig. 25C that for a space velocity of 18,657 $h^{-1}$, $C_{12}$ decreased at 20 mm, and depleted by 70 mm, while $CO_2$ formed at 20 mm and saturated at 70 mm. A shorter reaction region at a lower space velocity implies that due to the higher residence time the HC was completely oxidized at a smaller catalyst volume. A long-tail in the $CO_2$ concentration profile at the higher space velocity suggests a strong competition between HC trapping and oxidation.

The amplitude of the oscillating $CO_2$ concentration was larger at lower space velocities (higher residence time). This larger oscillating amplitude was probably due to the less effective external mass transfer that made it more difficult to release the HC from BEA. The higher HC concentration accumulated on the BEA triggered a more abrupt release, leading to a larger $CO_2$ oscillating amplitude. At higher space velocity the HC mass transfer was more effective, so that less HC was accumulated in the BEA, leading to a smaller oscillating amplitude. The spatiotemporal temperature measurements supported the concentration observations. Fig. 25D–F shows that the temperature rise was larger at a lower space velocity. The 3D spatiotemporal temperature profiles exhibits a synchronized periodic oscillating behavior, providing direct evidence of the competition between the HC trapping and oxidation.

#### 3.3.3.5 Mechanism of Multiple Hydrocarbon Light-Off During TPO

The occurrence of multiple light-off in the DOC/BEA catalyst during the TPO experiment is illustrated in Fig. 26. The HC prestored catalyst was

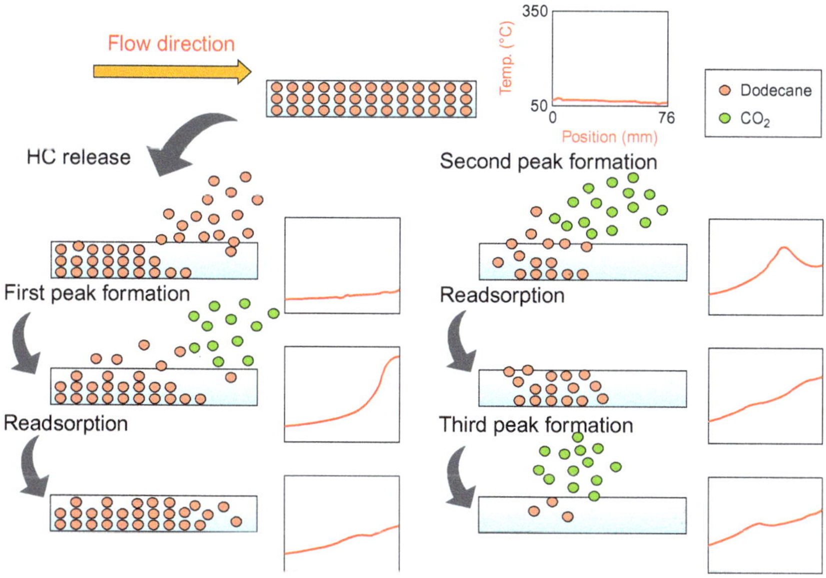


**Fig. 26** Illustration of the mechanism of the multiple-stage light-off during the TPO experiment.

subjected to a temperature ramp-up. The downstream temperature gradually increased when the furnace temperature exceeded the feed flow temperature. When the temperature exceeded the desorption temperature, the $C_{12}$ was first released in the downstream due to the high temperature. During the $C_{12}$ light-off in the downstream, the first $CO_2$ peak occurred and a sudden temperature rise appeared in the downstream. In the meanwhile, partially released $C_{12}$ from the midstream and upstream traveled to the downstream and was recaptured by the downstream sites emptied from the stored $C_{12}$. The temperature was not high enough to oxidize the upstream released $C_{12}$, so its temperature decreased due to insufficient reactant for a continuous reaction. This recapture caused a relocation of $C_{12}$ from upstream and midstream to downstream. When the midstream temperature became sufficiently high, another $C_{12}$ light-off occurred and the second peak appeared. The second light-off caused a temperature rise in the midstream section. The same scenario happened again for the third light-off until all the prestored HC exhausted. The powerful spatiotemporal measurements by SpaciMS and c-OFDR provided insight that led to a thorough understanding of this complicated coupling process.

## 4. CONCLUSIONS

c-OFDR is a sophisticated method that is capable of following the spatial–temporal temperature inside a reactor. The c-OFDR enables a nonintrusive measurement of the spatial temperature by shining a laser beam through an optical fiber and measuring the local reflected scattering. The local reflected scatter frequency is a function of the optical fiber local temperature as the local optical fiber density depends on the temperature. One can measure the spatial temperature along the optical fiber by referencing it with a reference beam and a reference sate. The c-OFDR has very good spatial and temporal resolution of 3 mm increment and 0.9 Hz frequency. The maximum operating temperature depends on the shielding material and is up to 800°C when using a gold coating. The main advantages of the c-OFDR are high noninvasive spatiotemporal resolution and a broad operating range. We successfully demonstrated the successful application of c-OFDR for ensuring steady state, transient, and cyclic behavior of different catalytic reaction systems. The c-OFDR can instantaneously measure the spatiotemporal temperature a feat that cannot be accomplished by a moving thermocouple.

The simultaneous measurements of spatiotemporal concentrations by a SpaciMS and temperature by c-OFDR, provide useful insight into the dynamic intricate features of transient behavior of a reaction system that cannot be obtained by use of just one of these measurements. A challenge for combined c-OFDR and SpaciMS measurements is the potential detrimental impact resulting from the invasiveness of these measurements. Although both sensors have a small cross-sectional area, an adequate operating condition (i.e., space velocity, monolith size, inlet temperature) should be seriously considered in case of the occurrence of the dis-match between the concentration and temperature profiles due to the blockage, suction, and transverse gradient effect. Also, in small monolith channels it may be better to place the two sensors in adjacent channels and not in the same one. However, in current circumstance, we still can rely on the temperature measurement by c-OFDR and the $\Delta Z$ map we created to decide the proper condition to use SpaciMS. We successfully demonstrated the successful application of SpaciMS and c-OFDR joint technique for verifying the steady state, lean/rich cycling, and temperature programed oxidation behavior among various catalysis systems.

## ACKNOWLEDGMENT

This work is funded by National Science Foundation (CBET-1067709).

## REFERENCES

Allison S, Gillies G: Remote thermometry with thermographic phosphors: instrumentation and applications, *Rev Sci Instrum* 68:2615–2650, 1997.

Ballinger TH, Manning WA, Lafyatis DS: *Hydrocarbon trap technology for the reduction of cold-start hydrocarbon emissions,* SAE Technical Paper, 1997.

Bethge H, Klaua M: Photo-electron emission microscopy of work function changes, *Ultramicroscopy* 11:207–214, 1983.

Bitsch-Larsen A, Degenstein N, Schmidt L: Effect of sulfur in catalytic partial oxidation of methane over Rh–Ce coated foam monoliths, *Appl Catal Environ* 78:364–370, 2008.

Brown J, D'Netto G, Schmitz R: *Spatial effects and oscillations in heterogeneous catalytic reactions temporal order,* Berlin, Heidelberg, 1985, Springer, pp 86–95.

Bugosh GS, Harold MP: Impact of zeolite beta on hydrocarbon trapping and light-off behavior on Pt/Pd/BEA/$Al_2O_3$ monolith catalysts, *Emission Control Sci Technol* 3(2):123–134, 2017.

Bugosh GS, Easterling VG, Rusakova IA, Harold MP: Anomalous steady-state and spatio-temporal features of methane oxidation on Pt/Pd/$Al_2O_3$ monolith spanning lean and rich conditions, *Appl Catal Environ* 165:68–78, 2015.

Chang H, Chen H, Koo K, Rieck J, Blakeman P: Gasoline cold start concept (gCSC™) technology for low temperature emission control, *SAE Int J Fuels Lubr* 7:480–488, 2014.

Chen K, Martirosyan K, Luss D: Wrong-way behavior of soot combustion in a planar diesel particulate filter, *Ind Eng Chem Res* 48:8451–8456, 2009.

Childers BA, Froggatt ME, Allison SG, et al: SPIE's 8th annual international symposium on smart, *Struct Mater:* 133–142, 2001.

Choi J-S, Partridge WP, Daw CS: Spatially resolved in situ measurements of transient species breakthrough during cyclic, low-temperature regeneration of a monolithic Pt/K/$Al_2O_3$ $NO_x$ storage-reduction catalyst, *Appl Catal Gen* 293:24–40, 2005.

Choi J-S, Partridge WP, Pihl JA, Daw CS: Sulfur and temperature effects on the spatial distribution of reactions inside a lean $NO_x$ trap and resulting changes in global performance, *Catal Today* 136:173–182, 2008.

Crisp J: *Introduction to fiber optics,* Amsterdam, The Netherlands, 2005, Elsevier.

Elangovan S, Ogura M, Zhang Y, Chino N, Okubo T: Silicoaluminophosphate molecular sieves as a hydrocarbon trap, *Appl Catal Environ* 57:31–36, 2005.

Elangovan S, Ogura M, Ernst S, et al: A comparative study of zeolites SSZ-33 and MCM-68 for hydrocarbon trap applications, *Microporous Mesoporous Mater* 96:210–215, 2006.

Engler B, Lindner D, Lox E, Ostgathe K, Schäfer-Sindlinger A, Müller W: *Reduction of exhaust gas emissions by using hydrocarbon adsorber systems,* SAE Technical Paper, 1993.

Froggatt M, Moore J: High-spatial-resolution distributed strain measurement in optical fiber with Rayleigh scatter, *Appl Optics* 37:1735–1740, 1998.

Glombitza U, Hoff H: Proceedings of 13th international conference on automatic fire detection 2004, pp 1–23.

Gossler H, Kee BL, Zhu H, Hettel M, Deutschmann O, Kee RJ: Flow and pressure characteristics in rectangular channels with internal cylindrical bodies, *Chem Eng Sci* 149:296–305, 2016.

Güemes A, Fernández-López A, Soller B: Optical fiber distributed sensing-physical principles and applications, *Struct Health Monit* 9:233–245, 2010.

Hartig H, Wicke E: Brennzonen in Kontaktschichten bei der katalytischen CO-oxydation, *Z Phys Chem* 38:265–284, 1963.

Hartog A: Distributed fibre-optic temperature sensors: technology and applications in the power industry, *Power Eng J* 9:114–120, 1995.

Heimrich MJ, Smith LR, Kitowski J: *Cold-start hydrocarbon collection for advanced exhaust emission control,* SAE Technical Paper, 1992.

Hettel M, Diehm C, Torkashvand B, Deutschmann O: Critical evaluation of in situ probe techniques for catalytic honeycomb monoliths, *Catal Today* 216:2–10, 2013.

Heyes A, Seefeldt S, Feist J: Two-colour phosphor thermometry for surface temperature measurement, *Opt Laser Technol* 38:257–265, 2006.

Hill KO, Meltz G: Fiber Bragg grating technology fundamentals and overview, *J Lightwave Technol* 15:1263–1276, 1997.

Il'in A, Luss D: Wrong-way behavior of packed-bed reactors: influence of reactant adsorption on support, *AIChE J* 38:1609–1617, 1992.

Iliyas A, Zahedi-Niaki M, Eić M, Kaliaguine S: Control of hydrocarbon cold-start emissions: a search for potential adsorbents, *Microporous Mesoporous Mater* 102:171–177, 2007.

Jaree A, Budman H, Hudgins RR, Silveston PL, Yakhnin V, Menzinger M: Temperature excursions in packed bed reactors with an axial variation of catalyst activity, *Catal Today* 69:137–146, 2001.

Kaiser EW, Siegl WO, Henig YI, Anderson RW, Trinker FH: Effect of fuel structure on emissions from a spark-ignited engine, *Environ Sci Technol* 25:2005–2012, 1991.

Kapička J, Marek M: Oscillations on individual catalytic pellets in a packed bed: CO oxidation on PtAl2O3, *J Catal* 119:508–511, 1989.

Kellow JC, Wolf EE: Infrared thermography and FTIR studies of catalyst preparation effects on surface reaction dynamics during CO and ethylene oxidation on Rh/SiO2 catalysts, *Chem Eng Sci* 45:2597–2602, 1990.

Kellow JC, Wolf EE: Propagation of oscillations during ethylene oxidation on a Rh/SiO2 catalyst, *AIChE J* 37:1844–1848, 1991.

Kersey AD: A review of recent developments in fiber optic sensor technology, *Opt Fiber Technol* 2:291–317, 1996.

Khan AA, Vrabie V, Mars JI, Girard A, D'Urso G: A source separation technique for processing of thermometric data from fiber-optic DTS measurements for water leakage identification in dikes, *IEEE Sensors J* 8:1118–1129, 2008.

Kobayashi M, Hanafusa H, Takada K, Noda J: Polarization-independent interferometric optical-time-domain reflectometer, *J Lightwave Technol* 9:623–628, 1991.

Kumar A, Harold MP, Balakotaiah V: Isotopic studies of NOx storage and reduction on $Pt/BaO/Al_2O_3$ catalyst using temporal analysis of products, *J Catal* 270:214–223, 2010.

Li H-X, Donohue J, Chrmier W, Chu Y: Application of zeolites as hydrocarbon traps in automotive emission controls, *Stud Surf Sci Catal* 158:1375–1382, 2005.

Li M, Easterling VG, Harold MP: Towards optimal operation of sequential $NO_x$ storage and reduction and selective catalytic reduction, *Appl Catal Environ* 184:364–380, 2016.

Lobban L, Luss D: Spatial temperature oscillations during hydrogen oxidation on a nickel foil, *J Phys Chem* 93:6530–6533, 1989.

Lobban L, Philippou G, Luss D: Standing temperature waves on electrically heated catalytic ribbons, *J Phys Chem* 93:733–736, 1989.

Mehta P, Sams W, Luss D: Wrong-way behavior of packed-bed reactors: 1. The pseudo-homogeneous model, *AIChE J* 27:234–246, 1981.

Meltz G, Morey WW, Glenn W: Formation of Bragg gratings in optical fibers by a transverse holographic method, *Opt Lett* 14:823–825, 1989.

Migliardini F, Iucolano F, Caputo D, Corbo P: MFI and FAU-type zeolites as trapping materials for light hydrocarbons emission control at low partial pressure and high temperature, *J Chem* 2015, 2015. https://www.hindawi.com/journals/jchem/2015/269694/abs/.

Morgan K, Touitou J, Choi J-S, et al: Evolution and enabling capabilities of spatially resolved techniques for the characterization of heterogeneously catalyzed reactions, *ACS Catal* 6:1356–1381, 2016.

Mukai K, Kanesaka H, Akama H, Ikeda T: *Adsorption and desorption characteristics of the adsorber to control the HC emission from a gasoline engine,* SAE Technical Paper, 2004.

Murakami K, Tominaga S, Hamada I, et al: *Development of a high performance catalyzed hydrocarbon trap using Ag-zeolite,* SAE Technical Paper, 2004.

Murphy KA, Gunther MF, Vengsarkar AM, Claus RO: Quadrature phase-shifted, extrinsic Fabry–Perot optical fiber sensors, *Opt Lett* 16:273–275, 1991.

Nakamura S, Morooka S, Kawasaki K: Conductor temperature monitoring system in underground power transmission XLPE cable joints, *IEEE Trans Power Delivery* 7:1688–1697, 1992.

Navlani-García M, Puértolas B, Lozano-Castelló D, Cazorla-Amorós D, Navarro MV, García T: CuH-ZSM-5 as hydrocarbon trap under cold start conditions, *Environ Sci Technol* 47:5851–5857, 2013a.

Navlani-García M, Varela-Gandía FJ, Bueno-López A, et al: BETA zeolite thin films supported on honeycomb monoliths with tunable properties as hydrocarbon traps under cold-start conditions, *ChemSusChem* 6:1467–1477, 2013b.

Nguyen H, Harold MP, Luss D: Optical frequency domain reflectometry measurements of spatio-temporal temperature inside catalytic reactors: applied to study wrong-way behavior, *Chem Eng J* 234:312–317, 2013.

Nguyen H, Harold MP, Luss D: Spatiotemporal behavior of Pt/Rh/$CeO_2$/BaO catalyst during lean–rich cycling, *Chem Eng J* 262:464–477, 2015.

Nguyen H, Peng PY, Luss D, Harold MP: Assessing intrusion by the capillary during spatially resolved mass spectrometry measurement, *Chem Eng J* 307:845–859, 2017.

Oberson P, Huttner B, Guinnard O, Guinnard L, Ribordy G, Gisin N: Optical frequency domain reflectometry with a narrow linewidth fiber laser, *IEEE Photon Technol Lett* 12:867–869, 2000.

Ogura M, Okubo T, Elangovan SP: Hydrocarbon reformer trap by use of transition metal oxide-incorporated beta zeolites, *Catal Lett* 118:72–78, 2007.

Partridge WP, Storey JM, Lewis SA, et al: *Time-resolved measurements of emission transients by mass spectrometry,* SAE Technical Paper, 2000.

Puertolas B, Navlani-García M, García T, Navarro MV, Lozano-Castelló D, Cazorla-Amorós D: Optimizing the performance of catalytic traps for hydrocarbon abatement during the cold-start of a gasoline engine, *J Hazard Mater* 279:527–536, 2014.

Puszyński J, Hlavacek V: Experimental study of ignition and extinction waves and oscillatory behavior of a tubular nonadiabatic fixed bed reactor for the oxidation of carbon monoxide, *Chem Eng Sci* 39:681–692, 1984.

Rotermund H, Engel W, Kordesch M, Ertl G: Imaging of spatio-temporal pattern evolution during carbon monoxide oxidation on platinum, *Nature* 343:355, 1990.

Russell A, Epling WS, Hess H, et al: Spatially-resolved temperature and gas species changes in a lean-burn engine emissions control catalyst, *Ind Eng Chem Res* 49:10311–10322, 2010.

Sá J, Fernandes DLA, Aiouache F, et al: SpaciMS: spatial and temporal operando resolution of reactions within catalytic monoliths, *Analyst* 135:2260–2272, 2010.

Serra RM, Miró EE, Bolcatto P, Boix AV: Experimental and theoretical studies about the adsorption of toluene on ZSM5 and mordenite zeolites modified with Cs, *Microporous Mesoporous Mater* 147:17–29, 2012.

Shwan S, Partridge W, Choi J-S, Olsson L: Kinetic modeling of NOx storage and reduction using spatially resolved MS measurements, *Appl Catal Environ* 147:1028–1041, 2014

Smolen J, Van der Spek A: *A DTS primer for oil & gas production,* 2003, Shell International Exploration and Production BV.

Soller BJ, Gifford DK, Wolfe MS, Froggatt ME: High resolution optical frequency domain reflectometry for characterization of components and assemblies, *Opt Express* 13:666–674, 2005.

Turing AM: The chemical basis of morphogenesis, *Bull Math Biol* 52:153–197, 1990.

van Sint Annaland M, Nijssen R: A novel reverse flow reactor coupling endothermic and exothermic reactions: an experimental study, *Chem Eng Sci* 57:4967–4985, 2002.

Westermann A, Azambre B, Finqueneisel G, Da Costa P, Can F: Evolution of unburnt hydrocarbons under "cold-start" conditions from adsorption/desorption to conversion: on the screening of zeolitic materials, *Appl Catal Environ* 158:48–59, 2014.

Yamamoto S, Matsushita K, Etoh S, Takaya M: *In-line hydrocarbon (HC) adsorber system for reducing cold-start emissions,* SAE Technical Paper, 2000.

Yoshino T, Kurosawa K, Itoh K, Ose T: Fiber-optic Fabry–Perot interferometer and its sensor applications, *IEEE Trans Microwave Theory Tech* 30:1612–1621, 1982.

Zhou D-P, Qin Z, Li W, Chen L, Bao X: Distributed vibration sensing with time-resolved optical frequency-domain reflectometry, *Opt Express* 20:13138–13145, 2012.

Zohour B, Noon D, Senkan S: New insights into the oxidative coupling of methane from spatially resolved concentration and temperature profiles, *ChemCatChem* 5:2809–2812, 2013.

Zones SI, Williams CM, Rufael TS, Burton AW: *Treatment of cold start engine exhaust,* 2013. Google Patents.

## FURTHER READING

Di Sante R: Fibre optic sensors for structural health monitoring of aircraft composite structures: recent advances and applications, *Sensors* 15:18666–18713, 2015.

CHAPTER THREE

# In Situ Spatially Resolved Techniques for the Investigation of Packed Bed Catalytic Reactors: Current Status and Future Outlook of Spaci-FB

**Alexandre Goguet*,[1], Caomhán Stewart*, Jamal Touitou†, Kevin Morgan*,[1]**

*School of Chemistry and Chemical Engineering, Queen's University Belfast, Belfast, United Kingdom
†School of Physics, Trinity College Dublin, College Green, Dublin 2, Ireland
[1]Corresponding authors: e-mail address: a.goguet@qub.ac.uk; kmorgan08@qub.ac.uk

## Contents



## Abstract

Over the past two decades a large effort has been dedicated to the development and optimization of techniques aimed at resolving the evolution of the chemistry within structured catalysts such as monoliths or foams under operando conditions. This approach, known as spatial resolution, is a powerful tool in the unraveling of complex reaction networks, as well as kinetic characterization. The most recent developments have been made at providing such spatial resolution at the early stages of the catalyst development, i.e., when the catalyst is still in its powdered form. Techniques aimed at providing spatial resolutions of the gas composition, of structural information, and of adsorbate distributions via X-ray and vibrational-based spectroscopies, respectively, are developing rapidly. The ability to simultaneously operate these spatially resolved

*Advances in Chemical Engineering*, Volume 50
ISSN 0065-2377
https://doi.org/10.1016/bs.ache.2017.05.001

approaches has the potential of providing a powerful platform for reaction mechanism investigations. Herein, we discuss and evaluate the development of a spatially resolved technique for powdered catalysts developed at Queen's University Belfast, namely the Spaci-FB. Furthermore, some aims and aspirations for further evolution of spatially resolved techniques are presented.

## 1. INTRODUCTION

Heterogeneous catalysts can be found in all sorts of shapes and forms depending on the stage of their development and final application. In academia and at the early stage of industrial research the most common form will be a powder since it is the easiest form to prepare. In a more advanced stage of research (pilot plant) or when commercialized, the catalysts' shape will preferably be that of pellets, tablets, rings, spheres, or monoliths (http://www.bioingenieur.eu/recherches.html, consulted on 4-11-2013). The development of these commercialized catalysts is the result of combined research in various fields including catalytic activity/selectivity measurements, kinetics characterization, reaction mechanism resolutions, materials properties, and process engineering developments. Optimization is possible only with a good understanding of the phenomena occurring at the surface of the catalyst and in the reactor.

The conventional test is simple, whereby the catalyst is loaded in a reactor within which the pressure and the temperature are controlled and the reactant gas mixture flows through the reactor. The gas is then sampled and analyzed at the outlet of the reactor to determine its composition using different techniques (for example, gas chromatography or mass spectrometry). This method of analysis is known as "tail-pipe end" or "catalyst effluent" analysis (Morgan et al., 2016). Frustratingly, such methods can be quite limited since there is no understanding of the chemical processes actually occurring *inside* the reactor as the only information that can be extracted is the difference of concentration between the inlet (known) and the outlet (measured). Unfortunately, the real point of interest in catalytic research is actually within the catalyst bed and traditional methods see the catalyst bed as a "black box/mystery box" with no actual access to the chemistry occurring within the catalyst bed (interaction between the reactants and the catalyst).

In an attempt to overcome such limitations, and to provide reaction and temperature profiles, computational models of reactions in the catalyst bed have also been studied. Most theoretical models have only been applicable for very specific conditions, and due to experimental limitations, it has been difficult in the past to conduct detailed experimental validation of such models (Stewart et al., 2014).

Borne from such frustrations has been the revolution in the field of spatial resolution, which has provided the opportunity to obtain concentration and temperature profiles/maps across a catalyst, and clear evolutions of such maps have been reported which can be directly attributed to conversions, yields, and exotherms. Several detailed discussions on imaging of spatial heterogeneities within catalysts and catalytic reactors have been reported (Akpa et al., 2005; Beale et al., 2014; Buurmans and Weckhuysen, 2012; Chupas et al., 2001; Clausen et al., 1991; Couves et al., 1991; Daniel et al., 2010; Eu et al., 2009; Figueroa and Newton, 2014; Gänzler et al., 2015; Geske et al., 2013; Gladden and Sederman, 2013; Gladden et al., 2006; Grunwaldt and Baiker, 2005; Grunwaldt et al., 2004, 2006; Hettel et al., 2014; Koptyug, 2014; Lee et al., 2011b; Margulies et al., 1999; Nic An tSaoir et al., 2011, 2012; O'Brien et al., 2011, 2012; Pistidda et al., 2015; Rasmussen et al., 2013; Schroer and Grunwaldt, 2013; Sederman et al., 2005; Tan et al., 2009; Thomas et al., 2017; Titze et al., 2015; Urakawa and Baiker, 2009; Urakawa et al., 2008; Weckhuysen, 2009; Wragg et al., 2012; Yuen et al., 2002; Zaera, 2014; Zellner et al., 2015; Zhang et al., 2015). The recent review by Morgan et al. (2016) has summarized the evolution and achievements of such spatial resolution methodologies.

Characterization of the catalyst is another way to collect information, where the data obtained focus on the chemical nature and structure of the catalyst and, more particularly, the active site(s). Numerous techniques have been developed and are available; these techniques are based on different principles and can provide a large variety of information on the catalyst. These techniques are, for instance, based on spectroscopy, microscopy, and probe molecules (Morgan et al., 2016). It is common practice to combine a number of the techniques in order to obtain as much characterization data as possible. In the case of the spectroscopic characterization, it is generally accepted that reaction environment and the population of the surface by adsorbed species can modify the structure of the catalyst. Most traditional characterization techniques are not conducted under *operando* conditions, giving rise to the "material gap" (Niemantsverdriet et al., 1999) and "pressure gap" (Freund et al., 2001), and so it can be risky to attempt to correlate reactive structure or mechanism observed under model conditions or with a model catalyst to real-life operation.

In the case of the spectroscopic characterization, an additional issue is present. Its impact and origin are closely interwoven with the kinetics characterization. During the catalytic process the population of adsorbed species present on the surface of the catalyst evolves with the reaction conditions and the catalyst structure. In return, it is today widely recognized that the structure of the catalyst can be modified by the nature of the reaction environment and the adsorbed species. If the spectroscopic reactor/cell is used in a "plug-flow" mode, two cases are possible:

- The reactor is under integral conditions, implying that the concentration of reactants and products changes significantly between the inlet and outlet of the reactor. This has two impacts, first the end-pipe data cannot be used to directly extract reaction rates. Second, if the catalyst structure and concentration of surface species depend on the chemical environment, they will change significantly from inlet to outlet. Consequently, unless a spatially resolved characterization is performed, it will not be possible to relate the spectroscopic data to activity/selectivity.
- If the reactor is used under differential conditions (low conversions), the gradient across the reactor is considered as being a straight line and therefore the reaction rate is the same all along. It is therefore possible to calculate reaction rates from activity data and the evolution rate of the adsorbates/structure extracted from the spectroscopy can be related to the end-pipe measurements. The main issue with such an approach is that the low conversion conditions are not representative of the true conditions that the catalyst will experience in a real-life process. Moreover, it is important to see that if the catalyst nature/concentration of surface species and the structure of the catalyst do depend on the chemical environment. Eventually, these differential conditions only represent a very narrow "range" of the "environment vs structure" curve and deriving structure/activity/selectivity relationship from such a narrow range can be dangerous.

If the reactor/cell is not following a plug-flow behavior, the reaction profile development across the catalyst can be very difficult to resolve and relating the spectroscopic data (which will likely be localized, i.e., only probe a small fraction of the catalyst placed in the reactor) will be often impossible.

Consequently, in order to study real catalysts under conditions as realistic as possible, there is near constant development of new *operando* and/or in situ methodologies, which can offer the same detailed understanding as traditional techniques, but which have increased relevance. Such methodologies have been recently appraised in the review of Weckhuysen (2009).

Ultimately, the combination of spatial resolution analysis along with *operando* characterization is seen as highly desirable. In this chapter, we focus on such an experimental application to powdered catalysts, known as the Spaci-FB, a method which provides spatially resolved profiles of gas concentration, temperature, and most recently structure across a fixed-bed powder catalytic reactor.

## 2. SPATIALLY RESOLVED CHARACTERIZATION TECHNIQUES FOR FIXED BEDS

Spatial resolution has allowed for the direct observation of the chemistry processes as they evolve within the catalyst bed without significant disruption/impact on *operando* conditions. As highlighted recently (Morgan et al., 2016), these have mostly focused on obtaining detailed information to understand phenomena occurring on the structured catalyst surface and in the catalytic reactor under *operando* conditions using a range of analytical methods including infrared (Aftab et al., 2008; Baiker and Bergougnan, 1985; Baiker and Epple, 1986; Daniel et al., 2010; Irani et al., 2009; Luo et al., 2011a,b; Russell et al., 2010, 2011; Shakir et al., 2009; Urakawa and Baiker, 2009; Urakawa et al., 2008), mass spectrometry (Bitsch-Larsen et al., 2008; Bosco and Vogel, 2006; Choi et al., 2005, 2006, 2007, 2008; Cong et al., 1999; Currier et al., 2000; Dalle Nogare et al., 2008, 2011; Donnazi et al., 2008, 2010; Geske et al., 2013; Horn, 2011; Horn et al., 2006a,b, 2007a,b, 2010; Johansson et al., 2004; Kopyscinski et al., 2010; Korup et al., 2011, 2013; Michael et al., 2010; Orschel et al., 1999; Parks et al., 2005; Partridge and Choi, 2009; Partridge et al., 2000, 2002, 2006; Roos et al., 2007, 2010; Sá et al., 2010; West et al., 2004), Raman spectroscopy (Korup et al., 2013), gas chromatography (Beretta et al., 2011; Bosco and Vogel, 2006; Donazzi et al., 2011a,b; Kopyscinski et al., 2010; Liu et al., 2006; Livio et al., 2011), and XAFS (Chupas et al., 2001; Clausen et al., 1991; Eu et al., 2009; Grunwaldt and Baiker, 2005; Grunwaldt et al., 2004; Margulies et al., 1999). There have also been reports of the comparison of some of these experimental results with simulations generated from mathematical/kinetic models (Baiker and Bergougnan, 1985; Baiker and Epple, 1986; Kopyscinski et al., 2010; Korup et al., 2013; Livio et al., 2011). Surprisingly, despite powders being the most studied textural form for laboratory-based catalysis research, there have been a limited number of reports of techniques which have been developed for noninvasive spatial resolution of fixed-catalyst beds.

## 2.1 Spatially Resolved Spectroscopy for Fixed Beds

Imaging of spatial heterogeneities within catalysts and catalytic reactors has been reported (Weckhuysen, 2009), and this growing field can be broadly distinguished by the technique used for analysis: optical spectroscopy, magnetic resonance, and X-ray-based techniques.

As an example, $NO_x$ storage–reduction (NSR) over a Pt/Ba/$CeO_2$ catalyst was studied and spatially and temporally resolved DRIFTS and Raman were reported (Urakawa and Baiker, 2009; Urakawa et al., 2008). The principle is similar to that reported for a single-channel structured catalyst (Daniel et al., 2010) in that an $x$–$y$–$z$ translational stage was used for the movement of the catalyst. Consequently, through the combination of space- and time-resolved spectra, dynamic surface and bulk processes were determined for NSR. There are also reports of obtaining spatial resolution using other analytical techniques, including transmission FTIR (Rasmussen et al., 2013; Tan et al., 2009; Thomas et al., 2017), Raman (Geske et al., 2013), magnetic resonance imaging (Gladden and Sederman, 2013; Gladden et al., 2006; Koptyug, 2014), near infrared diffused transmittance tomography (Nic An tSaoir et al., 2011, 2012), microimaging IR microscopy (Titze et al., 2015), and 2D spatially resolved planar laser-induced fluorescence (Zellner et al., 2015), and examples of these are discussed in greater detail elsewhere (Morgan et al., 2016).

Interestingly, there has also been significant development of in situ X-ray spectroscopy from a range of research groups, including spatially resolved methods coupled with various forms of synchrotron radiation as it has now become an extremely popular medium by which to develop such techniques (Chupas et al., 2001; Clausen et al., 1991; Couves et al., 1991; Eu et al., 2009; Figueroa and Newton, 2014; Gänzler et al., 2015; Grunwaldt and Baiker, 2005; Grunwaldt et al., 2004, 2006; Lee et al., 2011b; Margulies et al., 1999; Zhang et al., 2015). This has included the work by Weckhuysen (O'Brien et al., 2009, 2011) and combining multiple spectroscopic techniques together to achieve a spatially resolved profile for the reduction of molybdenum oxide. A further example is that of Grunwaldt et al. (Cavusoglu et al., 2016; Doronkin et al., 2014) developing spatially resolved X-ray reactors for water gas shift and $NO_x$ reduction (see Fig. 1). There are also reports of the combination of multiple spatially resolved spectroscopies (Gibson et al., 2013) and through utilization of noninvasive tomographic sampling methods (Vamvakeros et al., 2015). Recent reviews by Newton (Newton, 2017; Newton et al., 2014) and by

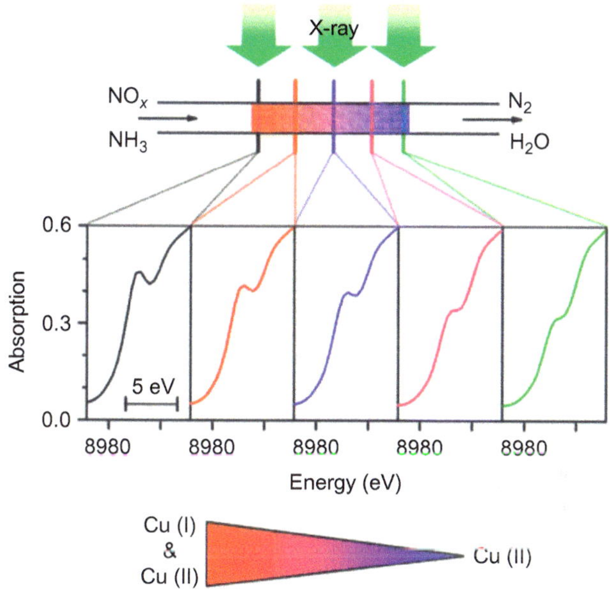


**Fig. 1** Spatially resolved X-ray absorption spectroscopy (XAS) during the selective catalytic reduction (SCR) of $NO_x$ by $NH_3$ along a catalyst bed consisting of Fe-containing BEA and ZSM-5 zeolites (Doronkin et al., 2014).

Grunwaldt (Grunwaldt et al., 2013) highlight some of the recent advances in greater detail, with numerous other resources also available for the application of X-ray spectroscopy available in more general terms for newcomers to the area (Beale et al., 2012, 2013, Nachtegaal et al., 2015, 2017). All this combined helps move toward the much vaunted catalytic dream machine proposed by Tinnemans et al. (2006).

## 2.2 Spatial Resolution of Fixed Catalyst Beds Using Physical Probes

While assessing the level of experimental complexity required to aid kinetic model development, Baiker et al. reported examples of spatially resolved temperature profiles (using thermocouples and IR analyzer probes) for the hydrogenation of toluene from a working packed bed reactor (Baiker and Bergougnan, 1985; Baiker and Epple, 1986). Gas concentration data were only obtained before (zero conversion) and after the catalyst bed (maximum conversion), which is tantamount to end-pipe sampling.

It can only be assumed that the size of the probes used (which were not provided) was very much larger 30 years ago compared to those currently available and it should be considered, therefore, that the bed and the flow will likely be disturbed due to the use of so many invasive probes. Even so, this work has been instrumental in advancing the field of spatial resolution, and with the development of smaller probes in intervening years, improvements have been achieved in terms of increased spatial and temporal resolution with the decreasing invasiveness of probes.

One such development was a setup for a fixed bed of millimeter-sized catalyst pellets and foam monoliths which provided gas concentration and temperature profiles using a single movable axially capped-end mounted capillary with a side wall gas sampling orifice (Bitsch-Larsen et al., 2008; Dalle Nogare et al., 2008, 2011; Donnazi et al., 2008; Geske et al., 2013; Horn, 2011; Horn et al., 2006a,b, 2007a,b, 2010; Korup et al., 2011, 2013; Michael et al., 2010) (see Fig. 2). Further improvements include a

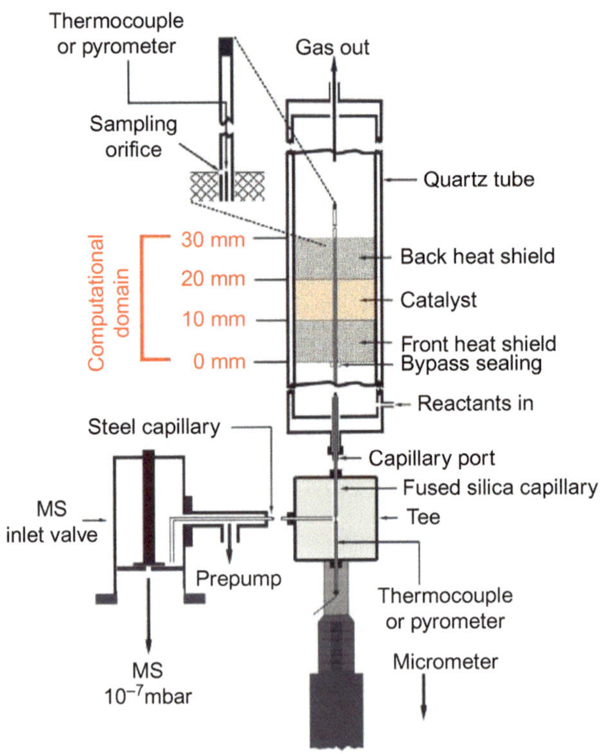


**Fig. 2** Schematic representation of the system for resolving species and temperature distributions within a foam-monolith catalyst showing the location of the different elements of the system (Horn et al., 2006b).

reactor design which can tolerate temperatures up to 1300°C and pressure up to 45 bar (Geske et al., 2013; Horn, 2011; Horn et al., 2010; Michael et al., 2010) and the ability to collect spectroscopic data regarding the catalyst surface (Geske et al., 2013; Korup et al., 2011).

## 2.3 First-Generation Spaci-FB

One of the better-known methods is the spatially resolved capillary-inlet mass spectrometer (Spaci-MS) (Sá et al., 2010) system, which can provide temporal and spatial (axial and radial) analytical information for monolith-supported catalysts. Originally developed by Oak Ridge National Laboratory (ORNL) (Partridge et al., 2000) and then further improved via collaboration between ORNL and Queen's University Belfast (QUB) (Sá et al., 2010), it was proposed that the knowledge developed from Spaci-MS could be applied to powdered catalyst beds.

Recently, Touitou et al. have reported a spatially resolved technique for fixed beds comprised of smaller particle-sized powders, the Spaci-FB (Touitou, 2014; Touitou et al., 2012, 2013). Initially, a manual linear transfer mechanism and fused silica gas sampling capillary were deployed (Fig. 3) and have served as a proof of concept of the Spaci-FB for use with fixed powdered catalyst beds (Touitou, 2014). Using some of the smallest equipment commercially available, the components selected were chosen to have negligible impact on the packed powdered catalyst bed. To reduce any possible effect of channeling or other disturbances to the fixed bed, it is recommended that the lower range of catalyst particle size be no smaller than the outer diameter of the capillary in order to maintain a similar tortuosity to the case when no capillary would be present. This first-generation system can be summarized as follows:

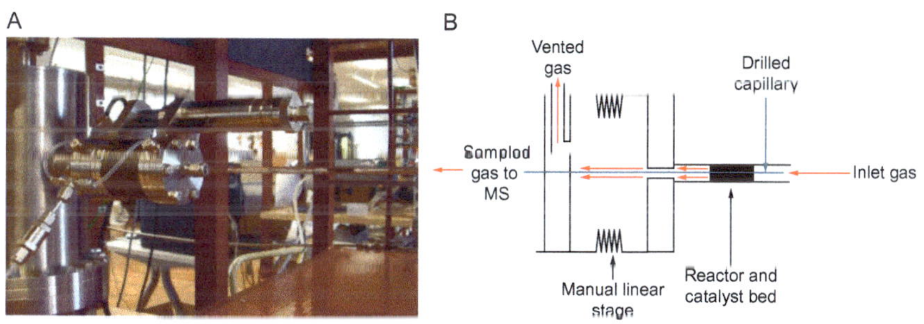


**Fig. 3** (A) Photograph of first-generation Spaci-FB (Touitou, 2014). (B) Representation of first-generation Spaci-FB (Touitou, 2014). The capillary in this case was fixed and the linear stage translates the reactor/packed bed across the sampling holes, thereby providing the spatially resolved gas concentrations.

- A sampling capillary (75 μm ID, 150 μm OD) was used.
- Upstream end of capillary sealed, and downstream end remains open.
- Three pairs of lateral gas sampling holes (20 μm diameter, 180° spacing) added via laser ablation.
- Thin sampling zone of 0.5 mm in length.
- Catalyst loaded into quartz reactor.
- The capillary loaded through the center of the fixed bed.
- The capillary outlet is sampled by quadrupole mass spectrometer.
- A hot air heater (Le Mini, Leister) to heat the reactor.

The operation of the technique is summarized in Fig. 4. The gas sampling capillary was fixed and the reactor/catalyst bed was moved via a linear transfer stage constituted of a bellow that can be extended or compressed. The spatial resolution was obtained by moving the reactor (and therefore the catalyst bed) relative to the sampling hole in the capillary which was placed perpendicular to the direction of travel of the catalyst bed. As the movement

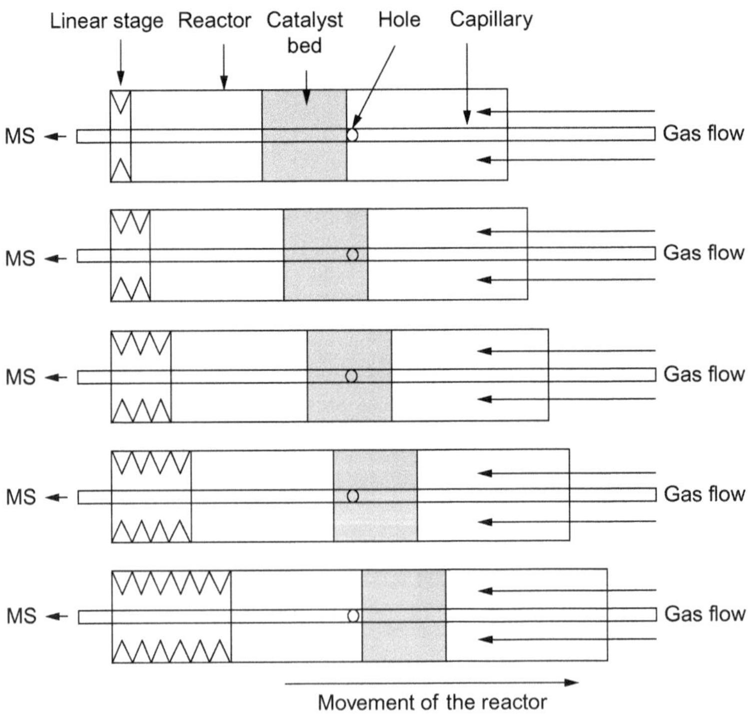


**Fig. 4** Principle of the sampling system (Touitou, 2014).

was achieved by manual adjustment of the linear transfer stage, the spatial resolution was maintained at 1 mm in order to provide greater assurances over reproducibility. The results obtained are an instantaneous sampling and a true representation of the gas concentrations at that particular point within the sampling bed.

CO oxidation over 1 wt% Pd/$Al_2O_3$ experiments were conducted to validate and provide the proof of concept of the first-generation Spaci-FB technique in terms of the provision of gas concentration profiles across a fixed powdered catalyst bed. In order to probe the sensitivity of Spaci-FB to variations in experimental conditions (i.e., temperature and feed concentrations), the effect of temperature change was studied, as was the addition of $H_2$ to the feed. As would be expected, there were clear promotional effects of increasing temperature, and some of the results are reported in Fig. 5.

It is from Fig. 5 that the significance of the capability of the Spaci-FB to provide greater insight can begin to be appreciated. At the end of the catalyst bed, 11 mm (which corresponds to traditional tail-pipe end measurements), there is no observable difference in conversion at 250°C and 300°C (both are at 100% conversion). There are, however, significant differences in the reaction profiles at 250°C and 300°C and this can be observed when considering the full conversion profile across the catalyst bed. Profiles for conversion of $O_2$ and the yield of $CO_2$ were also reported, while the promotional effect of the addition of $H_2$ (so-called PROX reaction) was also studied with the prototype Spaci-FB (Touitou et al., 2012). Based on the

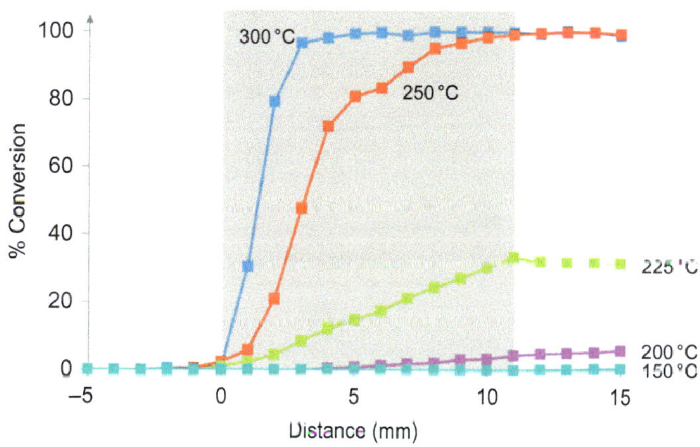


**Fig. 5** CO conversion in the catalyst bed at different temperatures. The catalyst bed is represented by the shaded area (Toutitou et al., 2012).

positive results obtained with the first-generation Spaci-FB, further technique development and improvements were implemented (Touitou et al., 2013).

## 2.4 Second-Generation Spaci-FB

As the first-generation Spaci-FB was manually operated, one of the main modifications for the second-generation Spaci-FB (Touitou et al., 2013) was to introduce an automatic linear transfer system (LNR50SE/M, Thorlabs), which would also allow for increased spatial resolution (see Figs. 6 and 7). In order to accommodate this and to reduce any interference of the automation on the catalyst bed a new frame containing two conflat flanges was designed to hold the reactor, while these flanges were directly connected to bellows, so as to ensure a smoother movement. As was the case for the first-generation Spaci-FB, in order to reduce strain on/breakages of the capillary the reactor was moved, while the capillary remained stationary and rigid. In order to reduce dead volumes and possible stagnation of flow the inert gas was introduced upstream of bellow 1, while the reactant gases were introduced directly downstream of bellow 1. This modification ultimately improved MS response time.

In a further improvement on the first-generation Spaci-FB, in order to improve the isothermal heating of the catalyst bed, the hot air blower

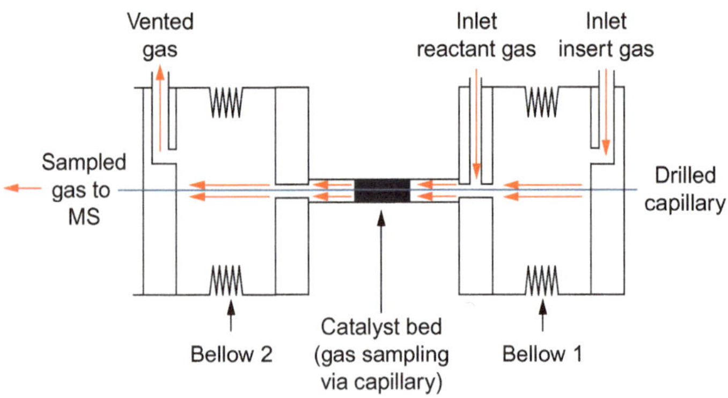


**Fig. 6** Schematic diagram of the second-generation automated Spaci-FB system (Touitou et al., 2013). The capillary is fixed and the automated stage translates the bellow system/reactor, passing the packed bed across the sampling holes, thereby providing the spatially resolved gas concentrations. A thermocouple is positioned inside the gas sampling capillary, thereby providing simultaneous spatially resolved temperature profiles.

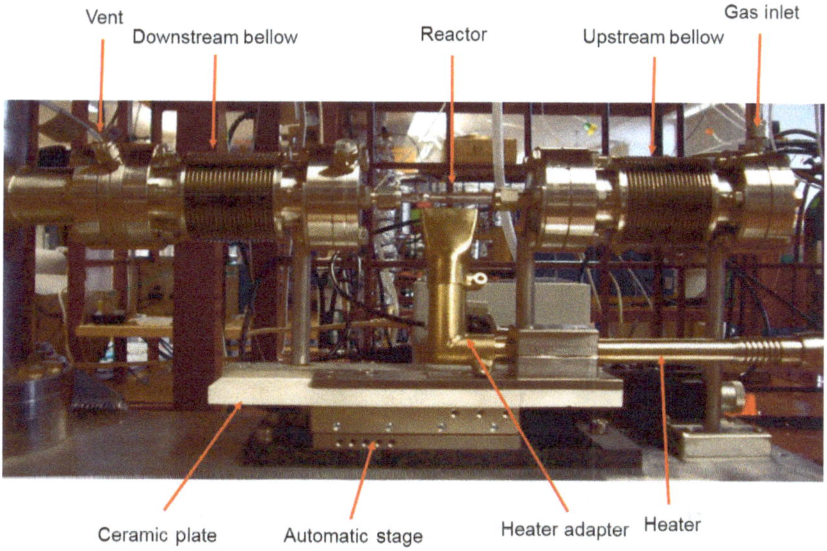


**Fig. 7** Photograph of the second-generation Spaci-FB (Touitou et al., 2013).

(Le Mini, Leister) was equipped with a specially designed nozzle (stainless steel, 8 mm × 30 mm), located at the extremity of the heater adaptor, as shown in Fig. 7. This nozzle improved the direction and focus of the hot air flow onto the reactor. Furthermore, to maintain the relative position of the heater compared with the reactor during the experiments, the whole heating assembly was incorporated onto the linear transfer mechanism.

The first-generation Spaci-FB could only provide gas concentration profiles across the catalyst bed. By way of improvement, it was decided to incorporate a thermocouple within the Spaci-FB system so that simultaneous spatially resolved temperature profiles could also be provided. To achieve this, a thermocouple (80 μm O.D., T33, Okazaki) was inserted into the drilled capillary and aligned with the gas sampling holes (see Fig. 8B). As such, in order to seal the gas sampling capillary, both the downstream (vacuum chamber side) and upstream ends (where the thermocouple is attached) are sealed with an adaptor with a compression ferrule. These adapters were attached to stationary posts which ensured the capillary was kept taut and stationary.

It was also noticed that, while preparing gas sampling capillaries for the first-generation Spaci-FB, the laser ablation method used was not well suited to provide mechanically resistant capillaries. As such, a xenon focused ion beam (Plasma-FIB) has since been used to provide the same orientation

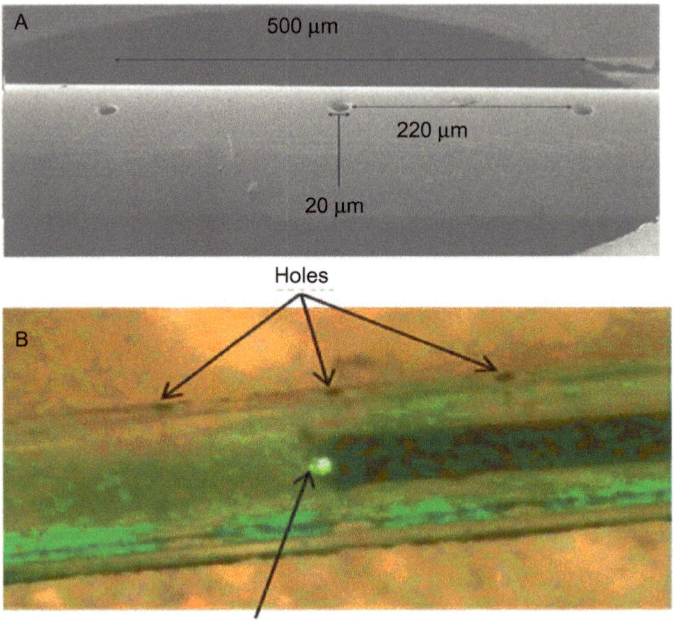


**Fig. 8** Sampling holes of the capillary. (A) Capillary drilled by plasma-FIB. (B) Picture of the capillary with the thermocouple inserted (picture obtained by optical microscopy) (Touitou et al., 2013).

and dimensions of gas sampling orifices, albeit with a higher accuracy/control of drilling and with a better alignment of sampling holes (see Fig. 8A).

When these modifications were complete, further validation tests were conducted which studied the effect of temperature and reactant concentration on the gas and temperature profiles (Touitou et al., 2013), an example of which is reported in Fig. 9. The inclusion of the automated linear stage allowed precise movement and an even higher spatial resolution, but the limiting factor in this instance is the length of the sampling zone itself (i.e., 0.5 mm).

An obvious advantage of using spatially resolved reactor data is the ability to determine *operando* kinetic parameters for use with catalyst models. With this in mind, data from the second-generation Spaci-FB have also been compared to a microkinetic model (Touitou et al., 2014), which was a modification of a previously reported CO oxidation kinetic model (Stewart et al., 2013). For this model, the following information was used to mathematically represent the catalyst bed and the reactor:

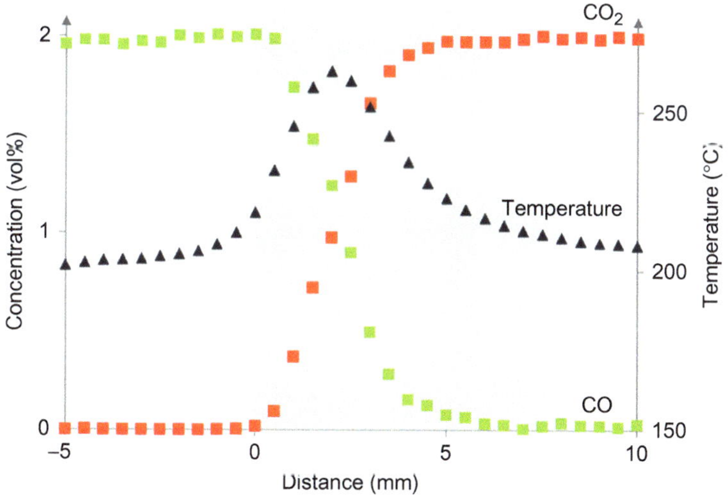


**Fig. 9** Gas concentration profiles of CO and $CO_2$; and temperature profile. The catalyst bed is represented by the shaded area (Toutitou et al., 2013).

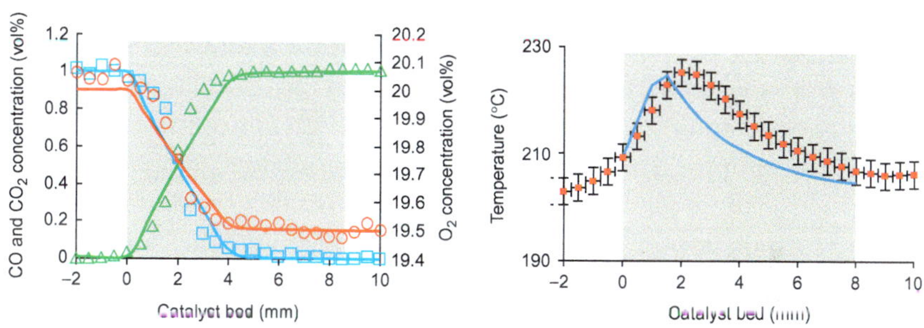


**Fig. 10** Comparison of experimental (*symbols*) and fully simulated (*solid lines*) results of gas concentration (CO in *blue*, $O_2$ in *red*, and $CO_2$ in *green*) and temperature profiles from within a packed powder catalyst bed (Touitou et al., 2014).

*Catalyst*: particle size, volume, surface area, precious metal loading, and precious metal dispersion.

*Reactor*: physical dimensions such as diameter and length (for both reactor and catalyst bed); the reactant concentrations; and total gas flow.

This information was utilized to model the reactor terms of heat and mass transfer in axial and radial directions. A fully simulated (concentration and temperature profiles) study and a semi-empirical (simulated concentrations, experimental temperature) study were conducted (Touitou et al., 2014), and the results are reported in Figs. 10 and 11. It is evident that the fully simulated model could not fully account for the exotherm that occurred

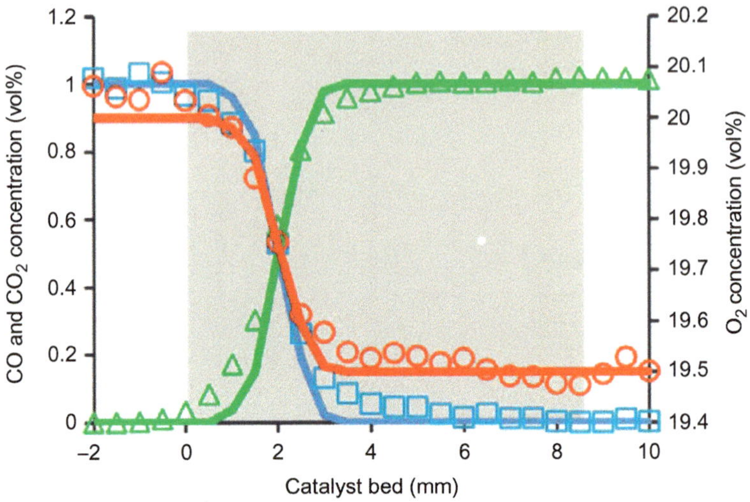


**Fig. 11** Comparison of experimental (*symbols*) and semi-empirical simulated (*solid lines*) results of gas concentration profiles (CO in *blue*, $O_2$ in *red*, and $CO_2$ in *green*) from within a packed powder catalyst bed (Touitou et al., 2014).

as a result of the chemical reaction as seen from the variance in the model and experimental temperatures in Fig. 10. This in turn caused some errors in the ability of the model to accurately simulate the gas concentration profiles. As the semi-empirical model utilized the experimental temperature, the full impact of the exotherm is accounted for and this resulted in a much improved simulation of the gas concentration profiles. The capabilities of the Spaci-FB have also been assessed in terms of performance and influence of the probe on the measurements and found to be minimally invasive (Touitou et al., 2014).

## 2.5 Coupling Second-Generation Spaci-FB With XAS

As discussed briefly in Section 2.1, catalytic in situ *operando* XAFS experiments have become more robust in bridging the existing gaps in analysis at nanoscale level on microreactors through continued development of spatially resolved methodologies. Cell design now focuses on being able to map the performance of the catalytic reactor over the reactor bed as opposed to previously over pelletized samples, through the use of spatial resolution methods in a spectroscopic environment. One such instance is the previously unreported spatial resolution investigation with Spaci-FB on diesel oxidation catalysis (DOC) over a $Pd/Al_2O_3$ catalyst. Herein we disclose the information which can be obtained, while also discussing the difficulties and challenges encountered when attempting to combine spatial resolution and spectroscopy, particularly under the pressures of premium synchrotron beam line experiments.

While the Spaci-FB technique was initially developed to monitor gas concentrations and temperature within powdered catalyst beds, enhancement of the method has been achieved through coupling with X-ray adsorption spectroscopy (XAS). In addition to concentration profiles, this coupling provides Spaci-FB with the additional feature of observing the evolution of the local structural environment. Three of the key critical challenges in using the equipment at the start were:

- Accommodation of a large piece of equipment in a relatively small environment.
- Ensuring the gas concentration and structural data obtained are from the same axial point, i.e., ensuring alignment of beamline with gas sampling zone of Spaci-FB.
- The sample environment and aspects related to changeover of samples.

Accommodation of the equipment was overcome through consultation of drawings of the synchrotron experimental hutch and of the second-generation Spaci-FB; and then subsequent planning for the incorporation of the Spaci-FB within the hutch (Fig. 12). As part of this planning and

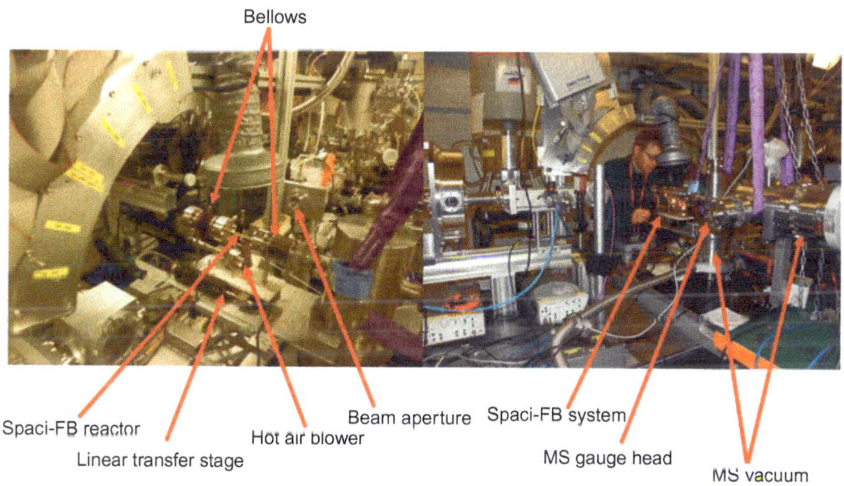


**Fig. 12** Second-generation Spaci-FB on beamline B18 Diamond Light Source.

incorporating stage, issues related to the alignment of the beamline with the gas sampling zone of the Spaci-FB were also addressed, and through the utilization of existing *x–y–z* translation stages within the hutch, fine tuning of this alignment is possible.

The greatest challenge encountered was in terms of the sample environment and the ability to change the sample. While the second-generation Spaci-FB was originally designed with routine operation within a catalyst research laboratory, this was not necessarily ideal for operation within the synchrotron hutch. With the greater level of access to the Spaci-FB within the catalyst research laboratory, and routine daily catalyst screening being the norm, there were no major time constraints for changing the catalyst sample, even on such a sensitive/fragile piece of equipment. In order to change catalyst samples in Spaci-FB (first and second generation) the ultra-high vacuum had to be shut down to remove the old sample and started up after the new sample was loaded followed by the dwell period to allow vacuum stabilization, all of which adds to experiment setup time.

It was determined that there were too many moving objects when trying to secure flanges and bellows to the vacuum chambers, which potentially compromise seals through copper gasket slipping, which could also break the gas sampling capillary. To improve on this, and to reduce sample changeover time, new flanges which sealed with clamps (rather than the threaded rods and nuts used previously) were used to form the seal between the vacuum chambers, flange, and the bellows. Additionally, new locks were added to the flanges which allowed the gaskets to be held in place (Fig. 13A and B), while new embedded prongs on flanges allowed connection to the bellows before attaching to the vacuum chambers (Fig. 13C). These modifications allowed for a reduction in challenges of sample changeover, including the system assembly (Fig. 13D–F) and potential for damage to the capillary.

These minor modifications allowed for the second-generation Spaci-FB to be coupled with XAS for the preliminary study of DOC over a $Pd/Al_2O_3$ catalyst with a pseudo exhaust gas mixture, at 260°C with reaction mix and water. As this was a preliminary study with strict time constraints to assess the feasibility of the coupling of Spaci-FB with XAS, simultaneous spatially resolved gas concentration and XAS profiles of the powdered catalyst were limited to three points, corresponding to the front (0.6 mm), middle (2 mm), and end (5 mm) of the reactor bed.

Fig. 14 reports the comparison of front, middle, and end of bed under DOC reaction conditions with reaction gas feed composed of 1000 ppm

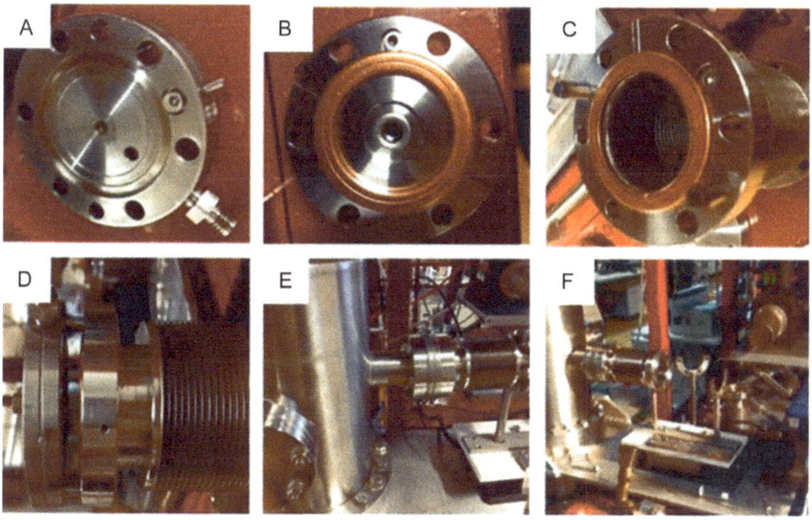


**Fig. 13** Second-generation Spaci-FB on vacuum development at QUB including the flanges (A–C), bellows (D), and construction (E, F).

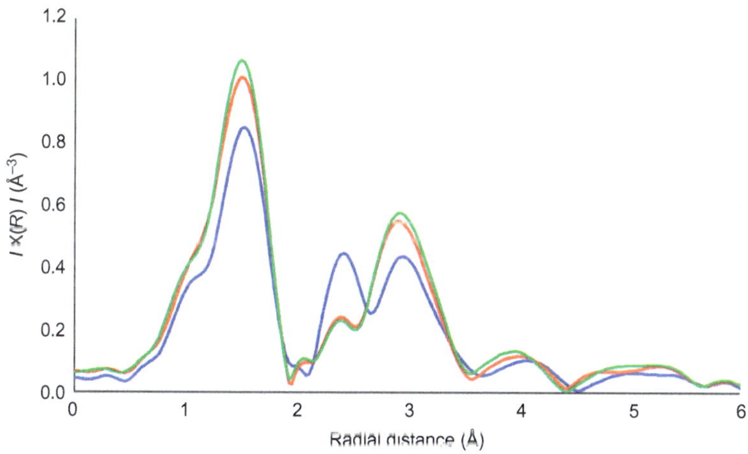


**Fig. 14** Fourier transforms of Chi data at front (*blue*), middle (*red*), and end (*green*) of the catalyst for DOC.

CO, 900 ppm $C_3H_6$, and 200 ppm NO, with 4.5% $H_2O$, 12% $O_2$, and balance with Ar. The XAS was analyzed using Artemis on the Demeter package (Ravel and Newville, 2005). There was a clear difference observed for the front compared to the other points, which indicated that there was a structural

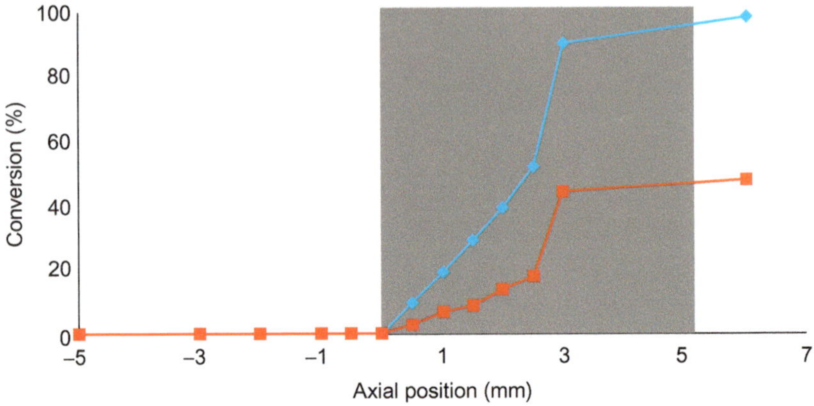


**Fig. 15** Spaci-FB conversion profiles for diesel oxidation catalysis; CO (*blue*) and propene (*red*).

change along the bed during the reaction, with the presence of a new shell at 2.5 Å observed for the front of the bed. Additionally, there was a small deviation for the first path (1.5 Å) in the case of the front compared with the rest of the bed.

Fig. 15 reports the spatially resolved conversion profiles for CO and propene during the DOC reaction. The front of the bed is found to be structurally different to the middle and end of the bed, and in this region, it is clear from Fig. 15 that there is no conversion of CO or propene, while reaction light-off has clearly been achieved in the middle (2 mm) of the bed, indicating the presence of the active phase of the catalyst in this region, which is structurally the same as the remainder of the fixed bed.

Despite these encouraging results, it should be noted that due to the cumbersome nature of fitting the thermocouple and strict time constraints, simultaneous spatially resolved temperature profiles were not obtained, a feature which will be addressed in future experiments. In addition to this, some other difficulties were encountered during the Spaci-FB synchrotron experiments and downtime was caused by three main factors:

- Sample changing and alignment for each individual sample.
- Heating and cooling system for water and heat tracing.
- Breaking and reestablishing of vacuum for MS and the subsequent stabilization time required.

In order to address these issues, beam line scientists have been consulted on the necessary modifications to produce the third-generation Spaci-FB,

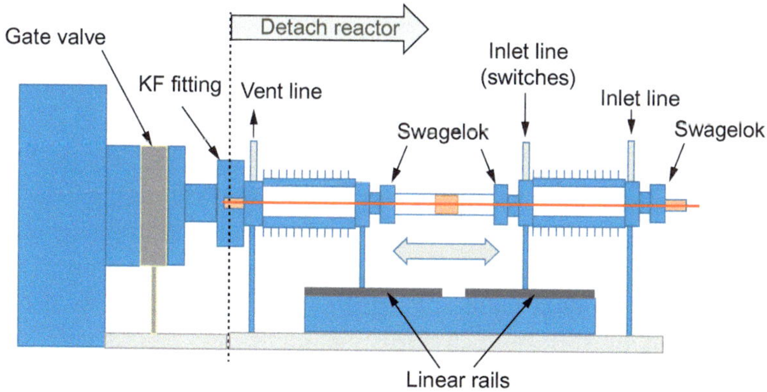


**Fig. 16** Schematic for third-generation Spaci-FB.

which will reduce downtime during the beam time for future studies. These modifications are represented in Fig. 16 and include:

- Smaller bellows for improved handling
- A lockable rail assembly for greater control of reactor loading, unloading, and adjustments
- A gate valve so that UHV for MS can be maintained during sample changeover
- Quick fit connections to reduce installation times
- A new plate assembly so the system can be decoupled so that the reactor and capillary can be worked on in a separate workspace as opposed to within the space constraints of the experimental table.

The modifications will reduce the major issues that occurred in this beam time for the future.

## 3. CONCLUSIONS AND PERSPECTIVE ON THE FUTURE OF SPATIALLY RESOLVED TECHNIQUES FOR PACKED BED CATALYTIC REACTORS

The Spaci-FB has been demonstrated as a valuable technology that can access the gas composition and temperature within a packed bed of catalyst for gas phase heterogeneously catalyzed reactions with minimal invasiveness. The quality of data has enabled its use for semi-empirical microkinetic modeling (Section 2.4). The incorporation of the experimental temperature measured with the spatially resolved technique into the model led to improved correlation in terms of gas concentration compared to the fully simulated model. This emphasized the importance of the technique since the combination of experimental data with simulations had improved model accuracy but also highlighted the limiting component of the model which constrained the accuracy.

These findings could be utilized to further develop the next generation of simulation models which can better predict the evolution of chemistry within catalysts beds and therefore provide a better understanding of catalysts and improved catalyst development. This will be particularly important as models become more complex in terms of reactions/feedstocks/processes which have to be simulated. A further consequence of this may lead to improved multiscale validations, and, as such, experimental methodologies such as Spaci-FB need to continue to be evolved to meet such demands. As has been previously proposed (Goguet et al., 2014; Morgan et al., 2016), this will be key to bridging the "knowledge gap" between traditional catalyst analysis and theories/models, ultimately improving the models and delivering improved catalytic processes. Part of such evolution may be to incorporate transient modes into spatial resolution methods, or indeed to combine existing transient techniques which can provide relevant kinetic information which can be used to assist in the modeling of spatially resolved profiles. One such method is the temporal analysis of products (TAP) (Gleaves et al., 1988), and it has been suggested that TAP experiments or other transients should be used to provide mechanistic insights and experimental data such as kinetics and diffusion coefficients for kinetic models (Morgan et al., 2016, 2017).

Further development of the integration of spectroscopic techniques with the Spaci-FB is also desirable. The preliminary work reported in Section 2.5 is promising. As discussed earlier in the manuscript, typical *operando* techniques thus far have tended to utilize a single method (e.g., MS, GC, IR, or XAS). Consequently, if one wishes to observe changes in gas concentrations as well as characterize the surface species, it was previously necessary to conduct a range of experiments followed by a correlation of the data from multiple sources. The ability to conduct simultaneous characterization (using a combination of techniques) under *operando* conditions must be the next stage for the spatial resolution methodologies. There is evidence that there is already movement on this aspect (Beale et al., 2012; Figueroa and Newton, 2014; Lee et al., 2011a), and it has been demonstrated that this is also the case with Spaci-FB (Section 2.5). Further work, as part of the newly created UK Catalysis Hub (http://www.ukcatalysishub.co.uk/themes/design, accessed 27th April 2017), is already addressing such challenges with further integrations on the coupling of vibrational spectroscopy to the Spaci-FB, work which will be reported in the near future.

The Spaci-FB work has highlighted an aspect transcending the modeling and structural characterization, i.e., that there is a problem of "connection" between the structure–activity and reaction modeling works. Indeed, the majority of modeling works infer that the intrinsic kinetic parameters are a unique set, valid for the whole catalyst present in the reactor irrespective of its position within the bed (or experienced reaction conditions). While this approach has had some success, a very common drawback is the limited predictive capabilities of the models outside the conditions used to calibrate them. It is easy to see that if there is catalyst structure variability with its position in the bed (as identified using the X-ray-based spectroscopy) and that the structure indeed affects the intrinsic catalyst activity, then the common approach to the setting and use of the kinetics parameters for the whole spectrum of reaction conditions is problematic. One can easily see that identifying the presence or absence of such spatial variabilities of the structure (and therefore activity) and the conditions and extent to which they prevail will be crucial with respect to the modeling approach and its complexity. Developing and refining the analysis approaches that will allow accessing such information is likely to be at the forefront of the research in this field for years to come.

Another more practical, but none less crucial, aspect is the importance of a multitude of technical details that can "make or break" a Spaci-FB experiment. The invasiveness aspect is a recurrent query and experience has demonstrated that keeping the capillary size close to the average catalyst particle diameter and the sampling rate low allows for minimal impact. A more detailed assessment/mapping of the "noninvasive" regimes depending on the relative size of the probe to that of the particle size and the space velocity, similar to that reported by Harold and coworkers for the Spaci-MS (Nguyen et al., 2017), would be desirable. Reactor geometry and design are also important aspects that should not be neglected. For example, the size of the capillary used was a balance between having sufficient strength to enable movement against the catalyst bed and to be drilled while being small enough to reduce any disturbance of the gas flow and catalyst bed. The latter is very important in order to prevent channeling of the gas and, therefore, significant bypass of the catalyst. When considering combination of Spaci-FB with spectroscopy, the reactor design is more complex, and material selection can become complicated for the nonspecialist. As an example, for XAS, the CXRO toolkit (http://henke.lbl.gov/optical_constants/

filter2.html, accessed 27th April 2017), a resource from the Centre for optics at Lawrence Berkeley National laboratory (http://www.cxro.lbl.gov/, accessed 27th April 2017), is a useful free access facility when designing a suitable reactor for a given energy range and therefore defining suitable build material.

While all these aspects can make Spaci-FB investigations challenging, we hope to have convinced the reader that the wealth of information that they provide certainly more than makes up for these.

## ACKNOWLEDGMENTS

A.G., K.M., and C.S. wish to thank EPSRC UK for funding under the First Grant Scheme (A.G.; EP/F026390/1), the CASTech project (EP/G02152X/1), and the UK Catalysis Hub project (EP/K014714/1). C.S. would like to thank the RSC for assistance via a Researcher Mobility Grant and DEL (NI) for funding of the studentship. The Diamond Light Source are thanked for awarding the beam time (Experiment number SP10225), and Christopher Hardacre, Cristina Stere, Peter Wells, and Emma Gibson are acknowledged for their contributions to XAS acquisition. Colin Colligan (QUB) and Phil Robbins (beamline B18 Diamond Light Source) are thanked for support with the technical modifications to the reactor setup.

## REFERENCES

Aftab K, Mandur J, Budman H, Currier NW, Yezerets A, Epling WS: Spatially-resolved calorimetry: using IR thermography to measure temperature and trapped NO(X) distributions on a NO(X) adsorber catalyst, *Catal Lett* 125:229–235, 2008.

Akpa BS, Mantle MD, Sederman AJ, Gladden LF: In situ $^{13}C$ DEPT-MRI as a tool to spatially resolve chemical conversion and selectivity of a heterogeneous catalytic reaction occurring in a fixed-bed reactor, *Chem Commun* (21):2741–2743, 2005.

Baiker A, Bergougnan M: Investigation of a fixed-bed pilot-plant reactor by dynamic experimentation, *Can J Chem Eng* 63:138–145, 1985.

Baiker A, Epple D: Simple method for the modelling of the behavior of a nonisothermal, nonadiabatic fixed-bed catalytic reactor, *Appl Catal* 22:55–69, 1986.

Beale AM, O'Brien MG, Weckhuysen BM: Techniques coupling for catalyst characterisation. In Che M, Védrinedrine JC, editors: Weinheim, Germany, 2012, Wiley-VCH Verlag GmbH & Co. KGaA, pp 1075–1117 Characterization of solid materials and heterogeneous catalysts: from structure to surface reactivity, (vols. 1 and 2).

Beale AM, Hofmann JP, Sankar M, van Schrojenstein-Lantman EM, Weckhuysen BM: Recent trends in operando and in situ characterization: techniques for rational design of catalysts. In Wilson K, Lee AF, editors: *Heterogeneous catalysts for clean technology: spectroscopy, design, and monitoring*, Weinheim, Germany, 2013, John Wiley & Sons, pp 365–411.

Beale AM, Jacques SDM, Gibson EK, Di Michiel M: Progress towards five dimensional diffraction imaging of functional materials under process conditions, *Coord Chem Rev* 277:208–223, 2014.

Beretta A, Donazzi A, Livio D, et al: Optimal design of a $CH_4$ CPO-reformer with honeycomb catalyst: combined effect of catalyst load and channel size on the surface temperature profile, *Catal Today* 171:79–83, 2011.

Bitsch-Larsen A, Horn R, Schmidt LD: Catalytic partial oxidation of methane on rhodium and platinum: spatial profiles at elevated pressure, *Appl Catal A* 348:165–172, 2008.
Bosco M, Vogel F: Optically accessible channel reactor for the kinetic investigation of hydrocarbon reforming reactions, *Catal Today* 116:348–353, 2006.
Buurmans ILC, Weckhuysen BM: Heterogeneities of individual catalyst particles in space and time as monitored by spectroscopy, *Nat Chem* 4:873–886, 2012.
Cavusoglu G, Dallmann F, Lichtenberg H, Goldbach A, Dittmeyer R, Grunwaldt J-D: In situ characterization of catalysts and membranes in a microchannel under high-temperature water gas shift reaction conditions, *J Phys Conf Ser* 712, 2016. 012054.
Choi J-S, Partridge WP, Daw CS: Spatially resolved in situ measurements of transient species breakthrough during cyclic, low-temperature regeneration of a monolithic Pt/K/$Al_2O_3$ NOx storage-reduction catalyst, *Appl Catal A* 293:24–40, 2005.
Choi J-S, Partridge WP, Epling WS, Currier NW, Yonushonis TM: Intra-channel evolution of carbon monoxide and its implication on the regeneration of a monolithic Pt/K/$Al_2O_3$ NOx storage-reduction catalyst, *Catal Today* 114:102–111, 2006.
Choi J-S, Partridge WP, Daw CS: Sulfur impact on NOx storage, oxygen storage, and ammonia breakthrough during cyclic lean/rich operation of a commercial lean NOx trap, *Appl Catal B* 77:145–156, 2007.
Choi J-S, Partridge WP, Pihl JA, Daw CS: Sulfur and temperature effects on the spatial distribution of reactions inside a lean NOx trap and resulting changes in global performance, *Catal Today* 136:173–182, 2008.
Chupas PJ, Ciraolo MF, Hanson JC, Grey CP: In situ X-ray diffraction and solid-state NMR study of the fluorination of $\gamma$-$Al_2O_3$ with $HCF_2Cl$, *J Am Chem Soc* 123:1694–1702, 2001.
Clausen BS, Steffensen G, Fabius B, Villadsen J, Feidehhans'l R, Topsoe H: In situ cell for combined XRD and on-line catalysis tests: studies of Cu-based water gas shift and methanol catalysts, *J Catal* 132:524–535, 1991.
Cong P, Doolen RD, Fan Q, et al: High-throughput synthesis and screening of combinatorial heterogeneous catalyst libraries, *Angew Chem Int Ed* (4):484–488, 1999.
Couves JW, Thomas JM, Waller D, et al: Tracing the conversion of aurichalcite to a copper catalyst by combined X-ray absorption and diffraction, *Nature* 354:465–468, 1991.
Currier NW, Cunningham MJ, Partridge WP, et al: In *Exhaust after treatment gas phase measurements in situ, 6th diesel engine emissions reduction workshop, DEER, August 23*. 2000.
Dalle Nogare D, Degenstein NJ, Horn R, Canu P, Schmidt LD: Modeling spatially resolved profiles of methane partial oxidation on a Rh foam catalyst with detailed chemistry, *J Catal* 258:131–142, 2008.
Dalle Nogare D, Degenstein NJ, Horn R, Canu P, Schmidt LD: Modeling spatially resolved data of methane catalytic partial oxidation on Rh foam catalyst at different inlet compositions and flowrates, *J Catal* 277:134–148, 2011.
Daniel C, Clarté M-O, Teh S-P, et al: Spatially resolved catalysis in microstructured reactors by IR spectroscopy: CO oxidation over mono- and bifunctional Pt catalysts, *J Catal* 272:55–64, 2010.
Donazzi A, Maestri M, Michael BC, et al: Microkinetic modelling of spatially resolved autothermal CH4 catalytic partial oxidation experiments over Rh-coated foams, *J Catal* 275:270–279, 2010.
Donazzi A, Livio D, Beretta A, Groppi G, Forzatti P: Surface temperature profiles in CH4 CPO over honeycomb supported Rh catalyst probed with in situ optical pyrometer, *Appl Catal A* 402:41–49, 2011a.
Donazzi A, Livio D, Maestri M, et al: Synergy of homogeneous and heterogeneous chemistry probed by in situ spatially resolved measurements of temperature and composition, *Angew Chem Int Ed* 50:3943–3946, 2011b.

Donnazi A, Michael BC, Schmidt LD: Chemical and geometric effects of Ce and washcoat addition on catalytic partial oxidation of $CH_4$ on Rh probed by spatially resolved measurements, *J Catal* 260:270–275, 2008.

Doronkin DE, Casapu M, Günter T, Müller O, Frahm R, Grunwaldt J-D: Operando spatially- and time-resolved XAS study on zeolite catalysts for selective catalytic reduction of NOx by $NH_3$, *J Phys Chem C* 118:10204–10212, 2014.

Eu WS, Cheung WH, Valix M: Design and application of a high-temperature microfurnace for an in situ X-ray diffraction study of phase transformation, *J Synchrotron Radiat* 16:842–848, 2009.

Figueroa SJA, Newton MA: What drives spontaneous oscillations during CO oxidation using O2 over supported $Rh/Al_2O_3$ catalysts? *J Catal* 312:69–77, 2014.

Freund H-J, Kuhlenbeck H, Libuda J, Rupprechter G, Bäumer M, Hamann H: Bridging the pressure and materials gaps between catalysis and surface science: clean and modified oxide surfaces, *Top Catal* 15:201, 2001.

Gänzler MA, Casapu M, Boubnov A, et al: Operando spatially and time-resolved X-ray absorption spectroscopy and infrared thermography during oscillatory CO oxidation, *J Catal* 328:216–224, 2015.

Geske M, Korup O, Horn R: Resolving kinetics and dynamics of a catalytic reaction inside a fixed bed reactor by combined kinetic and spectroscopic profiling, *Catal Sci Technol* 3:169–175, 2013.

Gibson EK, Zandbergen MW, Jacques SDM, et al: Noninvasive spatiotemporal profiling of the processes of impregnation and drying within $Mo/Al_2O_3$ catalyst bodies by a combination of X-ray absorption tomography and diagonal offset Raman spectroscopy, *ACS Catal* 3:339–347, 2013.

Gladden LF, Sederman AJ: Recent advances in flow MRI, *J Magn Reson* 229:2–11, 2013.

Gladden LF, Mantle MD, Sederman AJ: Magnetic resonance imaging of catalysts and catalytic processes, *Adv Catal* 50:1–75, 2006.

Gleaves JT, Ebner JR, Kuechler TC: Temporal analysis of products (TAP)—a unique catalyst evaluation system with submillisecond time resolution, *Catal Rev Sci Eng* 30:49, 1988.

Goguet A, Partridge WP, Aiouche F, et al: Comment on "The critical evaluation of in situ probe techniques for catalytic honeycomb monoliths" by Hettel et al, *Catal Today* 236:206–208, 2014.

Grunwaldt J-D, Baiker A: Axial variation of the oxidation state of $Pt\text{-}Rh/Al_2O_3$ during partial methane oxidation in a fixed-bed reactor: an in situ X-ray absorption spectroscopy study, *Catal Lett* 99:5–12, 2005.

Grunwaldt J-D, Caravati M, Hannemann S, Baiker A: X-ray absorption spectroscopy under reaction conditions: suitability of different reaction cells for combined catalyst characterization and time-resolved studies, *Phys Chem Phys* 6:3037–3047, 2004.

Grunwaldt J-D, Hannemann S, Schroer CG, Baiker A: 2D-mapping of the catalyst structure inside a catalytic microreactor at work: partial oxidation of methane over $Rh/Al_2O_3$, *J Phys Chem B* 110:8674–8680, 2006.

Grunwaldt JD, Wagner JB, Dunin-Borkowski RE: Imaging catalysts at work: a hierarchical approach from the macro- to the meso- and nano-scale, *ChemCatChem* 5:62–80, 2013.

Hettel M, Diehm C, Deutschmann O: Answer to the comment from Goguet et al. to the paper "The critical evaluation of in situ probe techniques for catalytic honeycomb monoliths" by Hettel et al. [1], *Catal Today* 236:209–213, 2014.

Horn, R., In-situ measurements of spatially resolved spectroscopic data within a reactor chamber of a reactor. International Patent WO 2011/072701 A1, 2011.

Horn R, Degenstein NJ, Williams KA, Schmidt LD: Spatial and temporal profiles in millisecond partial oxidation processes, *Catal Lett* 110:169–178, 2006a.

Horn R, Williams KA, Degenstein NJ, Schmidt LD: Syngas by catalytic partial oxidation of methane on rhodium: mechanistic conclusions from spatially resolved measurements and numerical simulations, *J Catal* 242:92–102, 2006b.

Horn R, Williams KA, Degenstein NJ, et al: Methane catalytic partial oxidation on autothermal Rh and Pt foam catalysts: oxidation and reforming zones, transport effects, and approach to thermodynamic equilibrium, *J Catal* 249:380–393, 2007a.

Horn R, Williams KA, Degenstein NJ, Schmidt LD: Mechanism of $H_2$ and CO formation in the catalytic partial oxidation of $CH_4$ on Rh probed by steady-state spatial profiles and spatially resolved transients, *Chem Eng Sci* 62:1298–1307, 2007b.

Horn R, Korup O, Geske M, Zavyalova U, Oprea I, Schlögl R: Reactor for in situ measurements of spatially resolved kinetic data in heterogeneous catalysis, *Rev Sci Instrum* 81:064102–064109, 2010.

Irani K, Epling WS, Blint R: Spatial resolution of reactant species consumption in diesel oxidation catalysts, *Top Catal* 52:1856–1859, 2009.

Johansson M, Hoffmann Jørgensen J, Chorkendorff I: Combined high-pressure cell—ultrahigh vacuum system for fast testing of model metal alloy catalysts using scanning mass spectrometry, *Rev Sci Instrum* 75:2082–2093, 2004.

Koptyug IV: Chapter 1: magnetic resonance imaging methods in heterogeneous catalysis, *Spectrosc Prop Inorg Organomet Compd* 45:1–42, 2014.

Kopyscinski J, Schildhauer TJ, Vogel F, Biollaz SMA, Wokaun AJ: Applying spatially resolved concentration and temperature measurements in a catalytic plate reactor for the kinetic study of CO methanation, *J Catal* 271:262–279, 2010.

Korup O, Mavlyankariev S, Geske M, Goldsmith CF, Horn R: Measurement and analysis of spatial reactor profiles in high temperature catalysis research, *Chem Eng Process Process Intens* 50:998–1009, 2011.

Korup O, Goldsmith CF, Weinberg G, et al: Catalytic partial oxidation of methane on platinum investigated by spatial reactor profiles, spatially resolved spectroscopy, and microkinetic modelling, *J Catal* 297:1–16, 2013.

Lee AF, Ellis CV, Naughton J, Newton MA, Parlett CMA, Wilson K: Reaction-driven surface restructuring and selectivity control in allylic alcohol catalytic aerobic oxidation over Pd, *J Am Chem Soc* 133:5724–5727, 2011a.

Lee S, Lee B, Seifert S, Vajda S, Winans RE: Simultaneous measurement of X-ray small angle scattering, absorption and reactivity: a continuous flow catalysis reactor, *Nucl Instrum Methods Phys Res A* 649:200–203, 2011b.

Liu S, Li W, Wang Y, Xu H: Catalytic partial oxidation of methane to syngas in a fixed-bed reactor with an O2-distributor: the axial temperature profile and species profile study, *Fuel Process Technol* 89:1345–1350, 2006.

Livio D, Donazzi A, Beretta A, Groppi G, Forzatti P: Optimal design of a CPO-reformer of light hydrocarbons with honeycomb catalyst: effect of frontal heat dispersions on the temperature profiles, *Top Catal* 54:866–872, 2011.

Luo J-Y, Al-Harbi M, Pang M, Epling WS: Spatially resolving LNT desulfation: re-adsorption induced by oxygen storage materials, *Appl Catal B* 106:664–671, 2011a.

Luo J-Y, Hou X, Wijayakoon P, Schmieg SJ, Li W, Epling WS: Spatially resolving SCR reactions over a Fe/zeolite catalyst, *Appl Catal B* 102:110–119, 2011b.

Margulies L, Kramer MJ, McCallum RW, et al: New high temperature furnace for structure refinement by powder diffraction in controlled atmospheres using synchrotron radiation, *Rev Sci Instrum* 70:3554–3561, 1999.

Michael BC, Nare DN, Schmidt LD: Catalytic partial oxidation of ethane to ethylene and syngas over Rh and Pt coated monoliths: spatial profiles of temperature and composition, *Chem Eng Sci* 65:3893–3902, 2010.

Morgan K, Touitou J, Choi J-S, et al: Evolution and enabling capabilities of spatially resolved techniques for the characterization of heterogeneously catalyzed reactions, *ACS Catal* 6(2):1356–1381, 2016.

Morgan K, Maguire N, Fushimi R, et al: Forty years of temporal analysis of products, *Catal Sci Technol* 7:2416–2439, 2017.

Nachtegaal M, Müller O, König C, Frahm R: QEXAFS: techniques and scientific applications for time-resolved XAS. In van Bokhoven JA, Lamberti C, editors: *X-ray absorption*

*and X-ray emission spectroscopy: theory and applications*, vols. 1 and 2, Chichester, UK, 2015, John Wiley & Sons, pp 155–183.
Nachtegaal M, Hartfelder UB, van Bokhoven JA: From spectator species to active site using X-ray absorption and emission spectroscopy under realistic conditions. In Frenken J, Groot I, editors: *Operando research in heterogeneous catalysis, Springer chemical physics series*, vol. 114, Berlin, Germany, 2017, Springer International Publishing, pp 89–110.
Newton MA: Time resolved operando X-ray techniques in catalysis, a case study: CO oxidation by $O_2$ over Pt surfaces and alumina supported Pt catalysts, *Catalysts* 7:58, 2017.
Newton MA, di Michiel M, Ferri D: Catalytic adventures in space and time using high energy X-rays, *Catal Surv Asia* 18:134–148, 2014.
Nguyen H, Peng PY, Luss D, Harold MP: Assessing intrusion by the capillary during spatially resolved mass spectrometry measurement, *Chem Eng J* 307:845, 2017.
Nic An tSaoir M, Fernandes DLA, Sá J, et al: Visualization of water vapour flow in a packed bed adsorber by near-infrared diffused transmittance tomography, *Chem Eng Sci* 66:6407–6423, 2011.
Nic An tSaoir M, Fernandes DLA, Sá J, Kitagawa K, Hardacre C, Aiouache F: Three-dimensional water vapor visualization in porous packing by near-infrared diffuse transmittance tomography, *Ind Eng Chem Res* 51:8875–8882, 2012.
Niemantsverdriet JW, Engelen AFP, de Jong AM, Wieldraaijer W, Kramer GJ: Realistic surface science models of industrial catalysts, *Appl Surf Sci* 144–145:366, 1999.
O'Brien MG, Beale AM, Jacques SDM, di Michiel M, Weckhuysen BM: Spatiotemporal multitechnique imaging of a catalytic solid in action: phase variation and volatilization during molybdenum oxide reduction, *ChemCatChem* 1:99–102, 2009.
O'Brien MG, Beale AM, Jacques SDM, Di Michiel M, Weckhuysen BM: Closing the operando gap: the application of high energy photons for studying catalytic solids at work, *Appl Catal A* 391:468–476, 2011.
O'Brien MG, Jacques SDM, Di Michiel M, Barnes P, Weckhuysen BM, Beale AM: Active phase evolution in single $Ni/Al_2O_3$ methanation catalyst bodies studied in real time using combined μ-XRD-CT and μ-absorption-CT, *Chem Sci* 3:509–523, 2012.
Orschel M, Klein J, Schmidt H-W, Maier WF: Detection of reaction selectivity on catalyst libraries by spatially resolved mass spectrometry, *Angew Chem Int Ed* 38:2791–2794, 1999.
Parks, J.; Huff, S.; Pihl, J.; Choi, J.; West, B., Nitrogen selectivity in lean NOx trap catalysis with diesel engine in-cylinder regeneration. SAE technical paper 2005, 2005-01-3876.
Partridge WP, Choi J-S: $NH_3$ formation and utilization in regeneration of $Pt/Ba/Al_2O_3$ NOx storage-reduction catalyst with $H_2$, *Appl Catal B* 91:144–151, 2009.
Partridge, W.P.; Storey, J.M.E.; Lewis, S.A.; et al, Time-resolved measurements of emission transients by mass spectrometry. SAE technical paper 2000, 2000-01-2952.
Partridge, W.P.; Lewis, S.A.; Ruth, M.J.; Muntean, G.G.; Smith, R.C.; Stang, J.H., Resolving EGR distribution and mixing. SAE technical paper 2002, 2002-01-2882.
Partridge WP, Toops TJ, Green JB, Armstrong TR: Intra-fuel cell stack measurements of transient concentration distributions, *J Power Sour* 160:454–461, 2006.
Pistidda C, Santoru A, Garroni S, et al: First direct study of the ammonolysis reaction in the most common alkaline and alkaline earth metal hydrides by in situ SR-PXD, *J Phys Chem C* 119:934–943, 2015.
Rasmussen SB, Perez-Ferreras S, Bañares MA, Bazin P, Daturi M: Does pelletizing catalysts influence the efficiency number of activity measurements? Spectrochemical engineering considerations for an accurate operando study, *ACS Catal* 3(1):86–94, 2013.
Ravel B, Newville B: ATHENA, ARTEMIS, HEPHAESTUS: data analysis for X-ray absorption spectroscopy using IFEFFIT, *J Synchrotron Radiat* 12:537–541, 2005.
Roos M, Kielbassa S, Schirling C, Häring T, Bansmann J, Behm RJ: Scanning mass spectrometer for quantitative reaction studies on catalytically active microstructures, *Rev Sci Instrum* 78, 2007. 084104.

Roos M, Bansmann J, Zhang D, Deutschmann O, Behm RJ: Product gas evolution above planar microstructured model catalysts—a combined scanning mass spectrometry, Monte Carlo, and computational fluid dynamics study, *J Chem Phys* 133, 2010. 094504.

Russell A, Epling WS, Hess H, et al: Spatially-resolved temperature and gas species changes in a lean-burn engine emissions control catalyst, *Ind Eng Chem Res* 49:10311–10322, 2010.

Russell A, Henry C, Currier NW, Yezerets A, Epling WS: Spatially resolved temperature and gas species concentration changes during $C_3H_6$ oxidation over a $Pt/Al_2O_3$ catalyst following sulfur exposure, *Appl Catal A* 397:272–284, 2011.

Sá J, Fernandes DL, Aiouache F, et al: SpaciMS: spatial and temporal operando resolution of reactions within catalytic monoliths, *Analyst* 135:2260–2272, 2010.

Schroer CG, Grunwaldt J-D: In Rodriguez JA, Chupas PJ, Hanson JC, editors: *Spatially resolved X-ray absorption spectroscopy in in-situ characterization of heterogeneous catalysts*, Hoboken, 2013, John Wiley & Sons, pp 49–73.

Sederman AJ, Mantle MD, Dunckley CP, Huang ZY, Gladden LF: In situ MRI study of 1-octene isomerisation and hydrogenation within a trickle-bed reactor, *Catal Lett* 103:1–8, 2005.

Shakir O, Yezerets A, Currier NW, Epling WS: Spatially resolving concentration and temperature gradients during the oxidation of propylene on $Pt/Al_2O_3$, *Appl Catal A* 365:301–308, 2009.

Stewart J, Douglas R, Goguet A: Integrating intrinsic and global kinetics as a dual kinetic model for automotive catalysis, *Proc IMechE Part D J Automob Eng* 228:285–294, 2013.

Stewart J, Douglas R, Goguet A, Stere C: Detailed validation of an auto catalysis model using spatially resolved measurements within the catalyst substrate, *Can J Chem Eng* 92(9):1535, 2014.

Tan CKC, Delgass WN, Baertsch CD: Spatially resolved in situ FTIR analysis of CO adsorption and reaction on $Pt/SiO_2$ in a silicon microreactor, *Appl Catal B* 93:66–74, 2009.

Thomas S, Marie O, Bazin P, et al: Modelling a reactor cell for operando IR studies: from qualitative to fully quantitative kinetic investigations, *Catal Today* 283:176–184, 2017.

Tinnemans SJ, Mesu JG, Kervinen K, et al: Combining operando techniques in one spectroscopic-reaction cell: new opportunities for elucidating the active site and related reaction mechanism in catalysis, *Catal Today* 113:3–15, 2006.

Titze T, Chmelik C, Kullmann J, et al: Microimaging of transient concentration profiles of reactant and product molecules during catalytic conversion in nanoporous materials, *Angew Chem Int Ed* 54:5060–5064, 2015.

Touitou J: *Development of an in situ spatially resolved technique to investigate catalysts in a plug flow reactor*, PhD thesis, 2014, Queen's University Belfast.

Touitou J, Morgan K, Burch R, Hardacre C, Goguet A: An in situ spatially resolved method to probe gas phase reactions through a fixed bed catalyst, *Catal Sci Technol* 2:1811–1813, 2012.

Touitou J, Burch R, Hardacre C, et al: An in situ spatially resolved analytical technique to simultaneously probe gas phase reactions and temperature within the packed bed of a plug flow reactor, *Analyst* 138:2858–2862, 2013.

Touitou J, Aiouache F, Burch R, et al: Evaluation of an in situ spatial resolution instrument for fixed beds through the assessment of the invasiveness of probes and a comparison with a micro-kinetic model, *J Catal* 319:239–246, 2014.

Urakawa A, Baiker A: Space-resolved profiling relevant in heterogeneous catalysis, *Top Catal* 52:1312–1322, 2009.

Urakawa A, Maeda N, Baiker A: Space- and time-resolved combined DRIFT and Raman spectroscopy: monitoring dynamic surface and bulk processes during NOx storage reduction, *Angew Chem Int Ed* 47:9256–9259, 2008.

Vamvakeros A, Jacques SDM, Middelkoop V, et al: Real time chemical imaging of a working catalytic membrane reactor during oxidative coupling of methane, *Chem Commun* 51, 2015. 12752.

Weckhuysen BM: Chemical imaging of spatial heterogeneities in catalytic solids at different length and time scale, *Angew Chem Int Ed* 48:4910–4943, 2009.

West, B.; Huff, S.; Parks, J.; et al, Assessing reductant chemistry during in-cylinder regeneration of diesel lean NOx traps. SAE technical paper 2004, 2004-01-3023.

Wragg DS, O'Brien MG, Bleken FL, Di Michiel M, Olsbye U, Fjellvåg H: Watching the methanol-to-olefin process with time- and space-resolved high-energy operando X-ray diffraction, *Angew Chem Int Ed* 51:7956–7959, 2012.

Yuen EHL, Sederman AJ, Gladden LF: In situ magnetic resonance visualisation of the spatial variation of catalytic conversion within a fixed-bed reactor, *Appl Catal A* 232:29–38, 2002.

Zaera F: New advances in the use of infrared absorption spectroscopy for the characterization of heterogeneous catalytic reactions, *Chem Soc Rev* 43:7624–7663, 2014.

Zellner A, Suntz R, Deutschmann O: Two-dimensional spatial resolution of concentration profiles in catalytic reactors by planar laser-induced fluorescence: NO reduction over diesel oxidation catalysts, *Angew Chem Int Ed* 54:2653–2655, 2015.

Zhang C, Gustafson J, Merte LR, et al: An in situ sample environment reaction cell for spatially resolved X-ray absorption spectroscopy studies of powders and small structured reactors, *Rev Sci Instrum* 86(033112):1–7, 2015.

CHAPTER FOUR

# Analysis of the Impact of Gas-Phase Chemistry in Adiabatic CPO Reactors by Axially Resolved Measurements

**Alessandra Beretta[1], Gianpiero Groppi, Andrea Carrera, Alessandro Donazzi**

LCCP—Laboratory of Catalysis and Catalytic Processes, Dipartimento di Energia, Politecnico di Milano, Milano, Italy

[1]Corresponding author: e-mail address: alessandra.beretta@polimi.it

## Contents



## Abstract

This chapter focuses on the contribution that the in situ probing of temperature and concentration profiles in high-temperature and short contact time partial oxidation processes has given to the understanding of the interplay between homogeneous and heterogeneous chemistry. Experimental evidence was collected in our laboratory on the gas-phase conversion of $C_{2+}$ fuels in adiabatic CPO reformers. The formation of short-chain hydrocarbons can promote the syngas performance, since the fed fuel is partly transformed into highly reactive species which further react on the surface producing CO and $H_2$; however, short olefins are soot/coke precursors and the surface condensation or breakthrough of such species is highly undesired. The experience from

*Advances in Chemical Engineering*, Volume 50
ISSN 0065-2377
https://doi.org/10.1016/bs.ache.2017.09.001

CPO reactors and other similar processes allows to understand the effects that pressure and dilution, the nature of the fuel and of the catalyst, and heat dispersions across the reactor have on the final fate of the cracking intermediates. The possibility to tune the relative contributions of the gas-phase and the surface chemistry in order to limit the catalyst hot spot temperature is also discussed.

## 1. INTRODUCTION

Small-scale fuel processing has been studied in the last 15–20 years as one of the possible solutions for the decentralized and the on-board production of $H_2$ for $H_2$-fuelled vehicles using fossil fuels (Armor, 1999; Basini, 2005; Villegas et al., 2014; Xu et al., 2013). Originally, hydrocarbon reforming represented a key factor of the so-called hydrogen economy. Nowadays, while the perspective of fuel cells-driven vehicles becomes more and more close, the use of fossil fuels for $H_2$-refuelling stations is conceived as a transition stage necessary for supporting the birth of a $H_2$ market before the development of technologies based on renewable energy (e.g., $H_2$ production by water electrolysis powered by solar energy or wind energy) is complete.

Additionally, the interest for developing compact small-scale processors using fossil or biofuels remains lively for several potential applications, from stationary combined heat and power systems (Galvagno et al., 2013; Kaltschmitt and Deutschmann, 2012; Kolb et al., 2007), to on-board $H_2$ generation for $H_2$-enhanced IC engines (Galloni and Minutillo, 2007; Mehra et al., 2017; Soberanis and Fernandez, 2010), regeneration of $NO_x$ traps on diesel vehicles (Mundschau et al., 2010), and fuelling of SOFC- or HTPEM-based auxiliary power units for cars, ships, and aircrafts (Fabiano et al., 2016; Pasel et al., 2004; Samsun et al., 2017).

The catalytic partial oxidation of hydrocarbons presents several potential advantages, highly attractive at this scope: easy scalability, flexibility to fuels, fast dynamic response, and, most importantly, autothermal character which greatly simplifies the processing system due to the absence of exo- and endothermic compartments. The autothermal operation of the CPO reformers has been proven at the lab scale in the so-called short contact time reactors, pioneered by Schmidt and his group at Minnesota, consisting of foam or honeycomb supports coated with a metal-containing washcoat (Hickman and Schmidt, 1992, 1993; Hickman et al., 1993). Those original results have been largely confirmed over the years, and Rh-supported catalysts have become the materials of choice for several proven proof-of-concept and demonstration units based on reforming or autothermal reforming of gaseous and liquid fuels because of high activity and reduced tendency to coking (Basile et al., 2003; Pasel et al., 2016; Specchia et al., 2012; Vita et al., 2017).

Although a direct partial oxidation route to CO and $H_2$ has been claimed in early studies, the kinetics of the CPO of several hydrocarbons (methane has been mostly studied, but also higher hydrocarbons) over $Rh/Al_2O_3$ and other catalysts can be well explained by a reaction scheme, consisting on one side of the consumption of $O_2$ in the deep oxidation of the fuel and the post-combustion of CO and $H_2$, and on the other side of the steam reforming of the hydrocarbon to CO and $H_2$; WGS (water–gas shift), RWGS (reverse water–gas shift), and $CO_X$ methanation can then affect the syngas composition, depending on the temperature and chemical potential of the reacting mixture. The relative rate of gas–solid fluxes and surface reactions determines the dominant chemical pathways and the product distribution; in fact, in a chemically controlled regime, deep oxidation is the primary reaction of the fuel (thus producing $CO_2$ and $H_2O$) and syngas is produced only consecutively to the complete conversion of $O_2$, while under a diffusion-controlled regime (where the concentration of $O_2$ at the wall is zero) steam reforming is active even in the presence of $O_2$ in the gas phase and is fully responsible for the fuel conversion, while postcombustions (mostly of $H_2$) are responsible for $O_2$ conversion (Maestri et al., 2009).

Under the operating conditions of the real application (adiabatic reactor, absence of feed dilution, wash-coated structured catalysts, millisecond contact times, gas solid mass, and heat transfer limitations) the oxidation/reforming scheme results in the reaching of very high temperatures at the very entrance of the reactor. This turns out to be the inherent weakness of the CPO process, since high temperatures can cause catalyst sintering, activity loss, and reactor instability due to an autocatalytic deactivation process, resulting in a progressive heating of the system (Beretta et al., 2009).

Thus, a very accurate characterization of the temperature distribution across the reactor and especially at the entrance, as well as a detailed characterization of the chemical processes which determine the temperature evolution is extremely important.

The development of axially resolved sampling techniques has largely contributed to finely monitor the axial profiles of temperature and composition in short contact time CPO reactors (Morgan et al., 2016). It must be emphasized that in CPO reactors the peak of temperature and of reaction rates extinguishes within few millimeters from the monolith entrance; thus, an extremely high resolution has been necessary to disclose the sharpness of temperature and concentration gradients. In our laboratory, these techniques have been applied to study the CPO of light hydrocarbons (methane and propane) and liquid hydrocarbons (iso- and normal octanes) over Rh. The information gained has supported the development and validation of a mathematical model of the adiabatic reactor that implements the reaction schemes derived from the independent kinetic investigations. We strongly believe in the need of a strict interplay between the operando investigation and the development of engineering tools of simulations, necessary on one side to rationalize the observed phenomenon (e.g., by understanding relative importance of reaction rate vs heat and mass transfer rates) and on the other side to develop improved solutions that can lead to the final application. In this respect, a very interesting phenomenon revealed by the internal probing of the CPO reactor has been the formation of short-chain hydrocarbons, which can hardly detected at the reactor outlet; the observation prompted the development of a more comprehensive reactor model incorporating both the heterogenous chemistry and the homogeneous chemistry, which in turn provided new elements for the reaction engineering analysis of the process.

In this chapter, we illustrate the phenomenology of the CPO process as it results from the in situ probing of temperature and concentration for various fuels; we then discuss the interpretation of the observed trends based on a modeling analysis. The understanding of the interaction between radical and surface reactions and an engineering analysis of the effect of operating parameters on such interaction offered additional tools for the optimal design of the CPO reformer, where the key issue is the minimization of the inlet hot spot temperature.

Finally, we discuss how the learning gained from the CPO of $C_{2+}$ hydrocarbons can provide a key for the interpretation of the performances observed in similar high temperatures and fuel-rich oxidation processes (e.g., methane oxidative coupling (MOC), ethane oxidative dehydrogenation (ODH) to ethylene, ethanol partial oxidation).

## 2. APPLICATION OF THE SPATIALLY RESOLVED SAMPLING TECHNIQUE TO THE CPO OF $C_{2+}$ HYDROCARBONS

This section presents the results obtained in our laboratory by the application of the sampling technique to the investigation of temperature and concentration profiles in the CPO of propane and the CPO of octane isomers over Rh. Experimental details are provided first.

## 2.1 Experimental Methods

### *2.1.1 Catalyst*

The CPO tests were performed over a 2 wt% Rh/$\alpha$-$Al_2O_3$ catalyst supported onto 400 cpsi cordierite honeycombs. The catalyst was prepared by dry impregnation of the $\alpha$-$Al_2O_3$ support (10 $m^2$/g surface area by BET measurement) with an aqueous solution of $Rh(NO_3)_3$. Following a standard procedure, the catalyst was deposited over the honeycomb support by dip-coating and blowing-off of the slurry obtained from the catalyst powders. Adhesion of the deposited layer was achieved by subsequent drying at 280°C in an oven for 15 min. The catalyst load (~600–800 mg for monolith lengths of 2.5–3.5 cm) was estimated by weight difference before and after coating the monolith. The washcoat thickness (15–20 μm) was then calculated assuming a density of the catalyst layer of 1.38 g/$cm^3$. The final void fraction of the coated monolith varied in between 0.70 and 0.75.

### *2.1.2 Apparatus and Operating Conditions*

The CPO experiments were carried out in a lab-scale testing rig placed under a fume hood for safety reasons. A diagram of the apparatus is presented in Fig. 1. Three lines fed gaseous reactants (air, $CH_4$ or $C_3H_8$, $N_2$), and one fed liquid fuels to the reactor. Gas flows were regulated by Brooks 5850S thermal mass flow controllers. A Gilson Minipuls 3 peristaltic pump delivered the liquid fuel (stocked in a 250 mL burette) into an evaporator, which consisted of a 40-cm-long ½ inch pipe, filled with quartz spheres (3 mm OD), heated by an annular Carbolite oven at a set point temperature of 170°C. Dilution nitrogen was fed to the evaporator, in order to ensure complete octane evaporation. In octane CPO tests, air is preheated as well, using two Watlow Starflow heating cartridges. Instead, CPO tests with light hydrocarbons were typically performed without any preheating of reactants.

Fuel and air were mixed in a Sulzer static mixer, placed immediately before the reactor. A K-type thermocouple was placed after the static mixer, in order to measure the reactants' temperature.

The reactor (Fig. 2) consisted of a stainless steel tube, externally insulated by wrapping a thick pad of quartz wool. To prevent C formation, a quartz liner was used for hosting the catalytic monolith in the reactor. Within this liner, the catalytic monolith was placed in between two inert blank monoliths, kept at a distance of ~1 cm, which acted as heat shields. This choice

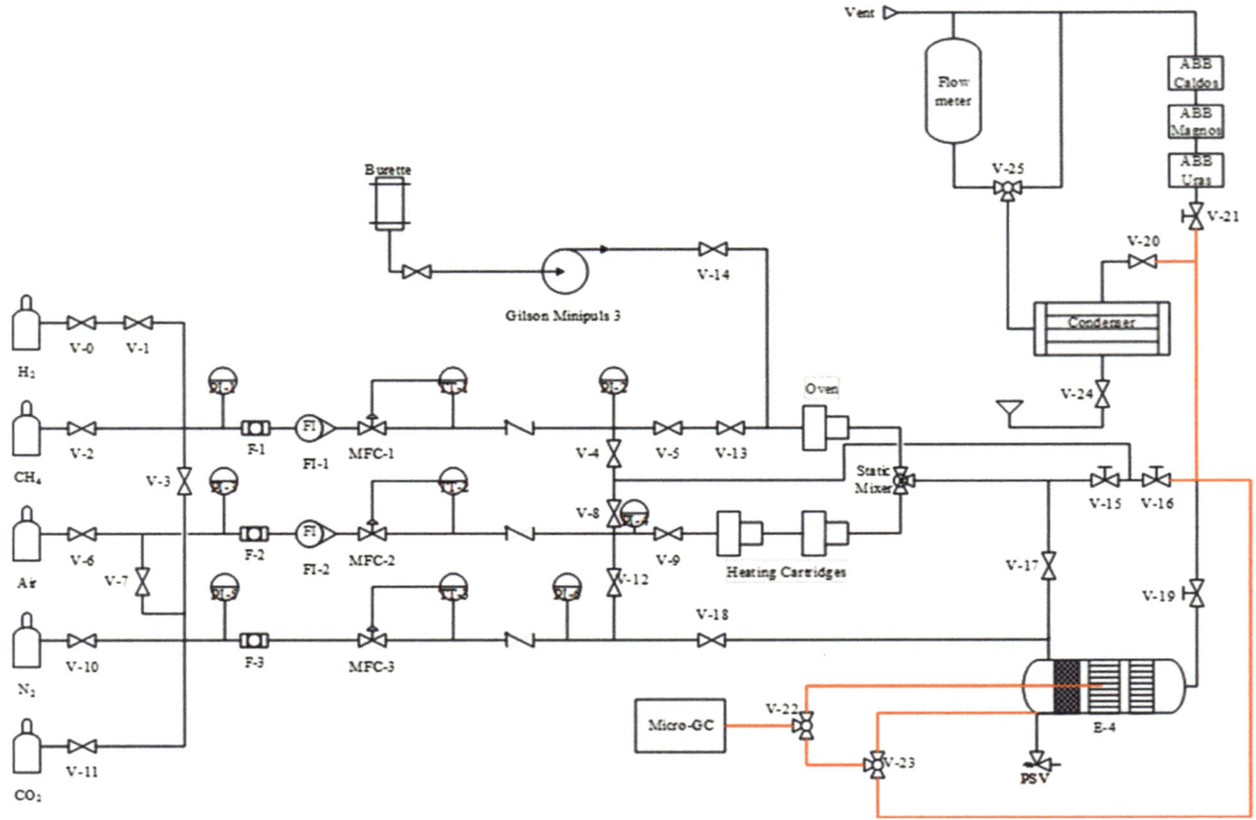


**Fig. 1** P&I diagram of the testing rig used in this work. Heated lines are colored in *red*.

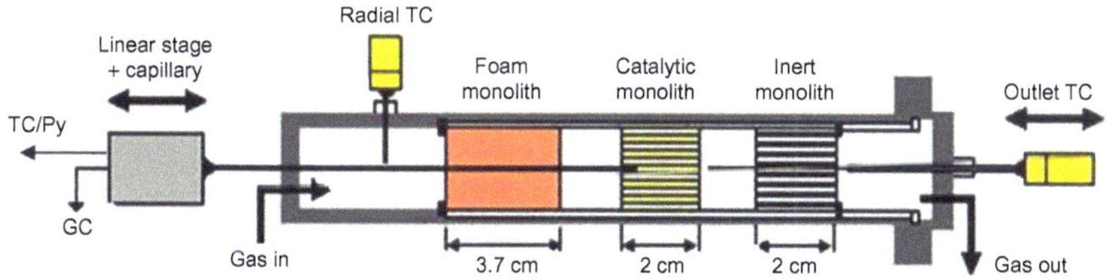


**Fig. 2** Simplified scheme of the configuration of the adiabatic reactor. *Adapted from Carrera A, Pelucchi M, Stagni A, Beretta A, Groppi G, et al: Catalytic partial oxidation of n-octane and iso-octane: experimental and modeling results,* Int J Hydrogen Energy, *2017, https://doi.org/10.1016/j.ijhydene.2017.08.020, with permission from Elsevier.*

increased the radiant heat dissipation at the edges of the catalytic monolith, thus preserving it from excessive heating, especially in the inlet portion (Livio et al., 2011).

The setup of the reactor was designed to carry out spatially resolved measurements up to 4 bar. The sealing system was improved by addition of a guard filled with high vacuum grease (Dow Corning), which allowed to move the capillary and to avoid gas leakage. The pressure of the reactor was measured at the inlet and was regulated with a control valve at the outlet.

For the measurement of the spatially resolved concentration profile of the gas species, a deactivated fused silica capillary (O.D. = 350 μm, I.D. = 200 μm) was inserted into the central channel of the honeycomb and externally connected to a loop purposely designed for tests under pressure. The sampled flow (2 Ncc/min) was carefully controlled by connecting the capillary to a regulation valve. After this valve, the sampled flow was admitted to an empty volume connected to a micro-GC (Agilent 3000). For the spatially resolved measurements of temperature, the reactor was equipped with a K-type thermocouple (250 μm O.D.) and a narrow band infrared pyrometer (Impac Infrared, IGA 5-LO) connected to an optical fiber (300 μm core diameter, 330 μm O.D.). A larger capillary (O.D. = 500 μm) sealed at one end was used to move the optical fiber and the thermocouple. The thermocouple measurements were taken as representative of the temperature of the gas phase, whereas the measurements of the pyrometer were representative of the temperature of the catalyst surface (Donazzi et al., 2011a).

A micro gas chromatograph analyzed the gas streams by two modules, each equipped with a TCD detector: a Molecular Sieve Column (10 m length), with PLOT U precolumn (3 m length), using Ar as carrier gas, and a PLOT U column (6 m length), using He as carrier gas. Column A worked at 70°C, thus separating $H_2$, $N_2$, $O_2$, $CH_4$, and CO. In experiments with liquid fuels, Column B was operated at two different temperatures, namely, 160°C and 60°C; the analysis at the highest temperature was performed to quantify iso-octane and *n*-octane concentrations, while the lowest temperature analysis separated light hydrocarbon species. Continuous gas analyzers also collected integral data at the reactor outlet.

The CPO tests were performed at 5–10 NL/min. Pressure was varied in between 1 and 4 bar for the $C_3H_8$ CPO tests, while atmospheric pressure was maintained in the campaign with octane fuels. Before each test, the catalyst was preheated with a 6% v/v hydrogen and 3% v/v oxygen in nitrogen mixture, with a flow rate of 5 NL/min. The catalytic combustion of hydrogen preheated the catalyst. As soon as the temperature measured by the rear thermocouple reached 200°C in case of propane or octane CPO or 240°C in case of methane CPO, the inlet mixture of the reactor was switched to the feed mixture for the desired CPO test. The results have been collected in steady-state conditions, at least 30′ after reaction ignition.

In the case of $C_3H_8$ CPO, the experiments were performed with propane/air mixtures with an $O_2$/C ratio between 0.56 and 0.60.

In the case of the liquid fuels, CPO tests were performed at varying dilution; most of the experiments were obtained at 2% $C_8H_{18}$ in the feed with a C/O ratio in between 0.83 and 0.90, which corresponds to a nitrogen dilution from 52% to 56%. Additional experiments were performed at decreasing dilution to verify the severity of the thermal behavior for a nondiluted fuel-rich mixture. The reactants were preheated at a temperature of 85°C, in order to avoid octane condensation.

## 2.2 Mathematical Model of the Reactor and Kinetic Scheme

The experiments were analyzed by a 1D, adiabatic, heterogeneous model of the honeycomb reactor, assuming single-channel approximation, i.e., all the channels are representative of the reactor performance. The model consists of mass and enthalpy balances for the gas phase and solid phase. The model also includes axial convection and diffusion terms, solid-phase conduction, and gas–solid transport terms, estimated with correlations for laminar flow in square channels. Heat conduction in the solid was taken into account with an effective axial thermal conductivity coefficient, corrected by addition of the radiation contribution. The complete set of the equations is reported in Table 1.

The model includes kinetic schemes for the solid phase and gas phase. The heterogeneous kinetic scheme was proposed in previous works starting from the development of a $C_1$ kinetic scheme (Donazzi et al., 2008a,b), and

**Table 1** Model Equations

*Gas phase*

Mass balance

$$\frac{\partial \omega_i}{\partial t} = -\frac{G}{\rho_g \varepsilon}\frac{\partial \omega_i}{\partial z} - \frac{a_v}{\varepsilon} k_{\text{mat},i}(\omega_i - \omega_{\text{wall},i}) + \frac{1}{\varepsilon}\frac{\partial}{\partial z}\left(\mathfrak{D}_i \frac{\partial \omega_i}{\partial z}\right) + \frac{MW_i}{\rho_g}\sum_{j=1}^{NR_s} \upsilon_{i,j} r_j^{\text{hom}} \quad (1)$$

Enthalpy balance

$$\frac{\partial T_g}{\partial t} = -\frac{G}{\rho_g \varepsilon}\frac{\partial T_g}{\partial z} - \frac{a_v h(T_g - T_s)}{\varepsilon \rho_g \hat{c}_{p,g}} + \frac{1}{\varepsilon \rho_g \hat{c}_{p,g}}\frac{\partial}{\partial z}\left(k_g \frac{\partial T_g}{\partial z}\right) - \frac{1}{\rho_g \hat{c}_{p,g}}\sum_{j=1}^{NR_s} \Delta H_{r,j} r_j^{\text{hom}} \quad (2)$$

*Solid phase*

Mass balance

$$0 = a_v \rho_g k_{\text{mat},i}(\omega_i - \omega_{\text{wall},i}) + MW_i \rho_w \alpha \sum_{j=1}^{NR_s} \upsilon_{i,j} r_j^{\text{het}} \quad (3)$$

Enthalpy balance

$$\frac{\partial T_s}{\partial t} = \frac{a_v h(T_g - T_s)}{(1-\varepsilon)\rho_s \hat{c}_{p,s}} + \frac{\frac{\partial}{\partial z}\left(k_{ax}^{\text{eff}} \frac{\partial T_s}{\partial z}\right)}{(1-\varepsilon)\rho_s \hat{c}_{p,s}} - \frac{\alpha \rho_w}{(1-\varepsilon)\rho_s \hat{c}_{p,s}}\sum_{j=1}^{NR_s} \Delta H_{r,j} r_j^{\text{het}} \quad (4)$$

*Boundary conditions*

Reactor inlet ($z = -0.5$ cm, any $t$)

$$\left.\frac{\partial T_s}{\partial z}\right|_{z=-0.5\,\text{cm}} = 0 \quad (5)$$

Catalyst inlet ($z = 0$, any $t$)

$$\omega_i|_{z=0} = \omega_{i,\text{feed}} + \left.\frac{D_i}{v}\frac{\partial \omega_i}{\partial z}\right|_{z=0};\ T_g|_{z=0} = T_{\text{feed}} + \left.\frac{k_g}{v \rho_g \hat{c}_{p,g}}\frac{\partial T_g}{\partial z}\right|_{z=0} \quad (6)$$

Catalyst outlet ($z = L$, any $t$)

$$\left.\frac{\partial \omega_i}{\partial z}\right|_{z=L} = 0;\ \left.\frac{\partial T_s}{\partial z}\right|_{z=L} = 0 \quad (7)$$

Initial conditions ($t = 0$, any $z$)

$$\omega_i(z,0) = 0;\ T_g(z,0) = T_{\text{feed}};\ T_s(z,0) = 650°\text{C} \quad (8)$$

progressively extended by the addition of lumped reactions for the oxidation and steam reforming of propane, iso-octane, and *n*-octane (Carrera et al., 2017a,b; Pagani et al., 2012). The whole scheme includes then the WGS reaction and its reverse RWGS, a CO methanation route, and the post-combustions of CO and $H_2$. The kinetic dependences for each reaction were studied over the years by dedicated experiments, where single reactions were "isolated" from the global CPO process with dedicated experiments and kinetic analysis. The whole scheme is reported in Table 2.

The kinetic scheme also incorporates the reaction of steam reforming of all the hydrocarbon species that are produced in the gas phase; these species include mainly propylene, ethylene, and ethane in the case of propane CPO, but include also several other $C_4$ and $C_{4+}$ species in the CPO of the octanes. As discussed in a following section, this further extension of the heterogeneous kinetic scheme was a key assumption to adequately describe the observed phenomenology.

Gas-phase reactions were described according to the schemes proposed by Ranzi and coworkers (Carrera et al., 2017b; Pelucchi et al., 2014; Ranzi et al., 2012; Stagni et al., 2016).

The numerical analysis herein reported is fully predictive, with no parameter adjustment. The only input data of the calculations were the catalyst amount, the Rh load and dispersion, the geometrical parameters, and the physical properties of the honeycomb support.

## 2.3 Detection of Gas-Phase Products in the Autothermal CPO of Propane

It is first instructive to analyze the behavior of the propane/$O_2$ reacting system in isothermal activity tests, as those performed in annular reactor. Fig. 3 reports the results of experiments at the space velocity of $2 \times 10^6$ L(NTP)/$kg_{cat}$/h, with feed composition of 1% $C_3H_8$ and $O_2$/C ratio = 0.56 in $N_2$ dilution. These data have been fully analyzed elsewhere (Pagani et al., 2012), and we herein focus on the evidence that the evolution of the product distribution is very well explained by a kinetic scheme wherein $CO_2$ and $H_2O$ are primary products of propane fuel-rich oxidation, while syngas is formed by a consecutive route of steam reforming and WGS; solid lines represent the prediction of a isothermal PFR model of the annular reactor incorporating the kinetic scheme reported in Table 2. Interestingly, $CH_4$ was also formed with a maximum (Fig. 3), coherently with the existence of a consecutive methanation reaction that approaches the thermodynamic equilibrium at high temperature. In other experiments at increasing flow rate,

**Table 2** Heterogeneous Kinetic Scheme

| Reaction | Rate$_i$ (mol $g_{cat}^{-1}$ $s^{-1}$) | $K_i^{873\,K}$ (mol $atm^{-1}$ $g_{cat}^{-1}$ $s^{-1}$) | $E_{ACT}$ (kJ $mol^{-1}$) |
|---|---|---|---|
| $i$-$C_8H_{18}$ total oxidation<br>$i\text{-}C_8H_{18} + \frac{25}{2}O_2 \rightarrow 8CO_2 + 9H_2O$ | $R_{Ox,\,i\text{-}C_8H_{18}} = \frac{k_{Ox,\,i\text{-}C_8} P_{i\text{-}C_8H_{18}}}{1 + K_{H_2O} P_{H_2O}} \sigma_{O_2}$ | 4.600E − 01 | 80.00 |
| $i$-$C_8H_{18}$ steam reforming<br>$i\text{-}C_8H_{18} + 8H_2O \rightarrow 8CO + 17H_2$ | $R_{SR,\,i\text{-}C_8H_{18}} = \frac{k_{SR,\,i\text{-}C_8} P_{i\text{-}C_8H_{18}}}{1 + K^{poison}_{i\text{-}C_8H_{18}} \frac{P_{i\text{-}C_8H_{18}}}{P_{H_2O}}} \cdot \left(1 - \eta^{SR}_{i\text{-}C_8H_{18}}\right) \cdot \sigma_{H_2O}$ | 7.500E − 02 | 69.00 |
| $n$-$C_8H_{18}$ total oxidation<br>$C_8H_{18} + \frac{25}{2}O_2 \rightarrow 8CO_2 + 9H_2O$ | $R_{Ox,\,n\text{-}C_8H_{18}} = \frac{k_{Ox,\,n\text{-}C_8} \cdot P_{n\text{-}C_8H_{18}}}{1 + K_{H_2O} P_{H_2O}} \cdot \sigma_{O_2}$ | 1.9E − 02 | 88 |
| $n$-$C_8H_{18}$ steam reforming<br>$C_8H_{18} + 8H_2O \rightarrow 8CO + 17H_2$ | $R_{SR,\,n\text{-}C_8H_{18}} = \frac{k_{SR,\,n\text{-}C_8} \cdot P_{n\text{-}C_8H_{18}}}{1 + K^{poison}_{n\text{-}C_8H_{18}} \frac{P_{i\text{-}C_8H_{18}}}{P_{H_2O}}} \cdot \left(1 - \eta^{SR}_{n\text{-}C_8H_{18}}\right) \cdot \sigma_{H_2O}$ | 7.5E − 02 | 69 |
| $C_3H_8$ total oxidation<br>$C_3H_8 + 5O_2 \rightarrow 3CO_2 + 4H_2O$ | $R_{Ox,\,C_3H_8} = \frac{k_{Ox,\,C_3} P_{C_3H_8}}{1 + K_{H_2O} P_{H_2O}} \cdot \sigma_{O_2}$ | 2.500E − 01 | 80.00 |
| $C_3H_8$ steam reforming<br>$C_3H_8 + 3H_2O \leftrightarrow 3CO + 7H_2$ | $R_{SR,\,C_3H_8} = \frac{k_{SR,\,C_3H_8} P_{C_3H_8}}{1 + K_{O_2} P_{O_2} + K_{CO} P_{CO}} \cdot \left(1 - \eta^{SR}_{C_3H_8}\right) \cdot \sigma_{H_2O}$ | 2.486E − 01 | 84.63 |
| $CH_4$ total oxidation<br>$CH_4 + 2O_2 \rightarrow CO_2 + 2H_2O$ | $R_{Ox,\,CH_4} = \frac{k_{Ox,\,CH_4} P_{CH_4}}{1 + K^{ads}_{H_2O} p_{H_2O}} \cdot \sigma_{O_2}$ | 1.030E − 01 | 91.96 |
| $CH_4$ steam reforming<br>$CH_4 + H_2O \leftrightarrow CO + 3H_2$ | $R_{SR,\,CH_4} = \frac{k_{SR,\,CH_4} P_{CH_4}}{1 + K_{O_2} P_{O_2} + K_{CO} P_{CO}} \cdot \left(1 - \eta^{SR}_{CH_4}\right) \cdot \sigma_{H_2O}$<br>$\eta^{SR}_{CH_4} < 1$ | 1.027E − 01 | 91.80 |
| $C_xH_y$ steam reforming<br>$C_xH_y + xH_2O \rightarrow xCO + \left(\frac{y}{2} + x\right)H_2$ | $r_{SR\,C_xH_y} = \frac{k_{SR\,C_xH_y} P_{C_xH_y} \left(1 - \eta_{SR\,C_xH_y}\right)}{1 + K_{CO} P_{CO} + K_{O_2} P_{O_2}} \sigma_{H_2O}$ | 2.486E − 01 | 84.63 |

*Continued*

**Table 2** Heterogeneous Kinetic Scheme—cont'd

| Reaction | Rate$_i$ (mol $g_{cat}^{-1}$ $s^{-1}$) | $K_i^{873K}$ (mol $atm^{-1}$ $g_{cat}^{-1}$ $s^{-1}$) | $E_{ACT}$ (kJ $mol^{-1}$) |
|---|---|---|---|
| Water–gas shift<br>$CO + H_2O \leftrightarrow CO_2 + H_2$ | $R_{WGS} = k_{WGS} P_{H_2O}(1 - \eta_{WGS})\sigma_{CO}$<br>$\eta_{WGS} < 1$ | 6.831E−03 | 74.83 |
| Reverse water–gas shift<br>$CO_2 + H_2 \leftrightarrow CO + H_2O$ | $R_{RWGS} = k_{RWGS} P_{CO_2}(1 - \eta_{RWGS})\sigma_{H_2}$<br>$\eta_{RWGS} < 1$ | 1.277E−02 | 62.37 |
| $H_2$ oxidation<br>$H_2 + \frac{1}{2}O_2 \rightarrow H_2O$ | $R_{OxH_2} = k_{OxH_2} P_{H_2} \sigma_{O_2}$ | 2.666E+03 | 61.65 |
| CO oxidation<br>$CO + \frac{1}{2}O_2 \rightarrow CO_2$ | $R_{OxCO} = k_{OxCO} P_{CO} \sigma_{O_2}$ | 1.937E+01 | 76.07 |
| CO methanation<br>$CO + 3H_2 \leftrightarrow CH_4 + H_2O$ | $R_{MetCO} = k_{MetCO} P_{H_2}(1 - \eta_{MetCO})\sigma_{CO}$<br>$\eta_{MetCO} < 1$ | 1.200E−03 | 88.02 |
| $r_{\text{chemical reaction}} = r_{\text{chemical reaction}}\big|_{873K} \exp\left[-\frac{E_{att,r_j}}{R}\left(\frac{1}{T} - \frac{1}{873}\right)\right]$ | | | |

| Surface Adsorption | $K_i^{873K}$ ($atm^{-1}$) | $\Delta H_{ADS}$ (kJ $mol^{-1}$) |
|---|---|---|
| $O_2$ | 5.461E+00 | −72.83 |
| CO | 2.114E+02 | −37.15 |
| $H_2O$ | 8.974E+00 | −57.48 |
| C-poisoning $i$-$C_8H_{18}$ | 6.0E+00 | −26.0 |
| C-poisoning $n$-$C_8H_{18}$ | 2.0E+01 | −35.5 |
| $K_j = K_j^0 \exp\left[-\frac{\Delta H_{ADS}}{R}\left(\frac{1}{T} - \frac{1}{873.}\right)\right]$ | | |

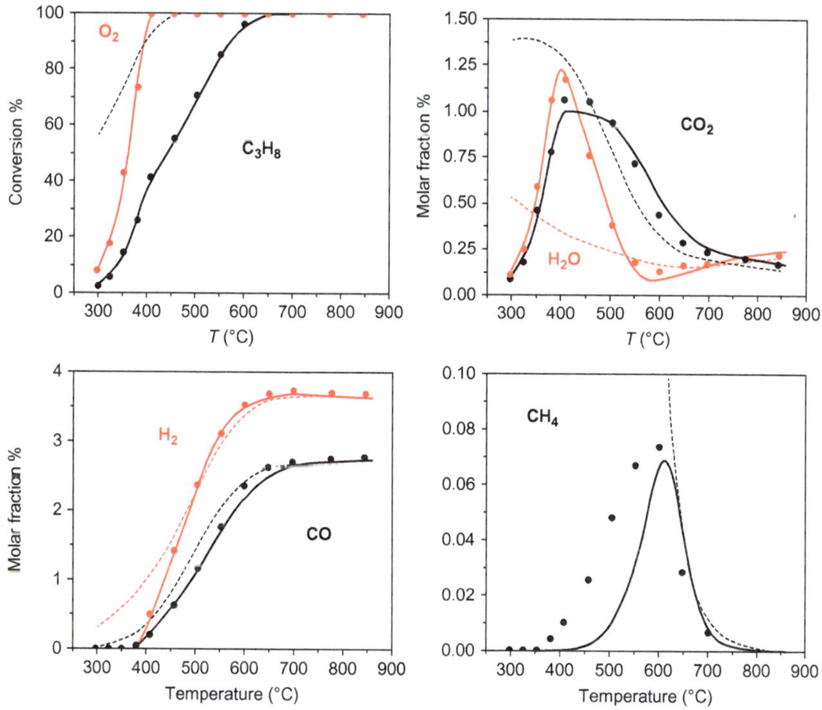


**Fig. 3** Effect of temperature on reactants' and products' molar fractions. Experimental (*symbols*), calculated (*solid lines*), and equilibrium (*dashed line*) composition. GHSV = $2 \times 10^6$ L(NTP)/$kg_{cat}$/h. Feed composition: $C_3H_8 = 1\%$ (v/v), $O_2/C = 0.56$, $N_2$ to balance; atmospheric pressure. *Reprinted from Pagani D, Livio D, Donazzi A, et al: A kinetic analysis of the partial oxidation of $C_3H_8$ over a 2% Rh/$Al_2O_3$ catalyst in annular microreactor,* Catal Today *197:265–280, 2012, with permission from Elsevier.*

partially reported in Fig. 4, we observed that at the highest space velocities ($4 \times 10^6$ and $6 \times 10^6$ L(NTP)/$kg_{cat}$/h) and above 700°C, small concentrations of methane, ethylene, and propylene were detected in the product mixture with an increasing trend with temperature (note that methane concentration grew against the thermodynamic equilibrium) and with flow rate; this clearly indicated that these species were produced in the gas phase by cracking reactions in the empty volume of the annular reactor downstream from the catalyst bed, and were not due to heterogeneous routes.

When the same process is performed in an adiabatic reactor, then the axial evolution of the process (consisting of highly exothermic and highly endothermic reactions) determines the interlinked evolution of reacting mixture composition and temperature.

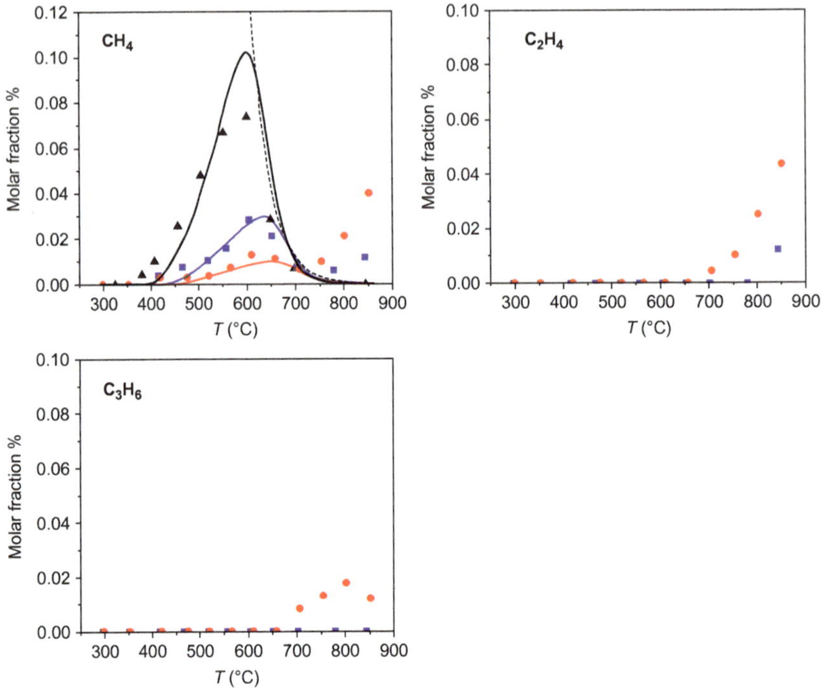


**Fig. 4** Effect of GHSV and temperature on hydrocarbon side products' molar fractions. Experimental (*symbols*) fractions are reported. (▲) GHSV = $2 \times 10^6$ L(NTP)/$kg_{cat}$/h; (■) GHSV = $4 \times 10^6$ L(NTP)/$kg_{cat}$/h; (●) GHSV = $6 \times 10^6$ L(NTP)/$kg_{cat}$/h; (*dashed line*) equilibrium. Feed composition: $C_3H_8$ = 1% (v/v), $O_2$/C = 0.56, $N_2$ to balance; atmospheric pressure. *Reprinted from Pagani D, Livio D, Donazzi A, et al: A kinetic analysis of the partial oxidation of* $C_3H_8$ *over a 2% Rh/*$Al_2O_3$ *catalyst in annular microreactor,* Catal Today *197:265–280, 2012, with permission from Elsevier.*

Fig. 5 reports the results of a propane CPO experiment that was performed at 5 L(NTP)/min by feeding a propane/air mixture with $O_2$/C ratio of 0.56 (Donazzi et al., 2011a,b). In the first 5 mm of the honeycomb, a sharp drop of $O_2$ and $C_3H_8$ concentration was observed, accompanied by the formation of total oxidation products ($CO_2$ and $H_2O$) and partial oxidation products ($H_2$ and CO). Correspondingly, a hot spot formed on the catalyst surface (980°C, measured by the pyrometer) and a steep rise was observed in the gas temperature (up to 945°C, measured by the thermocouple). In line with the occurrence of the endothermic steam reforming reaction, the evolution of $H_2O$ showed a maximum. As well, the temperatures of the solid surface and gas phase decreased toward the exit of the honeycomb. Qualitatively this is what our and other research groups have observed, also in the case of a $CH_4$-CPO experiment on Rh, and can be thus fully explained by

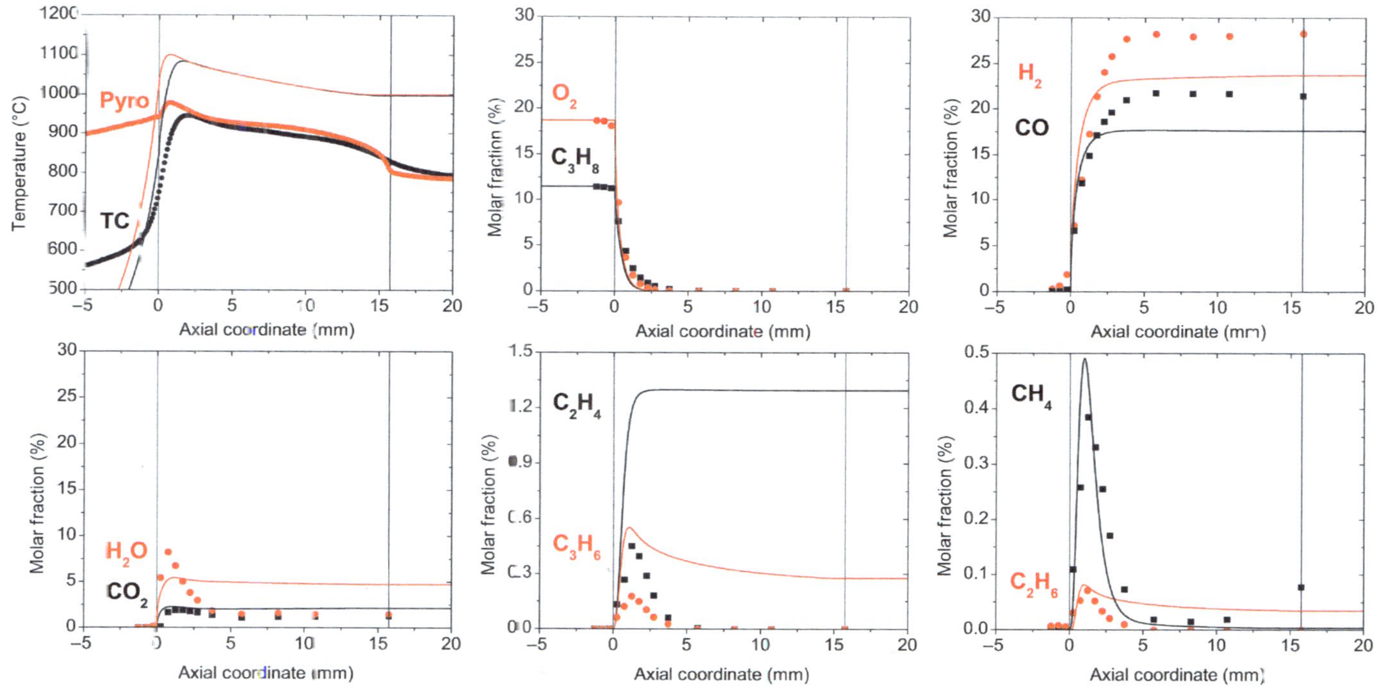


**Fig. 5** Spatially resolved profiles of temperature and composition for a $C_3H_8$ CPO experiment. Flow rate = 5 L min$^{-1}$ (STP), $P$ = 1 atm. Feed: $C_3H_8$/air, $C_3H_8$ = 11 vol%, C/O = 1.12. Experimental results in *symbols. Solid lines* are calculations obtained by neglecting the steam reforming reactions of cracking products. *Reprinted from Donazzi A, Livio D, Maestri M, et al: Synergy of homogeneous and heterogeneous chemistry probed by in situ spatially resolved measurements of temperature and composition*, Angew Chem Int Ed *50:3943–3946, 2011, with permission from Wiley.*

the CPO chemistry on Rh. Thus, integral measurements (i.e., measurements of temperature and composition collected exclusively at the reactor outlet) would have suggested the unique existence of a heterogeneous process. However, the spatially resolved measurements revealed for the first time that several other species, such as $C_2H_4$, $C_3H_6$, and $C_2H_6$, were formed at the very reactor inlet and were then completely consumed before the ending section of the catalyst, passing through maxima.

We also observed a complex evolution of $CH_4$ that, similarly to $C_2$ and $C_3$ species, was initially formed and then consumed within the first half of the reactor, but tended to be formed again in the ending portion of the reactor. Since hydrocarbon species like olefins cannot be explained by the syngas formation paths on Rh, this experimental finding provided evidence of gas-phase chemistry.

In line with the interpretation of the experiments, the adiabatic model of the honeycomb monolith was first extended to include the detailed gas-phase chemistry from Ranzi et al. (2012), thus combining heterogeneous and homogeneous routes of propane activation. The model (solid lines in Fig. 5) could nicely describe the rapid evolution of the reacting mixture in the very first millimeters of the catalyst bed, namely: the rapid consumption of the reactants, the formation of deep and partial oxidation products, and the formation of methane, ethylene, propylene, and ethane. However, the evolution of temperature and product distribution along the rest of the monolith was not correctly described. The overall production of $H_2$ and CO was largely underestimated, while the outlet concentration of hydrocarbon species was greatly overestimated. Also, the predicted temperature profiles were much higher than observed.

Clearly, the simple description of the onset of a gas-phase chemistry (that under fuel-rich conditions favors the formation of short olefins) could not explain the outlet-selective formation of syngas. Only in the case of methane, the formation of a maximum was correctly described by the model; this was due to the fact that the gas-phase production of $CH_4$ at the reactor entrance "activated" the surface reaction of methane steam reforming, already present in the $C_3$ heterogeneous scheme.

The kinetic model was thus revised by incorporating in the heterogeneous scheme oxidation and steam reforming reactions for all the hydrocarbon species, possibly produced in the gas phase. For simplicity, all these consecutive reactions were described by rate expressions and intrinsic parameters as those of propane.

The new model predictions are reported in Fig. 6 in solid lines and show that a very good description of the system was obtained. All the general

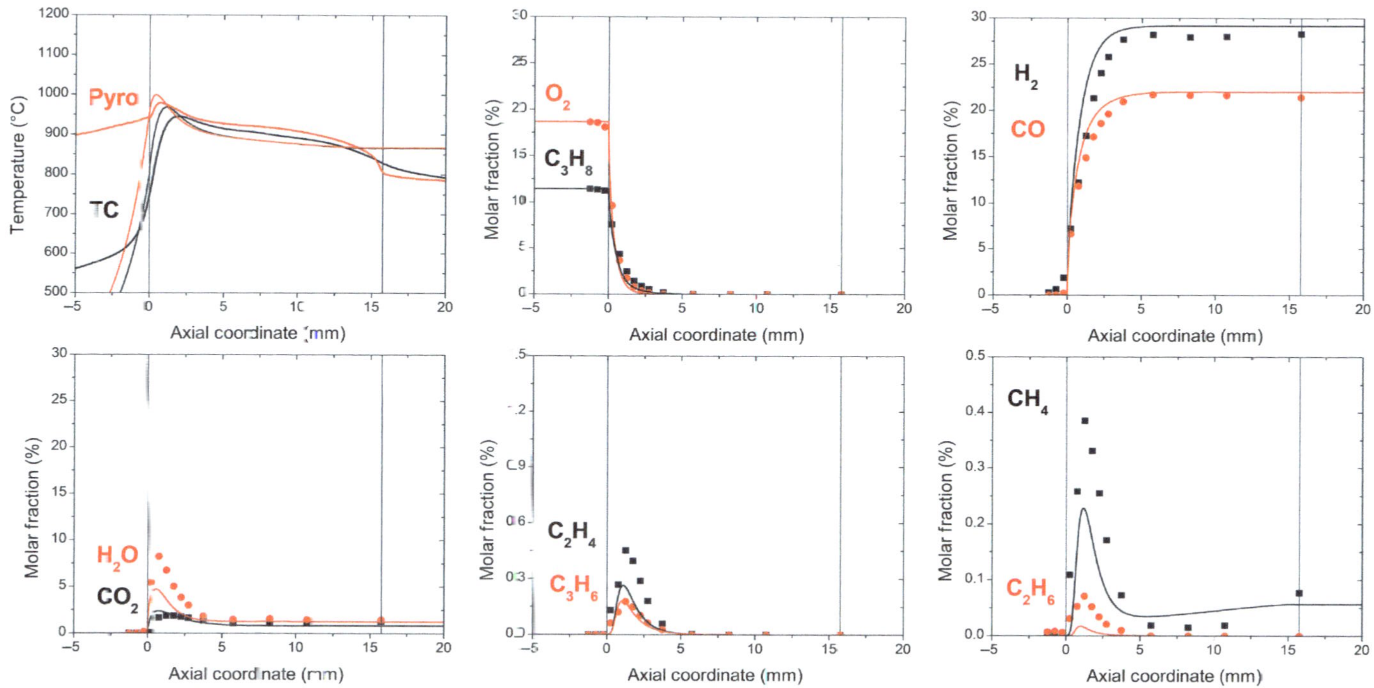


**Fig. 6** As in Fig. 5. *Solid lines are calculations obtained by accounting for the steam reforming reactions of cracking products. Reprinted from Donazzi A, Livio D, Maestri M, et al: Synergy of homogeneous and heterogeneous chemistry probed by in situ spatially resolved measurements of temperature and composition,* Angew Chem Int Ed *50:3943–3946, 2011, with permission from Wiley.*

trends of the product distribution were captured: the maxima of concentration for ethylene, ethane, and propylene, the production of synthesis gas, and the temperature profiles.

The analysis of the modeling results revealed that, in correspondence with the hot spot temperature, some gas-phase conversion of $C_3H_8$ occurs and is responsible for the production of all the hydrocarbon species observed at the reactor inlet: cracking of propane (1) explains an equimolar production of ethylene and methane, while dehydrogenation (2) explains the formation of propylene. Ethane, produced in much smaller amount, is the product of a consecutive hydrogenation of ethylene (3).

$$C_3H_8 \rightarrow CH_4 + C_2H_4 \tag{1}$$

$$C_3H_8 \rightarrow C_3H_6 + H_2 \tag{2}$$

$$C_2H_4 + H_2 \rightarrow C_2H_6 \tag{3}$$

According to the model, the consumption of $O_2$ is almost entirely due to surface oxidation reactions (mass transfer controlled), while the homogeneous consumption of oxygen in oxidative pyrolysis has a negligible role.

Once formed in the gas phase, the hydrocarbon species are then consumed on the catalytic surface via steam reforming with formation of syngas. The consecutive reactions of these species are in fact much slower in the gas phase than on Rh.

Concerning the complex evolution of $CH_4$, the model simulations showed that this can be very well explained by a series of contributions consisting of: (1) gas-phase formation via propane cracking, (2) catalytic consumption via steam reforming, and (3) catalytic formation via methanation (kinetically favored by the increasing partial pressures of $H_2$ and CO and thermodynamically favored by the decreasing temperatures).

The whole process can be visualized as the existence of a "by-pass" route detached from the prevailing heterogeneous scheme, wherein gas-phase reactions partly convert the propane feed, enriching it of olefins and methane, which then contribute to the reforming process to synthesis gas.

Fig. 7 summarizes this picture, and represents the CPO of propane on Rh as a process mainly "sticking" to the surface.

The application of the in situ probing techniques revealed in the case of propane CPO the existence of a concerted heterogeneous–homogeneous process that does not compromise the performance of the fuel reformer, in terms of fuel conversion and syngas yield. On the opposite, the short olefins are expected to promote the reactivity of the fuel.

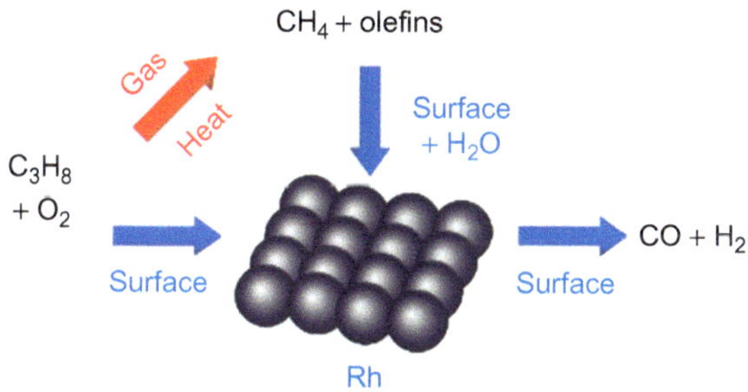


**Fig. 7** A schematic representation of the combined homogeneous–heterogeneous chemistry in the CPO of propane. *Reprinted from Donazzi A, Livio D, Maestri M, et al: Synergy of homogeneous and heterogeneous chemistry probed by in situ spatially resolved measurements of temperature and composition,* Angew Chem Int Ed *50:3943–3946, 2011, with permission from Wiley.*

One may wonder however which factors determine the fate of gas-phase species and the successful conversion to synthesis gas. An incomplete conversion of short olefins along the monolith could in fact represent a loss of selectivity, but also a risk for the reactor stability because of soot formation in the gas phase or catalyst deactivation by coking. All these aspects have a crucial impact and must be known and considered in designing small-scale hydrogen units.

In the following we explore on one side the effect of pressure, and on the other side the effect of the nature of the fuel.

## 2.4 Effect of Pressure in the Autothermal CPO of Propane

We studied the effect of pressure by experiments and modeling in the range 1–4 bar; simulations were then performed to explore the effect of pressure on the impact of gas-phase reactions at pressure larger than 4 bar (Donazzi et al., 2014).

### *2.4.1 Effect of Pressure up to 4 bar*

Figs. 8 and 9 show the results of spatially resolved experiments of $C_3H_8$ CPO carried out at 2, 3, and 4 bar. The molar flow rate was kept at 10 L(NTP)/min and the $O_2$/C ratio moderately increased from 0.56 at 2 bar to 0.60 at 4 bar. The experiments were simulated by using the same model above described where the homogeneous chemistry was included together with the catalytic routes of conversion of the hydrocarbon species. Model simulations are reported in solid lines.

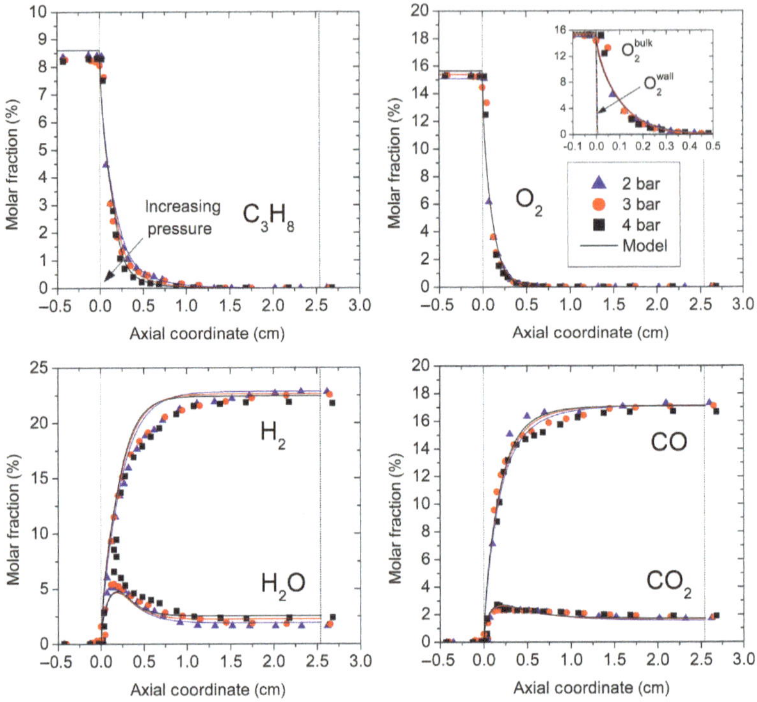


**Fig. 8** Spatially resolved temperature and composition profiles for an autothermal $C_3H_8$ CPO experiment at increasing pressure. Operating conditions: $C_3H_8=8.6\%$, $O_2/C=0.59–0.61$, flow rate = 10 NL/min, $P=2$ (▲), 3 (●), and 4 bar (■), $T_{in}=40°C$. Experimental measurements: *symbol* and *thick lines* in temperature profiles. *Thin lines*: model predictions. *Reprinted from Donazzi A, Livio D, Diehm C, Beretta A, Groppi G, Forzatti P: Effect of pressure in the autothermal catalytic partial oxidation of $CH_4$ and $C_3H_8$: spatially resolved temperature and composition profiles,* Appl Catal A Gen *469:52–64, 2014, with permission from Elsevier.*

The axial measurement of concentration profiles showed that at increasing pressure, the conversion of propane was promoted, while that of $O_2$ was unaffected. The production of $CO_2$, $H_2O$, CO, and $H_2$ was almost unchanged with a minor decrease of CO and $H_2$ selectivity at increasing pressure (Fig. 8); instead, a significant enrichment of the hydrocarbon species (ethylene, propylene, and $CH_4$) was observed (Fig. 9) while maintaining the same qualitative axial trend, characterized by the presence of a peak close to the entrance.

By comparing the conversion of propane to $CO_X$ and the conversion of propane to hydrocarbons (Fig. 10), it is easily understood that pressure uniquely promoted the gas-phase conversion of propane.

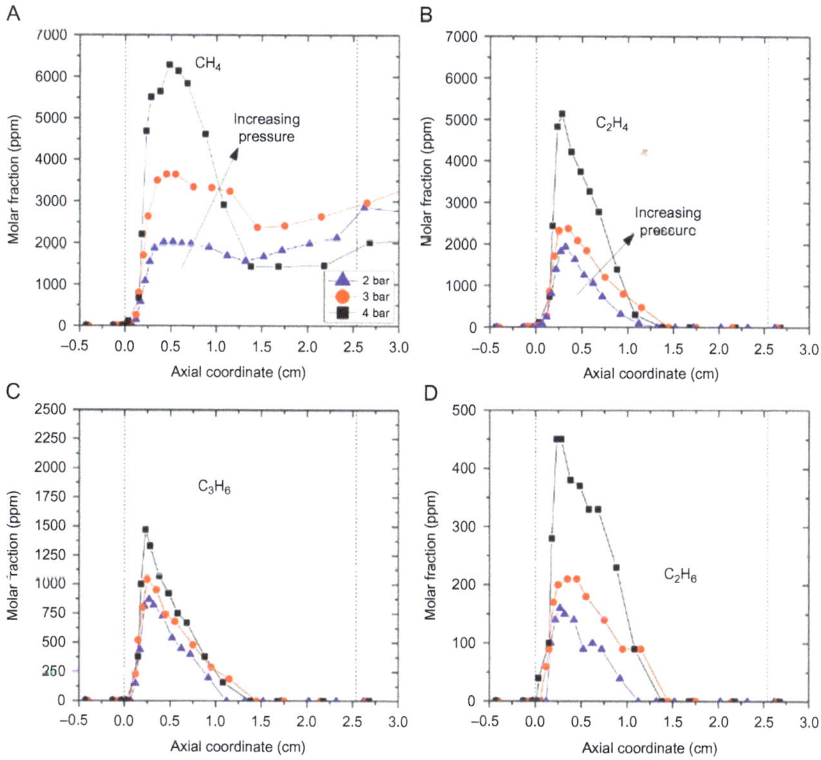


**Fig. 9** Spatially resolved composition profiles of methane and $C_{2+}$ products at increasing pressure from 2 to 4 bar. Conditions as in Fig. 8. Panels: $CH_4$ (A); $C_2H_4$ (B); $C_3H_6$ (C); $C_2H_6$ (D). *Reprinted from Donazzi A, Livio D, Diehm C, Beretta A, Groppi G, Forzatti P: Effect of pressure in the autothermal catalytic partial oxidation of $CH_4$ and $C_3H_8$: spatially resolved temperature and composition profiles,* Appl Catal A Gen *469:52–64, 2014, with permission from Elsevier.*

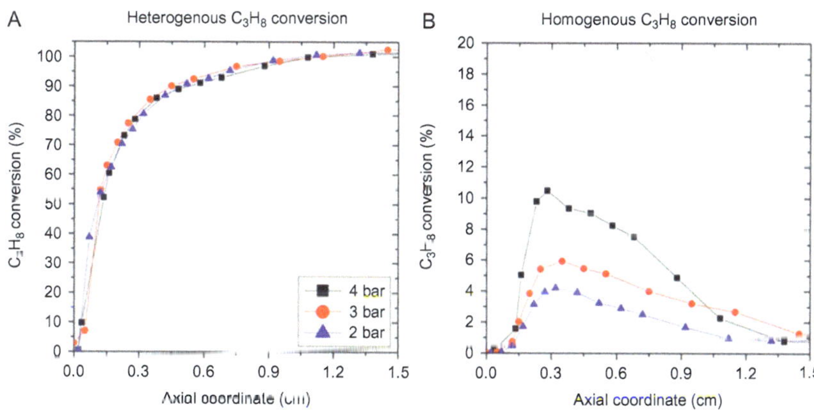


**Fig. 10** Calculated conversion of $C_3H_8$. Heterogeneous conversion to CO and $CO_2$ (A); homogenous conversion to methane and $C_{2+}$ products (B). *Reprinted from Donazzi A, Livio D, Diehm C, Beretta A, Groppi G, Forzatti P: Effect of pressure in the autothermal catalytic partial oxidation of $CH_4$ and $C_3H_8$: spatially resolved temperature and composition profiles,* Appl Catal A Gen *469:52–64, 2014, with permission from Elsevier.*

The modeling analysis of the CPO experiments was again used as the means for quantitatively interpreting the phenomena. Model predictions captured all the observed trends, although some inaccuracy was observed in the simulation of the hydrocarbon species, which is believed acceptable given the simplifying nature of the 1D model and not representative of any systematic error.

The model analysis (in particular the evaluation of the interphase concentration gradient and the evaluation of the Carberry number for the various species) revealed that all the heterogeneous reactions were fully controlled by mass transfer; thus not only the consumption of $O_2$ but also all the oxidation and steam reforming reactions are responsible for the conversion of propane and hydrocarbon species. Under mass transfer control, given the inverse dependence of molecular diffusivity on pressure, the process kinetics has a zeroth-order $P$ dependence which explains why pressure had no effect on the heterogeneous chemistry.

On the opposite, experimental results and model calculations showed that the maxima of the hydrocarbon peaks increased within increasing pressure and tended to shift toward the inlet of the channel (Fig. 11); this is explained by considering the promotion of the homogenous kinetics (which scale with $P^2$) and the reduction of the residence time of the gas flow. In the case of $C_{2+}$ species, at all the pressures investigated, the peak results from the balance between the global formation rate due to gas-phase cracking and the consumption rate due to surface reforming. $CH_4$ was still formed by cracking and consumed by reforming, but additional production by methanation occurs in the final part of the monolith, which is responsible for the presence of $CH_4$ among the reaction products.

A reaction path analysis for ethylene and methane is reported in Fig. 12 and illustrates the axial evolution of the gas phase and surface contribution to the formation and consumption of ethylene and methane at 2 and 4 bar pressure.

The heterogeneous reforming of gas-phase intermediates plays a key role in $C_3H_8$ CPO: the catalyst surface acts as a chemical quench, preventing the buildup of short hydrocarbons and their condensation to polycyclic aromatic species and possibly coke. Notably, although the rate of production in the gas phase grew considerably with pressure, no net breakthrough of olefins was observed up to 4 bar.

### 2.4.2 Effect of Pressure Above 4 bar

The effect of pressure at higher values was explored by extending the modeling analysis up to 15 bar. Fig. 13 reports the calculated axial profiles

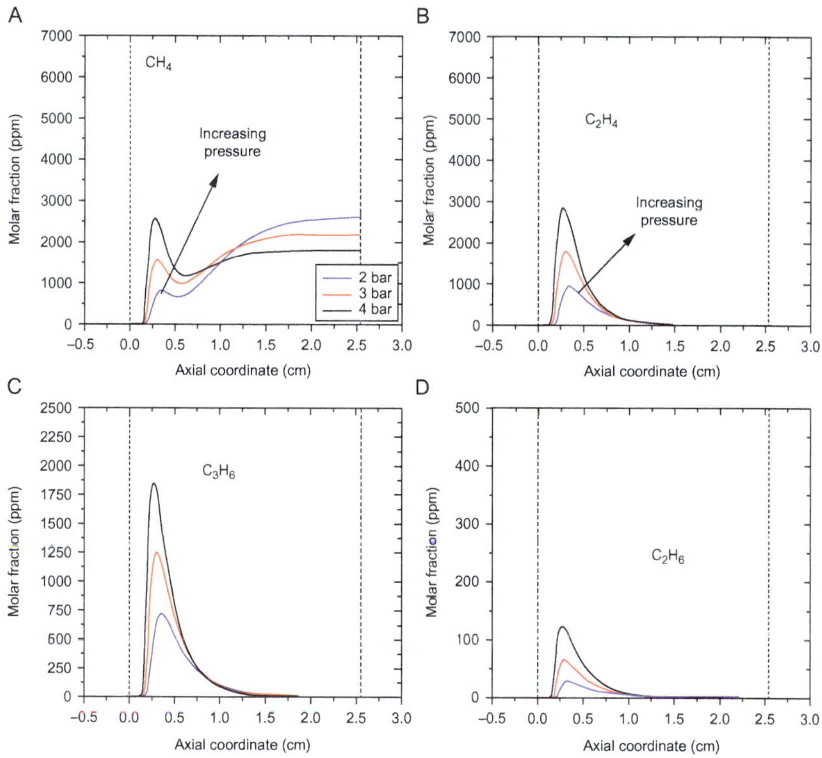


**Fig. 11** Calculated profiles for a $C_3H_8$ CPO experiment at increasing pressure. $P=2$ (*blue line*), 3 (*red line*), and 4 bar (*black line*). Panels: $CH_4$ (A); $C_2H_4$ (B); $C_3H_6$ (C); $C_2H_6$ (D). *Reprinted from Donazzi A, Livio D, Diehm C, Beretta A, Groppi G, Forzatti P: Effect of pressure in the autothermal catalytic partial oxidation of $CH_4$ and $C_3H_8$: spatially resolved temperature and composition profiles,* Appl Catal A Gen *469:52–64, 2014, with permission from Elsevier.*

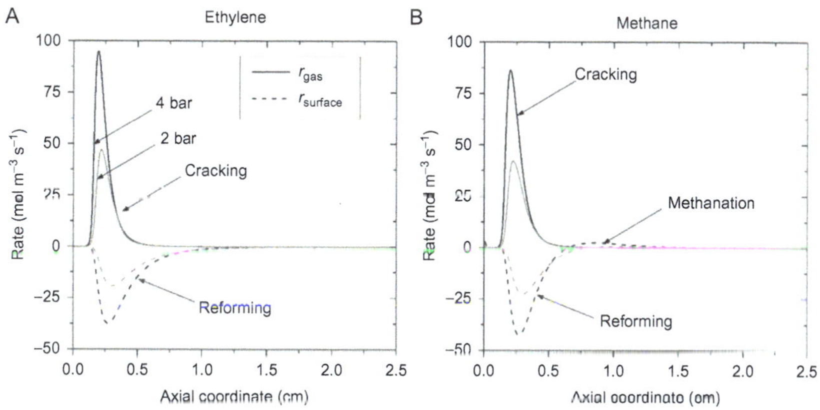


**Fig. 12** Calculated global formation and consumption rates for $C_2H_4$ (A) and $CH_4$ (B) along the axis of the catalyst. Conditions: $C_3H_8=8.6\%$, $O_2/C=0.59$, flow rate = 10 NL/min, $P=4$ bar. *Reprinted from Donazzi A, Livio D, Diehm C, Beretta A, Groppi G, Forzatti P: Effect of pressure in the autothermal catalytic partial oxidation of $CH_4$ and $C_3H_8$: spatially resolved temperature and composition profiles,* Appl Catal A Gen *469:52–64, 2014, with permission from Elsevier.*

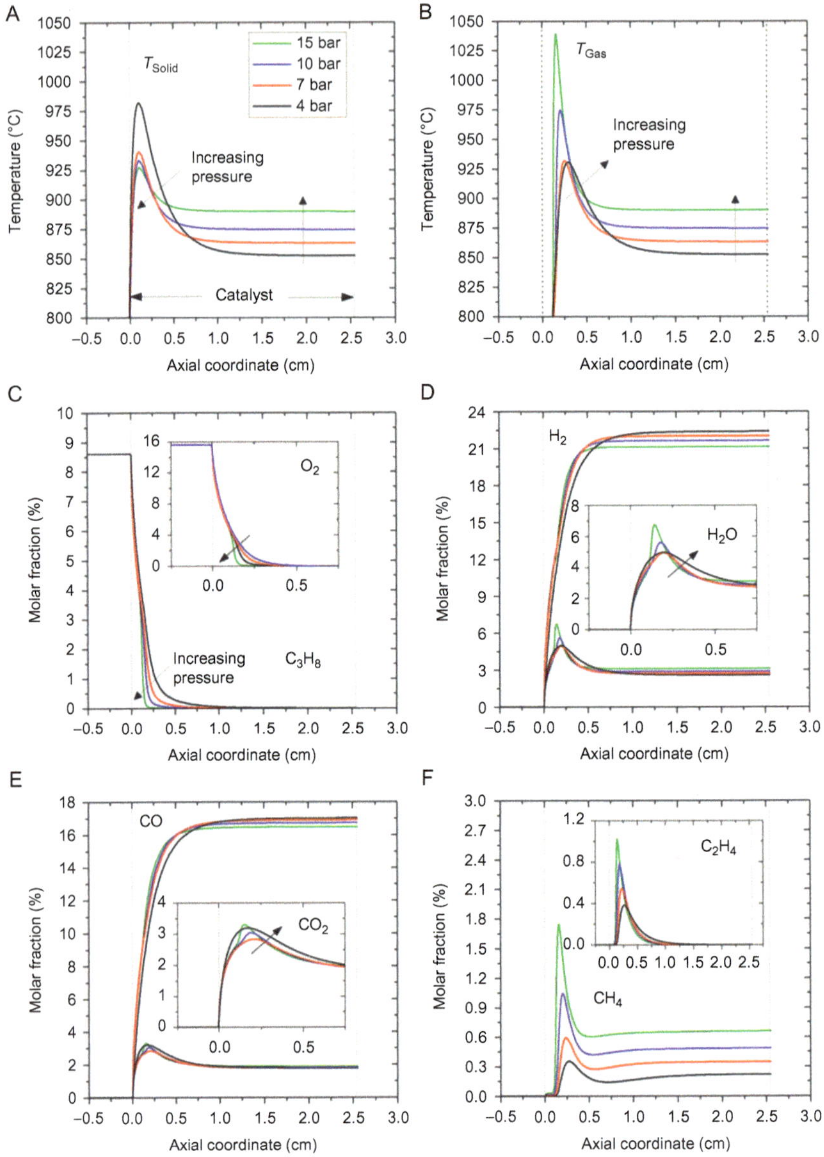


**Fig. 13** Calculated temperature and composition profiles for an autothermal $C_3H_8$ CPO experiment at increasing pressure from 4 to 15 bar. $C_3H_8 = 8.6\%$, $O_2/C = 0.59$, flow rate = 10 NL/min. Profiles of temperature (A and B), and molar fractions of reactants (C) and products (D–F). *Reprinted from Donazzi A, Livio D, Diehm C, Beretta A, Groppi G, Forzatti P: Effect of pressure in the autothermal catalytic partial oxidation of $CH_4$ and $C_3H_8$: spatially resolved temperature and composition profiles,* Appl Catal A Gen *469:52–64, 2014, with permission from Elsevier.*

of temperature and composition. Thermodynamics controls the reactor performance in the ending portion of the bed, and drives the CPO system toward increasing outlet temperatures and increasing selectivity of methane (with loss of CO and $H_2$ selectivity) at increasing pressure.

Kinetic factors govern instead the reactor performance in the first centimeter and model predictions suggest that at increasing pressure above 4 bar, the fraction of propane which is converted in the gas phase increases steadily; additionally, at increasing pressure, also $O_2$ is more and more consumed in the gas phase such that the gas-phase chemistry passes from a purely pyrolysis process at 1 bar to an oxidative pyrolysis process at 15 bar. The production of hydrocarbon species is predicted to grow steadily together with the production of $H_2O$, making sharper and sharper the hot spot in the gas-phase temperature profile.

A stunning result of such change of chemistry is that, because of the decreased radial flow of reactants to the wall and because of the lowering of $O_2$ to fuel ratio at the wall, the heat load at the catalyst surface decreases and the peak temperature on the surface is predicted to decrease (Fig. 13). It is thus envisaged that the homogeneous chemistry can be seen as a "degree of freedom" for tuning the thermal behavior of the reactor and moderate the temperature stress of the catalyst. Also at high pressure, the model predicted that all the gas-phase species were entirely consumed at the surface by steam reforming to synthesis gas; thus, the competition between surface process and gas-phase process does not affect the reactor performance, provided that the catalyst activity in steam reforming and the length of the catalyst bed are large enough to perfectly close and anchor to the surface the "by-pass" route.

## 2.5 Gas-Phase Chemistry in the Autothermal CPO of Octane Isomers

Another factor that in principle can influence the interplay between gas-phase chemistry and surface chemistry is the nature of the hydrocarbon fuel, this will affect the reactivity and the fragmentation pattern in the gas phase, but also the rate of gas–solid mass transfer and the surface reactivity.

We studied the CPO of iso-octane and normal octane with the general objective of analyzing the case of liquid model fuels of interest for the on-board application of the CPO process (Carrera et al., 2017a,b); with respect to the gas-phase/surface chemistry balance it is expected that long-chain hydrocarbons are more reactive in the gas phase and may represent a powerful source of soot/coke precursors, thus representing a challenge for the surface chemistry.

Most of the CPO tests were performed with $N_2$-diluted hydrocarbon/air feed mixtures (with 2% octane, $O_2/C$ ratio of 0.56) in order to avoid excessive heating of the monolith and minimizing the risk of deactivation by sintering. Fig. 14 shows the measured temperature profiles at 10 L(NTP)/min, while Fig. 15 reports the axial evolution of the reactants of the surface product species. The two fuels showed very close thermal behavior and syngas performance.

As expected, hydrocarbon species were also detected in GC analyses and their concentration profiles are reported in Fig. 16. All these species except methane (panel A) showed the same typical peak trend as observed in the CPO of propane, thus confirming that the mechanism of formation via gas-phase reactions and consumption via consecutive surface reforming have a general validity.

Methane concentration grew progressively along the catalyst length; however, its concentration in the first 4–5 mm from the catalyst entrance exceeded the local equilibrium of the methanation reaction, thus suggesting that cracking reactions were still responsible for its initial formation. On the other hand, downstream from the entrance zone catalytic methanation is thermodynamically allowed due to the parallel decrease of temperature and the increase of CO and $H_2$ molar fractions. As a result, methane

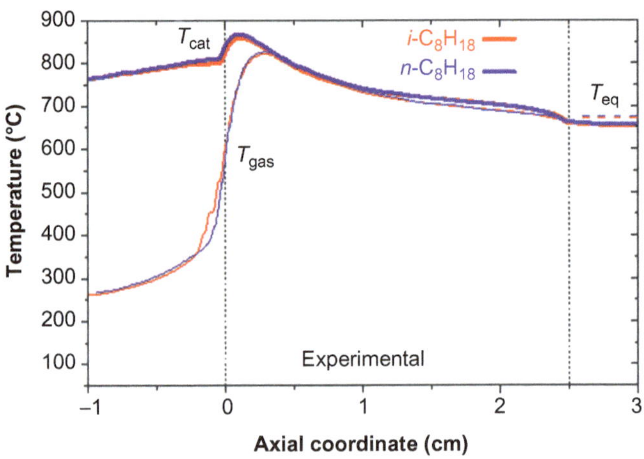


**Fig. 14** Measured temperature profiles in the CPO of iso-octane and *n*-octane; $C_8H_{18}$ = 1.9%, C/O = 0.83, $T_{in}$ = 85°C, flow rate = 10 NL/min. *Adapted from Carrera A, Pelucchi M, Stagni A, Beretta A, Groppi G, et al: Catalytic partial oxidation of n-octane and iso-octane: experimental and modeling results,* Int J Hydrogen Energy, *2017, https://doi.org/10.1016/j.ijhydene.2017.08.020, with permission from Elsevier.*

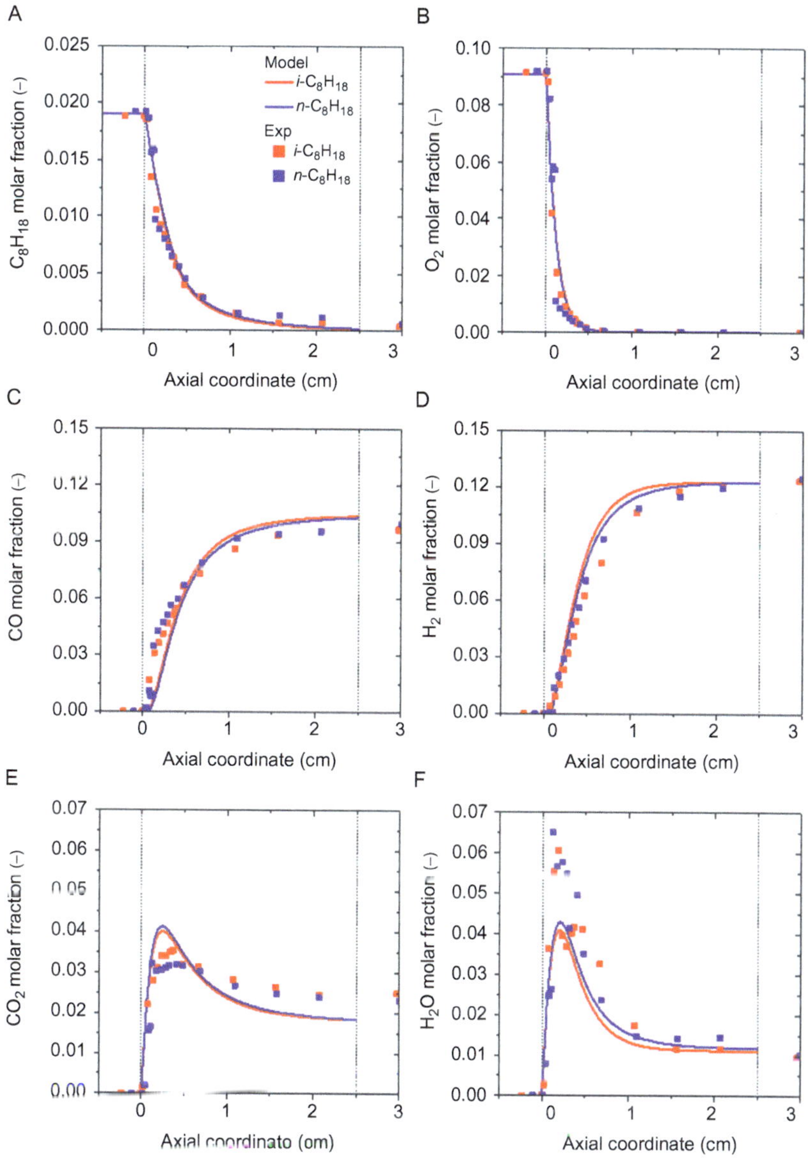


**Fig. 15** Experimental and calculated concentration profiles for the CPO of iso-octane and *n*-octane; $C_8H_{18} = 1.9\%$, C/O = 0.83, $T_{in} = 85°C$, flow rate = 10 NL/min. *Reprinted from Carrera A, Pelucchi M, Stagni A, Beretta A, Groppi G, et al: Catalytic partial oxidation of n-octane and iso-octane: experimental and modeling results,* Int J Hydrogen Energy, *2017, https://doi.org/10.1016/j.ijhydene.2017.08.020, with permission from Elsevier.*

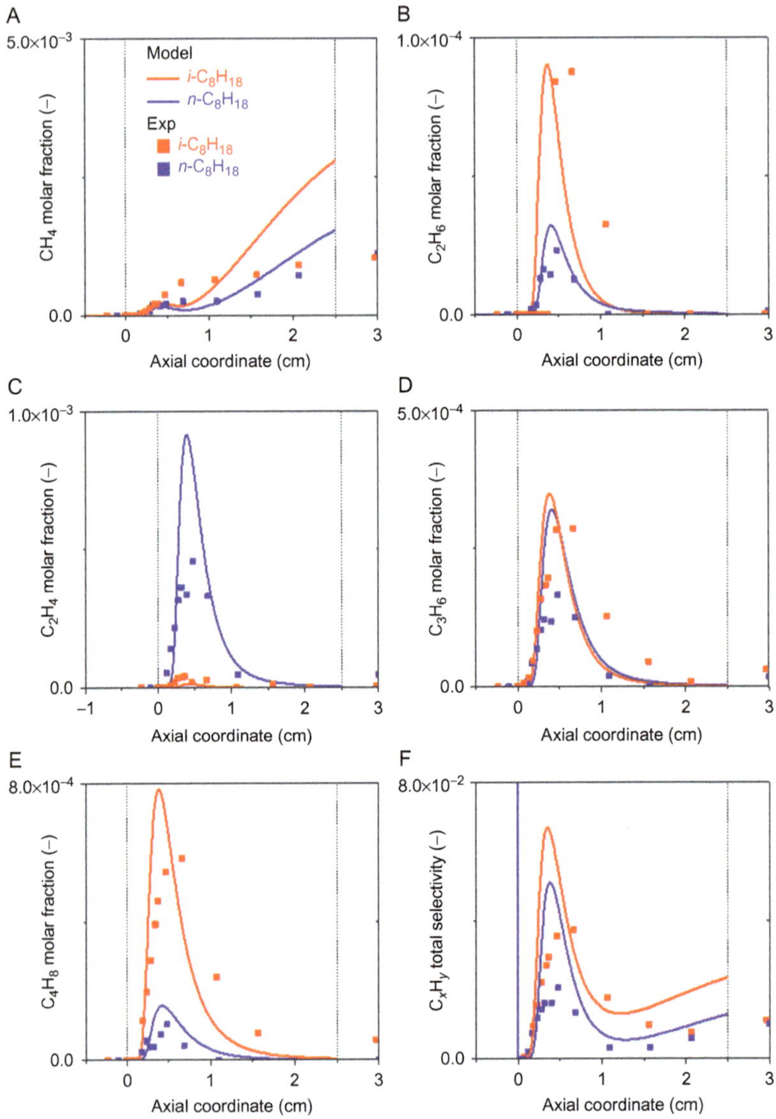


**Fig. 16** Experimental and calculated concentration profiles of light hydrocarbons for the CPO of iso-octane and *n*-octane; *n*-$C_8H_{18}$ = 2%, C/O = 0.83, $T_{in}$ = 85°C, flow rate = 10 NL/min. *Reprinted from Carrera A, Pelucchi M, Stagni A, Beretta A, Groppi G, et al: Catalytic partial oxidation of n-octane and iso-octane: experimental and modeling results,* Int J Hydrogen Energy, *2017, https://doi.org/10.1016/j.ijhydene.2017.08.020, with permission from Elsevier.*

concentration progressively increased, although to a very limited extent compared with the thermodynamic driving force.

Differently from the case of propane CPO, the breakthrough of hydrocarbon $C_{2+}$ species was observed. Their concentration approached zero at the catalyst outlet, and raised slightly after the catalyst end, likely due to cup-mix phenomena. In fact, due to peripheral heat dissipation, outer channels can operate at lower temperatures than the central one, thus resulting in slower reforming kinetics and hydrocarbon slip. Noteworthy the overall selectivity to such hydrocarbon species was kept below 4% all along the catalyst length, due to diluted test conditions adopted in this work, which maintain the temperature of the gas phase below 850°C.

Nevertheless, major differences can be observed in the productivity and distribution of gas-phase species. The total selectivity toward hydrocarbon species indicates that iso-octane was more reactive in the gas phase, with respect to *n*-octane. Besides, cracking of iso-octane was more selective toward iso-butylene and propylene (and methane), while cracking of *n*-octane was more selective toward ethylene. Such differences are in line with previous literature findings (Panuccio et al., 2006) and are due to the different gas-phase fragmentation mechanisms of the two fuels. Indeed model simulations well captured the differences between the two isomers showing that the thermal cracking of iso-octane is more reactive and selective toward iso-butylene, propylene, and methane (Curran et al., 2002), while *n*-octane, similarly to other linear alkanes (Curran et al., 1998), produces mainly ethylene. Noteworthy model simulations also captured the initial trend of $CH_4$ molar fraction, associated with thermal cracking in the gas phase, but markedly overestimated the $CH_4$ formation rate via methanation reaction in the downstream zone of the catalyst, suggesting that this surface reaction was partially inhibited in the last part of the reactor.

The single-channel model predicted a complete conversion of the $C_{2+}$ olefins and hydrocarbons before the end of the catalyst; thus, the slip of hydrocarbons from the peripheral channels could not be described. This experimental evidence however is an important warning that the CPO of liquid fuels is a less "clean" process than the CPO of light hydrocarbons; the increased reactivity in the gas phase and the enrichment in short olefins can actually promote not only soot formation through consecutive gas-phase reactions but also coking reactions on the surface (presently not included in the heterogeneous scheme). A certain hindering of the reforming/methanation routes could in fact explain the breakthrough of hydrocarbons from the colder peripheral channels and the overestimation of kinetics of surface methanation from synthesis gas.

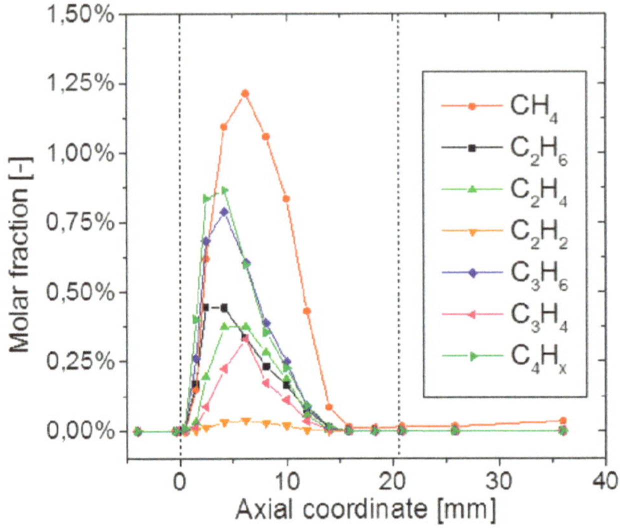


**Fig. 17** Measured molar fraction of hydrocarbon species in a *i*-C8 CPO experiment at 4% fuel concentration. Total flow rate = 10 L(NTP)/min, $O_2/C = 0.56$; atmospheric pressure.

In the absence of $N_2$ dilution, the temperature of the CPO reformer increases to a large extent. With almost stoichiometric *i*-octane/air feeds (with *i*-$C_8$ concentration of 4% and $O_2/C$ ratio of 0.56) the measured T-profile in the gas phase peaked at about 1000°C and reached about 850°C at the reactor outlet and the concentration of hydrocarbon species increased by almost 1 order of magnitude, as shown in Fig. 17.

At such very high temperatures, the surface chemistry was extremely effective and the chemical quench of the cracking chemistry was complete, with no hydrocarbon slip from the monolith, except for the traces of methane.

This result further confirms that the interaction between the two chemistries is extremely sensitive to both the nature of the fuel and the operating conditions.

## 3. ROLE OF HOMOGENEOUS CHEMISTRY IN OTHER FUEL-RICH OXIDATION PROCESSES

The understanding of the concerted homogenous/heterogeneous chemistries gained through the combined application of in situ sampling technique and reactor modeling can support the interpretation of the phenomenology in other complex cases of high-temperature partial oxidation processes.

## 3.1 ODH of Ethane Over Pt and Rh

An interesting case is represented by the conversion of ethane/$O_2$ and propane/$O_2$ mixtures to the corresponding short olefins over Pt-coated monoliths. Pioneered by Schmidt and his group, the relative roles of homogenous and heterogeneous chemistry in high-temperature ODH have been widely debated in the literature. Although Huff and Schmidt (1993) had originally proposed a purely heterogeneous mechanism of formation of ethylene, based on the evidence that the product selectivity was strongly affected by the nature of noble metal (being CO and $H_2$ the major reaction products on Rh), other research groups have instead proposed the homogeneous nature of ethylene production (Basini and Guarinoni, 2013; Beretta et al., 2001; Håkonsen et al., 2010; Lange et al., 2005; Minh et al., 2008; Zerkle et al., 2000). To explain why at comparably high reaction temperatures, Pt-coated monoliths were reported by several authors to produce mostly ethylene and water, while Rh yielded mostly CO and $H_2$, Beretta and Forzatti (2001) compared the intrinsic kinetics of the two metals in steam reforming experiments finding that Rh was more active than Pt and proposed that the different behavior of the adiabatic reactors depended on the peculiar reforming activity of Rh, responsible for the consecutive conversion of the cracking products to synthesis gas.

This proposal has been confirmed by Michael et al. (2010) who applied the spatially resolved sampling technique to measure temperature and composition profiles in ethane partial oxidation over Pt- and Rh-coated foams. A comparison of ODH experiments is reported in Fig. 18.

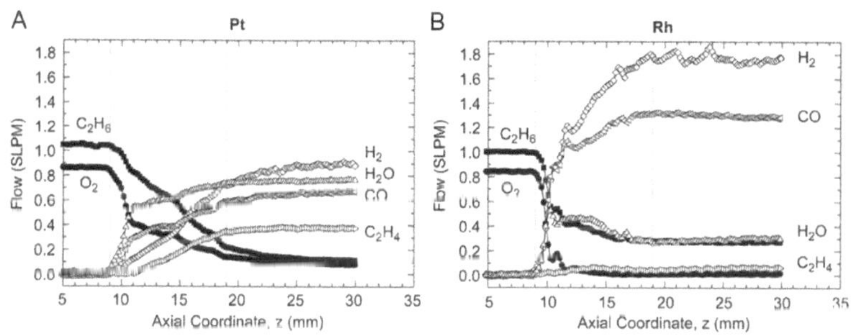


**Fig. 18** Spatial profile of major species flows for C/O = 1.2 over 4.7 wt% Pt/45 PPI α-$Al_2O_3$ (A) and 4.7 wt% Rh/45 PPI α-$Al_2O_3$ (B) for inlet flow of 5 SLPM, $T_{in}$ = 100°C. The catalytic monolith is located within $9 < z < 19$. *Reprinted from Michael BC, Nare DN, Schmidt LD: Catalytic partial oxidation of ethane to ethylene and syngas over Rh and Pt coated monoliths: spatial profiles of temperature and composition,* Chem Eng Sci *65:3893–3902, 2010, with permission from Elsevier.*

While on Rh synthesis gas was produced selectively from the very entrance of the bed (in line with the results above illustrated on the CPO of propane on Rh monoliths), on Pt a significant amount of ethylene was observed starting from a short distance from the monolith entrance (Fig. 19). By performing experiments at varying C/O ratios and temperature, the authors found that the appearance of ethylene among the product stream corresponded with the reaching of a threshold temperature of about 760°C in the gas phase; this was associated with the onset of gas-phase ODH reactions, thermally driven by the exothermic oxidation of ethane to $CO_X$ and $H_2O$ on the catalyst surface. In the absence of a sufficiently high rate of steam reforming on Pt, the production of olefin proceeded along the whole monolith length.

A possible explanation for the apparent inactivity of Pt on secondary reforming reactions has can be found in the studies performed by Horn and coworkers (Korup et al., 2012, 2013); by applying Raman spectroscopy to the inspection of Pt-coated monoliths in the CPO of methane, they proposed that Pt–C species were formed and reduced the surface activity. A modification of the surface state could justify the results by Michael et al. (2010); as shown in Fig. 18, a net transition was observed from the entrance zone (about 2 mm long, where ethane and $O_2$ were rapidly consumed to CO and $H_2O$) to the rest of the foam, where the conversion of reactants followed a more sluggish trend and gas-phase reactions occurred. The decreased reactivity of the surface possibly related to the formation of C species was likely responsible for the observed buildup of ethylene in the product stream.

## 3.2 Oxidative Coupling of Methane

Horn and coworkers have largely contributed to the development of axially resolved measurements of temperature and compositions in high-temperature oxidation processes in foams. In several studies, they have addressed the numerical simulation of the in situ measurements as a means for the comprehension of the governing kinetic phenomena (e.g., role of mass transfer), the discrimination among microkinetic schemes, the validation of gas-phase models, and the comparison among catalysts (Horn et al., 2007; Korup et al., 2011, 2012, 2013).

In relation with the role of homogeneous chemistry in high-temperature partial oxidations, Korup et al. (2011) have performed gas-phase measurements of MOC; the idea behind the experiments was the need of building a

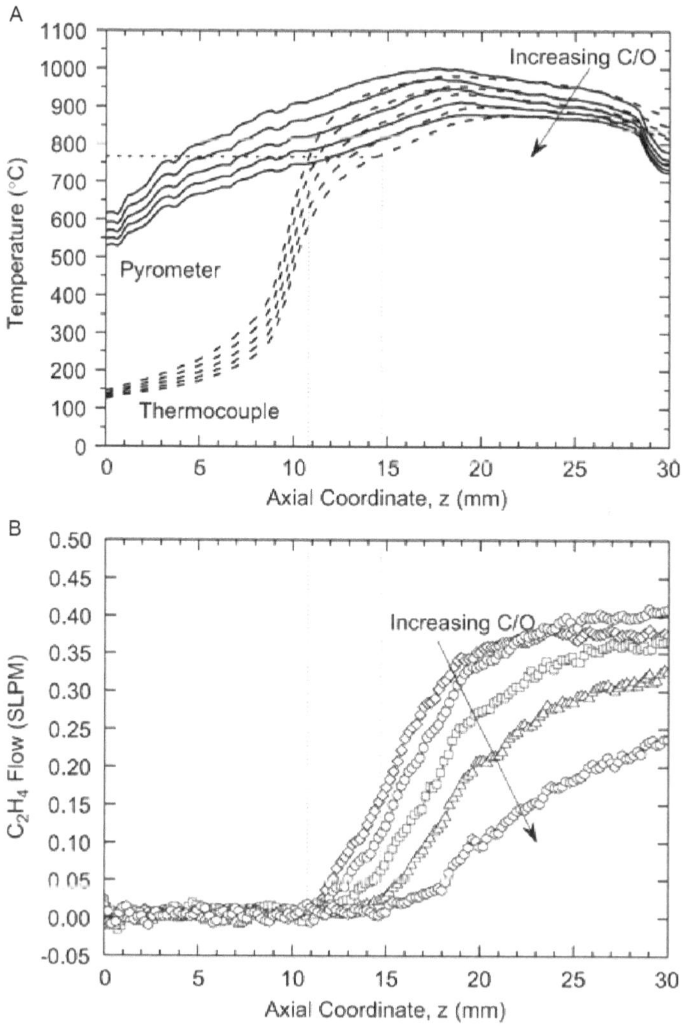


**Fig. 19** Comparison of reactor temperature profiles (A), measured by pyrometer (*solid lines*) and thermocouple (*dashed lines*), with $C_2H_4$ formation (B) for multiple C/O (*arrows* indicate C/O ranging from 1.2 to 2.0 by 0.2 increments) over the 4.7 wt% Pt/45 PPI α-$Al_2O_3$ catalyst. *Dotted lines* indicate isotherm/position relationship for $C_2H_4$ formation. *Reprinted from Michael BC, Nare DN, Schmidt LD: Catalytic partial oxidation of ethane to ethylene and syngas over Rh and Pt coated monoliths: spatial profiles of temperature and composition,* Chem Eng Sci *65:3893–3902, 2010, with permission from Elsevier.*

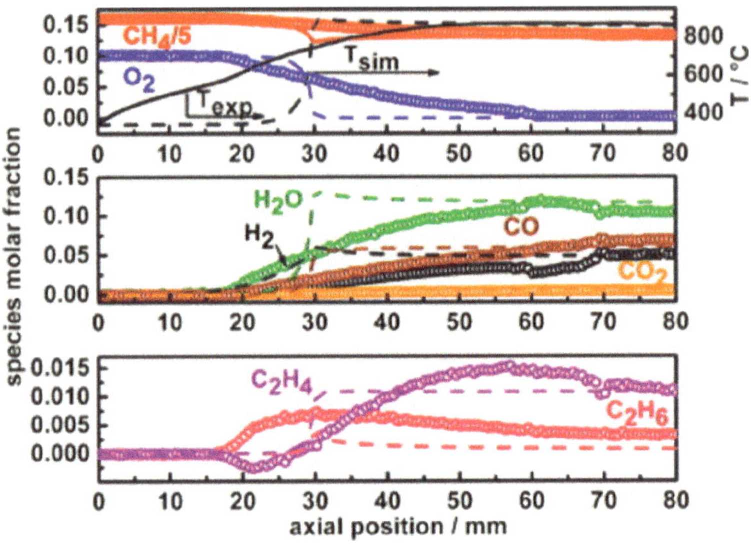


**Fig. 20** Experimental reactor profiles (*scatter*) and boundary layer numerical simulations (*dashed lines*) of gas-phase methane oxidative coupling. Reactor pressure $P=8$ bar, reactant stoichiometry C/O = 4.0, $CH_4$ flow = 3200 mln min$^{-1}$, $O_2$ flow = 400 mln min$^{-1}$, Ar flow = 400 mln min$^{-1}$, total flow = 4000 mln min$^{-1}$. *Reprinted from Korup O, Mavlyankariev S, Geske M, Goldsmith CF, Horn R: Measurement and analysis of spatial reactor profiles in high temperature catalysis research,* Chem Eng Process Process Intensif *50(10):998–1009, 2011, with permission from Elsevier.*

data set useful for verifying and possibly tuning gas-phase mechanisms for the design of MOC reactors, where the role of radical chemistry has been largely recognized.

Fig. 20 shows the axially resolved measurements obtained by the authors in an empty tube, fed with a methane/$O_2$ mixture with C/O = 4.0 and a total flow rate of 4000 mln min$^{-1}$. The dashed lines represent the predictions a boundary layer code implemented in CHEMKIN Pro; the model applies the OCM gas-phase kinetic model, developed by Zanthoff and Baerns (1990), comprising 33 species and 192 elementary reactions. It was found that, despite the model described with good accuracy the outlet product composition, the species development inside the reactor was not correctly described.

In particular the authors noted that the predicted gradients were too steep and not in quantitative agreement with the experimental data. Also important qualitative features of the experimental profiles such as the crossing of the $C_2H_6$ and the $C_2H_4$ profiles were not reproduced by the model.

It was thus concluded that microkinetic schemes can benefit from the information gained through spatial reactor data; these are more stringent than traditional integral data and can be used for validation, fine-tuning of parameters, and even identification of missing steps.

## 3.3 Partial Oxidation of Ethanol

Livio et al. (2013) have applied the in situ sampling technique to investigate the chemical pathways involved in partial oxidation of ethanol over rhodium-/alumina-coated monoliths. Fig. 21 reports the results obtained at 80% $N_2$ dilution, 4 NL/min total flow rate, C/O ratio = 0.4, and by preheating the reacting mixture at 363 K. The investigations clearly showed two zones inside the catalytic channel, an oxy-reforming zone as long as oxygen is present, with total oxidation and steam reforming as prevalent reactions, and a reforming zone, with steam reforming as the dominating reaction. Moreover, homogeneous gas-phase reactions, leading mainly to acetaldehyde and ethylene, were observed in front of the catalyst.

Interestingly, the same by-products were observed within the catalytic bed with the same bell-shaped concentration profile as illustrated above

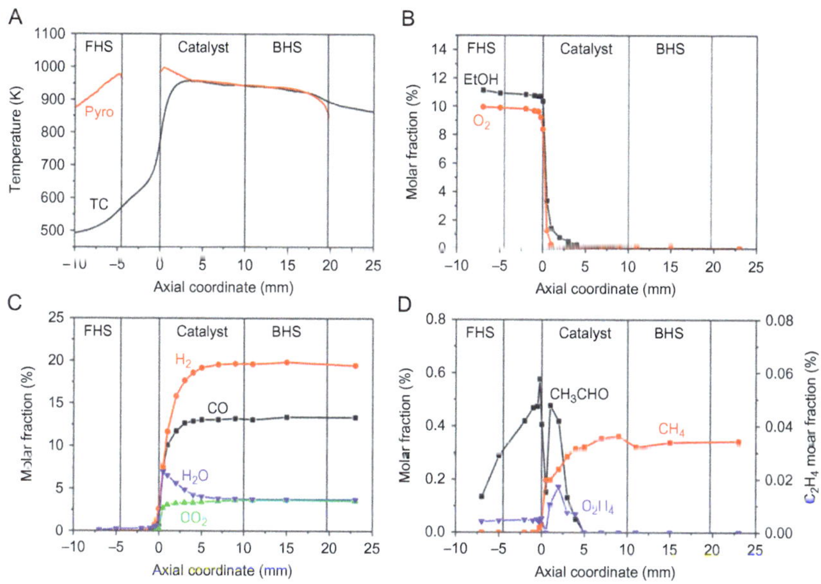


**Fig. 21** Ethanol partial oxidation. Temperature of the gas phase (TC) and surface (Pyro) as a function of the axial coordinate (A). Molar fraction of reactants (B), main products (C), and by-products (D) as a function of the axial coordinate at C/O = 0.75. *Reprinted from Livio D, Diehm C, Donazzi A, Beretta A, Deutschmann O: Catalytic partial oxidation of ethanol over Rh/Al$_2$O$_3$: spatially resolved temperature and concentration profiles,* Appl Catal A Gen *467:530–541, 2013, with permission from Elsevier.*

for the CPO of propane and octanes; this was interpreted as the evidence of gas-phase reactions proceeding inside the monolith channels, in combination with the reforming of the gas-phase products on the catalyst surface. The homogeneous nature of acetaldehyde and ethylene was confirmed by blank experiments, which showed that conversion of ethanol by ODH and dehydration may occur in the absence of a catalyst at temperatures higher than 640 K.

Also methane was formed as a by-product, but its formation could be well explained by a heterogeneous methanation route.

The authors extended the in situ probing to channels at different radial coordinate and, remarkably, they found that in the peripheral channels, where heat dispersions decreased the temperature, the same by-products were formed in larger amount (Fig. 22).

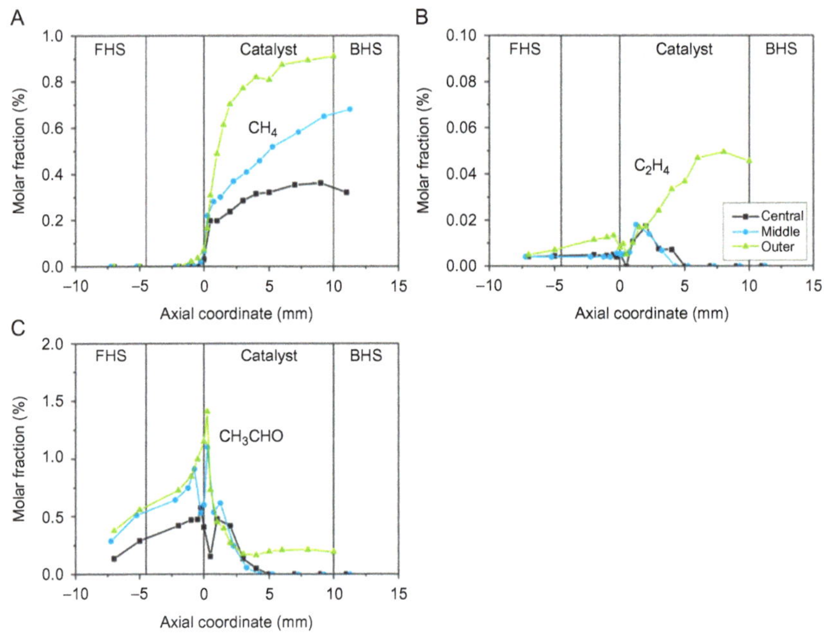


**Fig. 22** Ethanol partial oxidation. Molar fraction of the by-products methane (A), ethylene (B), and acetaldehyde (C) as a function of the axial coordinate at C/O = 0.75 in three differently positioned channels. One central channel (referred to as *central*), one channel at an intermediate position between the center and the edge of the monolith (*middle*), and one channel close to the edge of the monolith (*outer*). *Reprinted from Livio D, Diehm C, Donazzi A, Beretta A, Deutschmann O: Catalytic partial oxidation of ethanol over Rh/ $Al_2O_3$: spatially resolved temperature and concentration profiles,* Appl Catal A Gen *467:530–541, 2013, with permission from Elsevier.*

In the case of methane, this trend can be explained by the thermodynamic driving force on methanation; in the case of ethylene and acetaldehyde, the breakthrough from the colder peripheral channels could be ascribed to the kinetic control on steam reforming, responsible for a less efficient chemical quench of the gas-phase products on the catalyst surface.

Notably, the same evidence was indirectly collected when studying the CPO of *n*-octanes in $N_2$-diluted experiments; although the in situ probing showed complete consumption of the hydrocarbon fragments, a nonnegligible amount was present in the product mixture downstream of the monolith, likely due to cup-mixing from the whole section. A generalization can thus be drawn on the role of heat dispersions which disfavor the catalytic conversion of gas-phase products such that a breakthrough occurs. This may represent a critical issue, since gas-phase intermediates such as ethylene are coke and soot precursors and could favor both catalyst deactivation and soot deposition downstream from the catalyst bed.

## 4. CONCLUSIONS

Spatially resolved measurements of temperature and composition in CPO reformers with Rh-based honeycomb catalysts have revealed the existence of a concerted homogeneous–heterogeneous mechanism where homogeneous reactions (thermally activated in correspondence with the reactor hot spot) contribute to the conversion of reactants into short-chain hydrocarbons which are then converted into syngas via diffusion and reaction on the catalyst wall.

The formation and consumption of cracking products was observed in our laboratory in the CPO of propane and octanes, and reported in the literature in the CPO of ethane and ethanol.

Experiments and modeling have shown that:

- The overall syngas yield is unaffected by this additional route of fuel conversion through intermediate cracking species: in fact the molecular diffusion and the intrinsic reactivity of the short-chain hydrocarbons are even higher than those of the starting fuel.
- The measured evolution of composition and temperature can be correctly described only by accounting for the chemical quench via steam reforming of all the hydrocarbon species formed in the gas phase. In the absence of such consecutive surface process, high-temperature partial oxidation reactors are predicted to produce short olefins as well as syngas;

interestingly, this was actually observed in the partial oxidation of ethane on Pt, which further confirms that the final product distribution in the short contact time partial fuel-rich oxidation processes strictly depends on the reforming activity of the catalyst.

- Short olefins are coke and soot precursors; conditions have been reported where nonzero concentration of such species was measured in the reactor outlet. In particular, breakthrough was reported in $N_2$-diluted experiments of octane or ethanol partial oxidation, where the reactor was not fully adiabatic and colder peripheral zones of the reactor were present where the steam reforming activity was not insufficient for closing what we called the "by-pass" route.
- Pressure has little influence on the heterogeneous chemistry of high-temperature partial oxidation due to the important role of gas–solid mass transfer limitations; instead, pressure has a large effect on homogeneous reactions. We observed and modeled the promoting effect of pressure on the formation of short-chain hydrocarbons in the CPO of propane and predicted that at increasing pressure the onset of an oxy-cracking gas-phase chemistry can contribute to decrease the surface heat load, and consequently the catalyst hot spot temperature, due to a decreased $O_2$ flux at the wall.
- The axially resolved probing of concentration and temperature represents a powerful means for developing detailed heterogeneous/homogeneous kinetic schemes, allowing for very focused pieces of information useful for a fine-tuning or even identification of missing key reaction steps.

## REFERENCES


Armor JN: The multiple roles for catalysis in the production of $H_2$, *Appl Catal A Gen* 176:159–176, 1999.

Basile F, Fornasari G, Gazzano M, Kiennemann A, Vaccari A: Preparation and characterisation of a stable Rh catalyst for the partial oxidation of methane, *J Catal* 217:245–252, 2003.

Basini L: Issues in $H_2$ and synthesis gas technologies for refinery, GTL and small and distributed industrial needs, *Catal Today* 106:34–40, 2005.

Basini LE, Guarinoni A: Short contact time catalytic partial oxidation (SCT-CPO) for synthesis gas processes and olefins production, *Ind Eng Chem Res* 52:17023–17037, 2013.

Beretta A, Forzatti P: Partial oxidation of light paraffins to synthesis gas in short contact-time reactors, *Chem Eng J* 99:219–226, 2004.

Beretta A, Ranzi E, Forzatti P: Oxidative dehydrogenation of light paraffins in novel short contact time reactors. Experimental and theoretical investigation, *Chem Eng Sci* 56: 779–787, 2001.

Beretta A, Groppi G, Lualdi M, Tavazzi I, Forzatti P: Experimental and modeling analysis of methane partial oxidation: transient and steady-state behavior of Rh-coated honeycomb monoliths, *Ind Eng Chem Res* 48:3825–3836, 2009.

Carrera A, Beretta A, Groppi G: Catalytic partial oxidation of iso-octane over Rh/$\alpha$-$Al_2O_3$ in an adiabatic reactor: an experimental and modeling study, *Ind Eng Chem Res* 56:4911–4919, 2017a.

Carrera A, Pelucchi M, Stagni A, Beretta A, Groppi G: Catalytic partial oxidation of n-octane and iso-octane: experimental and modeling results, *Int J Hydrogen Energy* 42:24675–24688, 2017b. https://doi.org/10.1016/j.ijhydene.2017.08.020.

Curran HJ, Gaffuri P, Pitz WJ, Westbrook CK: A comprehensive modeling study of n-heptane oxidation, *Combust Flame* 114:149–177, 1998.

Curran HJ, Gaffuri P, Pitz WJ, Westbrook CK: A comprehensive modeling study of iso-octane oxidation, *Combust Flame* 129:253–280, 2002.

Donazzi A, Beretta A, Groppi G, Forzatti P: Catalytic partial oxidation of methane over a 4% Rh/$\alpha$-$Al_2O_3$ catalyst: part I: kinetic study in annular reactor, *J Catal* 255:241–258, 2008a.

Donazzi A, Beretta A, Groppi G, Forzatti P: Catalytic partial oxidation of methane over a 4% Rh/$\alpha$-$Al_2O_3$ catalyst: part II: role of $CO_2$ reforming, *J Catal* 255:259–268, 2008b.

Donazzi A, Livio D, Beretta A, Groppi G, Forzatti P: Surface temperature profiles in $CH_4$ CPO over honeycomb supported Rh catalyst probed with in situ optical pyrometer, *Appl Catal A Gen* 402:41–49, 2011a.

Donazzi A, Livio D, Maestri M, et al. Synergy of homogeneous and heterogeneous chemistry probed by in situ spatially resolved measurements of temperature and composition, *Angew Chem Int Ed* 50:3943–3946, 2011b.

Donazzi A, Livio D, Diehm C, Beretta A, Groppi G, Forzatti P: Effect of pressure in the autothermal catalytic partial oxidation of $CH_4$ and $C_3H_8$: spatially resolved temperature and composition profiles, *Appl Catal A Gen* 469:52–64, 2014.

Fabiano C, Italiano C, Vita A, Pino L, Laganà M, Recupero V: Performance of 1.5 $Nm^3/h$ hydrogen generator by steam reforming of n-dodecane for naval applications, *Int J Hydrogen Energy* 41:19475–19483, 2016.

Galloni E, Minutillo M: Performance of a spark ignition engine fuelled with reformate gas produced on-board vehicle, *Int J Hydrogen Energy* 32:2532–2538, 2007.

Galvagno A, Chiodo V, Urbani F, Freni F: Biogas as hydrogen source for fuel cell applications, *Int J Hydrogen Energy* 38:3913–3920, 2013.

Håkonsen SF, Walmsley JC, Holmen A: Ethene production by oxidative dehydrogenation of ethane at short contact times over Pt-Sn coated monoliths, *Appl Catal A Gen* 378:1–10, 2010.

Hickman DA, Schmidt LD: Synthesis gas formation by direct oxidation of methane over Pt monoliths, *J Catal* 138:267–282, 1992.

Hickman DA, Schmidt LD: Production of syngas by direct catalytic oxidation of methane, *Science* 259(5093):343–346, 1993.

Hickman DA, Haupfear EA, Schmidt LD: Synthesis gas formation by direct oxidation of methane over Rh monoliths, *Catal Lett* 17:223–237, 1993.

Horn R, Williams KA, Degenstein NJ, et al: Methane catalytic partial oxidation on autothermal Rh and Pt foam catalysts: oxidation and reforming zones, transport effects, and approach to thermodynamic equilibrium, *J Catal* 249:380–393, 2007.

Huff M, Schmidt LD: Ethylene formation by oxidative dehydrogenation of ethane over monoliths at very short contact times, *J Phys Chem* 97:11815–11822, 1993.

Kaltschmitt T, Deutschmann O: Fuel processing for fuel cells, *Adv Chem Eng* 41:1–64, 2012.

Kolb G, Schürer J, Tiemann D, et al: Fuel processing in integrated micro-structured heat-exchanger reactors, *J Power Sources* 171:198–204, 2007.

Korup O, Mavlyankariev S, Geske M, Goldsmith CF, Horn R: Measurement and analysis of spatial reactor profiles in high temperature catalysis research, *Chem Eng Process Process Intensif* 50:998–1009, 2011.

Korup O, Schlögl R, Horn R: Carbon formation in catalytic partial oxidation of methane on platinum: model studies on a polycrystalline Pt foil, *Catal Today* 181: 177–183, 2012.

Korup O, Goldsmith CF, Weinberg G, et al: Catalytic partial oxidation of methane on platinum investigated by spatial reactor profiles, spatially resolved spectroscopy, and microkinetic modeling, *J Catal* 297:1–16, 2013.
Lange J-P, Schoonebeek RJ, Mercera P, van Breukelen FW: Oxycracking of hydrocarbons: chemistry, technology and economic potential, *Appl Catal A Gen* 283:243–253, 2005.
Livio D, Donazzi A, Beretta A, Groppi G, Forzatti P: Optimal design of a CPO-reformer of light hydrocarbons with honeycomb catalyst: effect of frontal heat dispersions on the temperature profiles, *Topics Catal* 54:866–872, 2011.
Livio D, Diehm C, Donazzi A, Beretta A, Deutschmann O: Catalytic partial oxidation of ethanol over $Rh/Al_2O_3$: spatially resolved temperature and concentration profiles, *Appl Catal A Gen* 467:530–541, 2013.
Maestri M, Vlachos DG, Beretta A, Forzatti P, Groppi G, Tronconi E: Dominant reaction pathways in the catalytic partial oxidation of CH 4 on Rh, *Topics Catal* 52:1983–1988, 2009.
Mehra RK, Duan H, Juknelevičius R, Ma F, Li J: Progress in hydrogen enriched compressed natural gas (HCNG) internal combustion engines—a comprehensive review, *Renew Sustain Energy Rev* 80:1458–1498, 2017.
Michael BC, Nare DN, Schmidt LD: Catalytic partial oxidation of ethane to ethylene and syngas over Rh and Pt coated monoliths: spatial profiles of temperature and composition, *Chem Eng Sci* 65:3893–3902, 2010.
Minh HD, Bock HG, Tischer S, Deutschmann O: Optimization of two-dimensional flows with homogeneous and heterogeneously catalyzed gas-phase reactions, *AIChE J* 54:2432–2440, 2008.
Morgan K, Touitou J, Choi J-S, et al: Evolution and enabling capabilities of spatially resolved techniques for the characterization of heterogeneously catalyzed reactions, *ACS Catal* 6:1356–1381, 2016.
Mundschau MV, Gribble DA Jr, Plassmeyer PN, Henton LM, Jentzsch NR: Dry catalytic partial oxidation of diesel-fuel distillates into syngas, *Fuel* 89:1202–1211, 2010.
Pagani D, Livio D, Donazzi A, et al: A kinetic analysis of the partial oxidation of $C_3H_8$ over a 2% $Rh/Al_2O_3$ catalyst in annular microreactor, *Catal Today* 197:265–280, 2012.
Panuccio GJ, Williams KA, Schmidt LD: Contributions of heterogeneous and homogeneous chemistry in the catalytic partial oxidation of octane isomers and mixtures on rhodium coated foams, *Chem Eng Sci* 61:4207–4219, 2006.
Pasel J, Meißner J, Porš Z, et al: Hydrogen production via autothermal reforming of diesel fuel, *Fuel Cells* 4:225–230, 2004.
Pasel J, Wohlrab S, Kreft S, et al: Routes for deactivation of different autothermal reforming catalysts, *J Power Sources* 325:51–63, 2016.
Pelucchi M, Bissoli M, Cavallotti C, et al: Improved kinetic model of the low-temperature oxidation of n-heptane, *Energy Fuel* 28:7178–7193, 2014.
Ranzi E, Frassoldati A, Grana R, et al: Hierarchical and comparative kinetic modeling of laminar flame speeds of hydrocarbon and oxygenated fuels, *Prog Energy Combust Sci* 38:468–501, 2012.
Samsun RC, Krekel D, Pasel J, Prawitz M, Peters R, Stolten D: A diesel fuel processor for fuel-cell-based auxiliary power unit applications, *J Power Sources* 355:44–52, 2017.
Soberanis MAE, Fernandez AM: A review on the technical adaptations for internal combustion engines to operate with gas/hydrogen mixtures, *Int J Hydrogen Energy* 35:12134–12140, 2010.
Specchia S, Vella LD, Montini T, Fornasiero P: Syngas production by short contact time catalytic partial oxidation of methane. In Honnery DR, Moriarty P, editors: *Hydrogen production: prospects and processes*, Hauppauge, NY, 2012, Nova Science Publishers Inc., pp 95–139.

Stagni A, Frassoldati A, Cuoci A, Faravelli T, Ranzi E: Skeletal mechanism reduction through species-targeted sensitivity analysis, *Combust Flame* 163:382–393, 2016.

Villegas L, Guilhaume N, Mirodatos C: Autothermal syngas production from model gasoline over Ni, Rh and Ni-Rh/$Al_2O_3$ monolithic catalysts, *Int J Hydrogen Energy* 39:5772–5780, 2014.

Vita A, Italiano C, Pino L, Laganà M, Recupero V: Hydrogen-rich gas production by steam reforming of n-dodecane. Part II: stability, regenerability and sulfur poisoning of low loading Rh-based catalyst, *Appl Catal B Environ* 218:317–326, 2017.

Xu X, Li P, Shen Y: Small-scale reforming of diesel and jet fuels to make hydrogen and syngas for fuel cells: a review, *Appl Energy* 108:202–217, 2013.

Zanthoff H, Baerns M: Oxidative coupling of methane in the gas phase. Kinetic simulation and experimental verification, *Ind Eng Chem Res* 29:2–10, 1990.

Zerkle DK, Allendorf MD, Deutschmann O: Understanding homogeneous and heterogeneous contributions to the platinum-catalyzed partial oxidation of ethane in a short-contact-time reactor, *J Catal* (1):18–39, 2000.

CHAPTER FIVE

# Imaging Gas Flow in Gas–Solid Catalytic Systems by Near-Infrared Tomography

**Farid Aiouache**[1]
Lancaster University, Lancaster, United Kingdom
[1]Corresponding author: e-mail address: f.aiouache@lancaster.ac.uk

## Contents



## Abstract

Near-infrared imaging of 2D and 3D spatial resolutions allowed access to behavior of flowing gases and catalyst activity by measuring spatial distribution of composition and temperature in packed beds of low light scattering, subject to wall effects, and non-isothermal conditions. Water vapor was used as tracing species owing to high absorption coefficient in the near-infrared spectrum. Heterogeneous distribution of activity, flow maldistribution, and dynamic lags between the local mixing zones were observed and quantified. The uneven distribution of concentration and temperature highlighted the importance of mass and heat transport at gas–solid boundaries. The spatial resolution of near-infrared imaging was of importance to give reliable profiles of physical data, and dispersion coefficients and local activity were cited as examples of acceptable validation. The results were a proven stage for further developments, particularly those anticipated with infrared technology. However, key to that would be the development of affordable tunable lasers, optical accessories, and focal planar array sensors of similar sensitivities to those available in the near-infrared spectrum, allowing the technique to be applied to packed beds and other nontransparent reactive media.

*Advances in Chemical Engineering*, Volume 50
ISSN 0065-2377
https://doi.org/10.1016/bs.ache.2017.07.002

## 1. INTRODUCTION

In a continuous chemical process involving a packed bed, heterogeneity at different space scales and time scales lead to gradients of concentration and temperature that can be encountered in both bulk fluids and packing materials. The measurement of this heterogeneity is important, not only because it would provide information that helps understanding the underlying physical phenomena resulting from interactions of various types, but would anticipate development of models, particularly those of insufficient maturity (Korup et al., 2011). This heterogeneity is commonly investigated by a procedure of multiple probes that are placed at different locations as shown in Fig. 1. Discrete measuring points, however, lead to a limited number of sampling ports and would result in gradients between the data being too great to ascertain the changes (Niemantsverdriet et al., 1999). Alternative methods based on spatially resolved techniques under *operando* conditions, which are gaining popularity, are still not enough mature compared with conventional techniques for efficient and logical design of sustainable packed bed systems. Early spatially resolved methods relied on a continuous translation of sampling tools along a given direction of spatial coordinates for data transfer of species and temperature. The spatially resolved capillary inlet mass spectrometry system (SpaciMS) was successfully used to map trends of species and temperature in a multichanneled reactor of structured catalyst under stationary and transient operations as

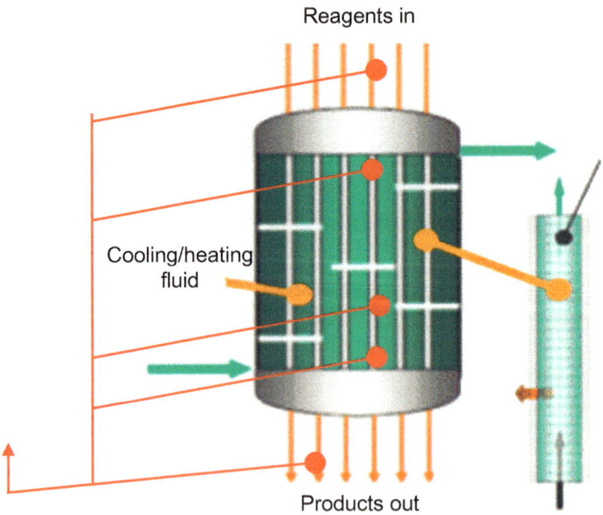


**Fig. 1** Multiple probes at different locations throughout the packed-bed.

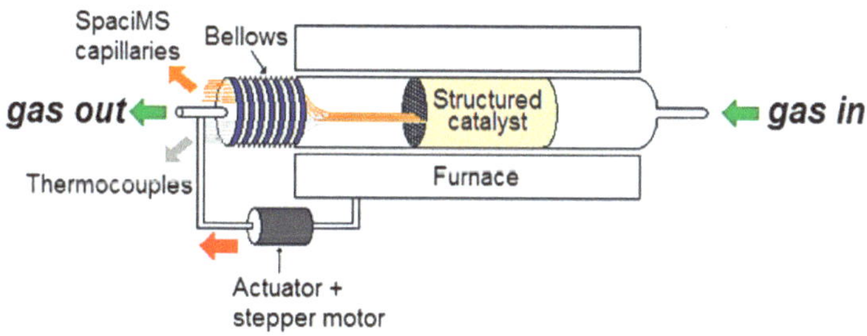


**Fig. 2** Sketch of SpaciMS instrument (Sa et al., 2010). *Reproduced with permission from the Royal Society of Chemistry.*

shown in Fig. 2 (Sa et al., 2010). Data by the SpaciMS capillary probe, which were further validated by three-dimensional computer fluid dynamics (CFD) modeling, demonstrated a promising sensitivity of the technique to measure spatiotemporal behavior of catalysts and were proven useful in providing reaction maps in monolithic channels and packed beds under reaction conditions. Kinetic oscillations were accessible for CO oxidation over a Pt/Rh on alumina catalyst supported on a cordierite monolith. It should be noted that the kinetic oscillations were only accessible inside the structured monolith but were not observed in a crushed catalyst, demonstrating the significance of *operando* experimental techniques. The capillaries did show, however, some invasiveness on the flow as the pressure drops and temperature varied when the probes were added (Sa et al., 2010; Horn et al., 2010). The use of capillary sampling techniques can prove problematic with unstructured catalysts; aside from the logistics of the technique is the staggered sampling, when the capillary is moved through the entire length of the packing, it will not measure everything axially at each step; heterogeneous distribution occurs regularly in unstructured packing and pockets of high or low concentration or hotspots are frequent. Capillary-based techniques and other sampling techniques might not sufficiently account for this. A third approach for getting access to spatially resolved information in packed beds is by using predictive simulations. With ongoing development of powerful computational tools, however, the simulation of flow and transports in packed beds is gaining substantial recognition to address a range of problems associated with experimental techniques. Simulations of packed beds of two-dimensional (2D) and three-dimensional (3D) geometries of discrete particles are successfully used as alternatives to mimic actual laboratory experiments, permitting prediction of physical phenomena inside a packing consisting of a few thousands of particles (Freund et al., 2005). These simulations, although they have reached rapid success, remain incapable to reach modeling of packed beds of industrial scale at highly spatially resolved resolution. Existing modeling is therefore insufficient for plant design but convenient for investigation of complex interaction between the flow, heat, and mass transports on a formerly unreachable level of detail, helping thus to establish cross-links of knowledge between this approach and simplified modeling approaches. It is therefore primordial to screen insights in local information of flow, mass, and heat by both experimental tools that rely on spatially resolved techniques as prerequisites for validation of performance of predictive simulations.

In this chapter, recent advances on spatially resolved techniques based on optical near-infrared (NIR) imaging are presented in order to acquire more comprehensive information of behavior of heat and mass through temperature and composition fields in the interior of gas–solid packed beds. Techniques that use planar 2D and tomographic 3D NIR imaging are applied to water vapor flow in heterogeneous systems of low aspect ratios (ARs), relevant capabilities are demonstrated by validation of advanced modeling codes, which often are not associated with experimental confirmation at high spatially resolved resolutions. In Section 2, a brief introduction to NIR imaging is given and followed by its application to assessment of the two domains of field inside gas–solid catalytic processes. The first domain is the solid domain by catalyst activity, and second domain is the fluid domain by mass and heat transports, and relevant interactions with the former solid domain in a fluidized bed reactor. This noninvasive technique uses a source of a broadband NIR light, an affordable quartz-based reactor, and two interference filters centered to absorbing wavelengths of water vapor. In Section 3, the 2D NIR imaging is extended to 3D NIR imaging and applied to diffuse transmittance tomography. Phenomena of mixing and interactions with catalytic processes are examined by observation of uneven distribution of these under dynamic operations inside gas–solid packed beds low light scattering, including adsorbers and reactors. Both 2D and 3D NIR imaging techniques are discussed along with validation of 2D and 3D models that simulate behavior of fluid flow, mass, and transfers in packed bed systems.

# 2. SPECIATION IN REACTORS BY 2D OPTICAL NIR IMAGING TECHNIQUE

## 2.1 Introduction to NIR Imaging

NIR imaging is a nondestructive technique that allows simultaneous access to quantitative information of multiconstituents at highly spatially resolved resolution. It relies on NIR spectroscopy with a wavelength ranging from the mid-infrared (IR) to the visible region. NIR spectroscopy region was first discovered by Herschel in the early 19th century when he noticed a substantial increase in temperature associating the electromagnetic spectrum beyond the red region (Reich, 2005). The technique has, however, seen moderate advances before the second half of the 20th century owing to characteristic weak absorption compared with competitive absorptions in the mid-IR spectrum. Throughout the last decades, the technology associated with emission sources and detectors in NIR region has seen substantial developments, gaining wide applications to various sectors, particularly biological organ function diagnostics, raw material testing, product quality control, and process monitoring. The NIR region of the electromagnetic spectrum is defined as the wavelength ranging from 780 to 2526 nm (ASTM, 2016), and absorption bands occurring in this region are driven by anharmonicity and Fermi resonance, which are related to overtones and combinations of fundamental mid-IR vibrations, but are of 10–100 times weaker in intensity as shown in Fig. 3. They include particularly –CH, –OH, –NH, and –SH functional groups, and typically require promotion of the sensitivity of probing tool by using data processing techniques. NIR, however, allows a penetration depth in scattering media that cannot be achieved in the mid-IR owing to weak absorption, making the technique convenient for nondestructive analysis in transmittance, diffusive, or reflectance modes, if careful attention is taken regarding the pretreatment and calibration of the medium of interest. NIR imaging is generally composed of a light source, spatially resolved detector, and optical accessories for light processing between the light source, the sample of interest, and the detector. Nowadays, detectors are based on focal planar array (FPA) technology, particularly of indium gallium arsenide (InGaAs) type which are popular compared to silicon and lead sulfide, as they combine the speed and size characteristics of the silicon detector with the wavelength range of the PbS detector (Reich, 2005). An example of detailed layout of optical components is shown in Fig. 4, where a series of lenses and mirrors direct the light to the surface area of observation and then to the FPA detector. The spectral information is provided either by a number of static or tunable filters, or an imaging interferometer. The exceptional feature of the FPA detector is its large number of small infrared detectors arranged in a rectangular grid pattern that allows simultaneous collection of spatial and spectral information for each pixel without any moving parts as shown in Fig. 4A. The data can therefore be thought of as 3D blocks that span either

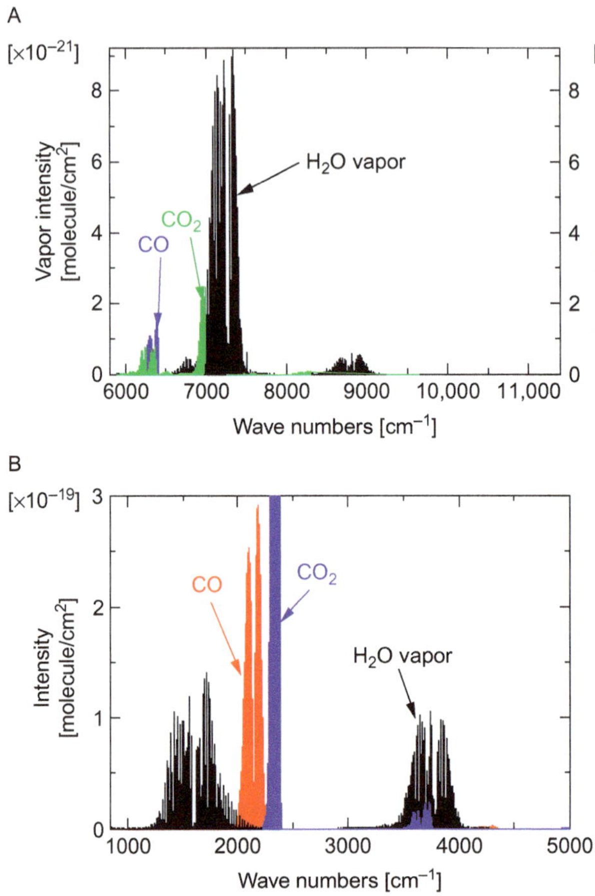


**Fig. 3** Spectra of $H_2O_v$, CO, and $CO_2$ at total pressure $1.013 \times 10^5$ Pa, $T = 913$ K, and path length = 0.02 m from Hitran data base (Rothman et al., 2009).

one spatial dimension with multispectral profiles (Fig. 4B) or two spatial dimensions at a selected frequency as shown in Fig. 4C, giving the opportunity to visualize the spatial distribution of a chemical species throughout the samples of interest.

## 2.2 Application of 2D NIR Imaging to Catalyst Activity and Fluid Flow Inside Gas–Solid Catalytic Reactors

Techniques for sensing species by 2D planar distribution have emerged in the recent years owing to developments in the laser technology for the telecommunication sector and imaging sensors of 2D array detectors, allowing

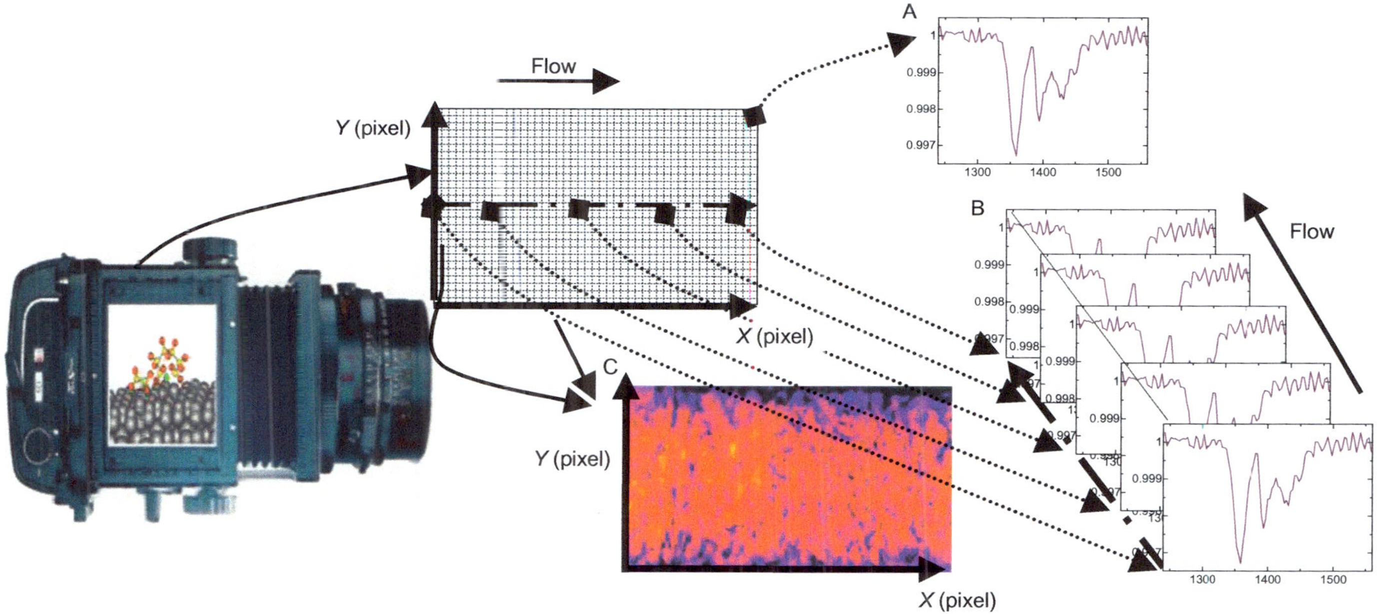


**Fig. 4** Focal planar array (FPA) detector and reactor coordinates: (A) Integral data of a single absorbance spectrum. (B) Differential data of one-dimensional hyperspectrum absorbance of a linear array along axial coordinate (C) 2D absorbance.

to run at spatiotemporal resolutions of submillimeters and subseconds. The principle of the 2D NIR imaging consists of a plane that is illuminated, and the attenuated light transmitted is captured by 2D array detectors such as CCD cameras. Application of NIR imaging to catalytic reactors emerged after sufficient maturity of comparative methods such as Raman imaging and IR imaging. For instance, spatially resolved Raman spectroscopy and optical fibers were used to extract information on the film transport limitations and the reaction pathways (Horn et al., 2010; Partridge et al., 2000). Raman spectroscopy was of significance for principles of spatially resolved spectroscopy with a fiber probe inside a circulated sampling capillary. An $Ar^+$ laser (300 mW, 488 nm) was coupled to a bare fluorine-doped fused silica fiber, and the scattered light was collected by the same fiber, transferred into a fiber bundle consisting of 19 fluorine-doped fused silica fibers arranged in a circular closed package at the reactor side, and analyzed by means of a triple filter Raman spectrometer with a CCD camera as detector operated in subtractive mode. This method provided valuable spectroscopic information that allowed measurements of concentration of species, temperature, and flow rates along the length of the reactor. The analysis of spatial reactor profiles for the extracted mechanistic information was outlined along with the validation of microkinetic models. Similarly, Nadeau et al. (1996) was able to reach a time resolution of milliseconds for measurements of residence time distributions in a plasma reactor. The resolution was made possible by directing an IR laser light to the reactor where methane was used as tracer probe. The residence time distribution was measured by running step changes in concentration of methane and which was then assessed over time by directing the laser light to inlet and outlet of the reactor. The subsecond resolution of response time allowed observation of rapid local maldistributions and radial dispersions. The measurement device consisted of an infrared He—Ne laser beam (wavelength of 3392 nm) passing through the diameter of the reactor through sapphire windows and was measured with a lead selenide detector. Although the technique provided a valuable insight into the flow, it did not differentiate the varying concentrations of the different species present and did not provide associated temperature data. We extended application of 2D NIR imaging to catalyst activity (Aiouache et al., 2006) and fluid flow in gas–solid catalytic reactors (Aiouache et al., 2011), and assessed how sensitive is the method to both gaseous and solid phases, allowing opportunity to investigate interactions of the two phases under *operando* conditions.

### 2.2.1 High-Throughput Screening of Catalyst Activity by NIR Imaging

The characteristic of NIR imaging in term of thermal and concentration sensitivity was first employed by Wolf's group (Qin & Wolf, 1996), who applied the concept of optical filtering of IR broadband radiations to detect surface catalytic species (Fig. 5A). Wolf used planar emissions of spectral vibrations of CO to discriminate relevant linear coverage of rhodium/silica catalyst. Owing to interference of thermal radiations with the background of spectral emissions, the results were most valid at low temperatures and pseudoisothermal conditions (Fig. 5B). Along with optical filtering of IR radiations, interferometric techniques such as those combined with Fourier transformed infrared red and an FPA detector were successfully used to quantitatively visualize emissions of $CO_2$ produced from CO oxidation by

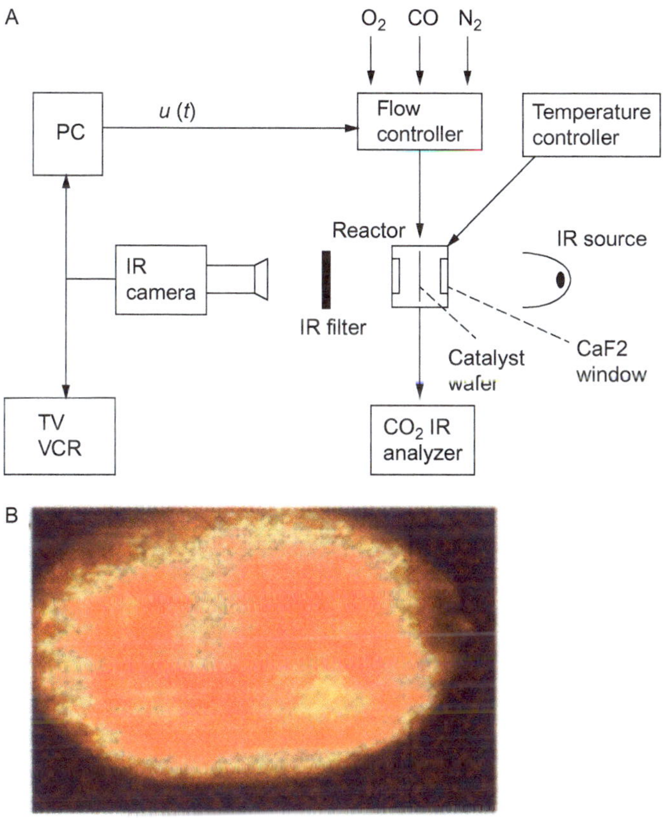


**Fig. 5** Diagram of the experimental system (A) and CO coverage of Rh/$SiO_2$ catalyst (B) (Qin and Wolf, 1996). *With permission of Springer.*

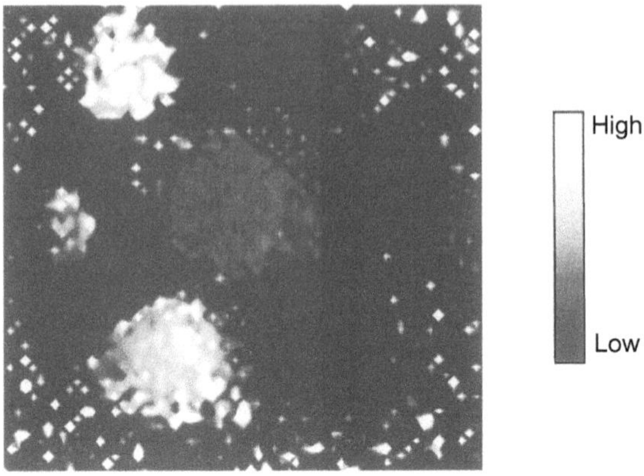


**Fig. 6** Absorbance distribution of CO adsorbed on $Pt/SiO_2$ relevant to location of three $Pt/SiO_2$ pellets (Snively et al., 2000). *Reprinted with permission from Elsevier.*

Lauterbach's group (Snively et al., 2000), as shown in Fig. 6. A multispectral FT-IR imager was used to determine light-off temperature of CO oxidation over a commercial catalyst monolith. Qualitative visualization of $CO_2$ composition was achieved by averaging absorbance of light transmitted over the entire catalyst. We developed a high-throughput diffuse reflectance NIR imaging technique, as alternative approach to IR imaging, to assess activity of catalyst of different types (Aiouache et al., 2006). A combination of thermal and spectroscopic imaging was used along with principles of combinatorial chemistry for screening activity of "multisample at a time" in a library of catalysts. Discrimination of activity of catalysts by thermal screening was popular but has been always uncertain when probing activity of reactions that are associated with weakly thermal exchanges. The added value of the technique was mapping species adsorbed on catalyst at realistic conditions. The case, that was investigated, was the atmospheric adsorption of molecules with high vibration bands in the NIR range, such as the bridged-bond or geminal hydroxyls overtones on reducible cerium oxide during methane steam reforming. Methane steam reforming at temperatures of 673 and 823 K was selected because two different mechanisms via formate intermediates or carbonate intermediates prevail. The high-throughput imaging experiments were carried out in a transparent quartz cell and included 25 catalyst wells on a ceramic template, which was supported by a stainless steel plate, allowing a uniform heat transfer throughout the ceramic template as shown in Fig. 7. The analysis was carried out by using a broadband halogen source of light, a multichannel fiber probe spectrometer equipped with InGaAs linear image sensor (Jasco, Tokyo, Japan), a NIR Vidicon-type camera and two band-pass interference filters, centered at 1381 and 1369 nm. The experiments were carried out under

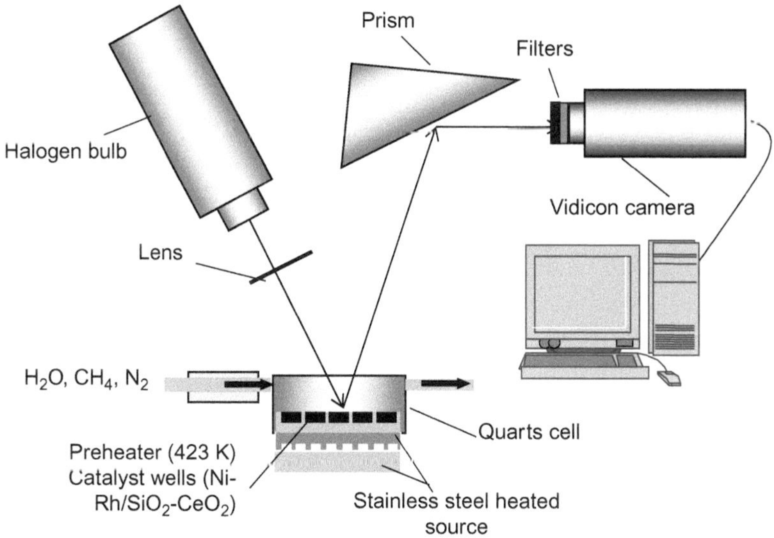


**Fig. 7** Experimental setup for high-throughput NIR imaging (Aiouache et al., 2006). *Reprinted by permission of John Wiley & Sons, Inc.*

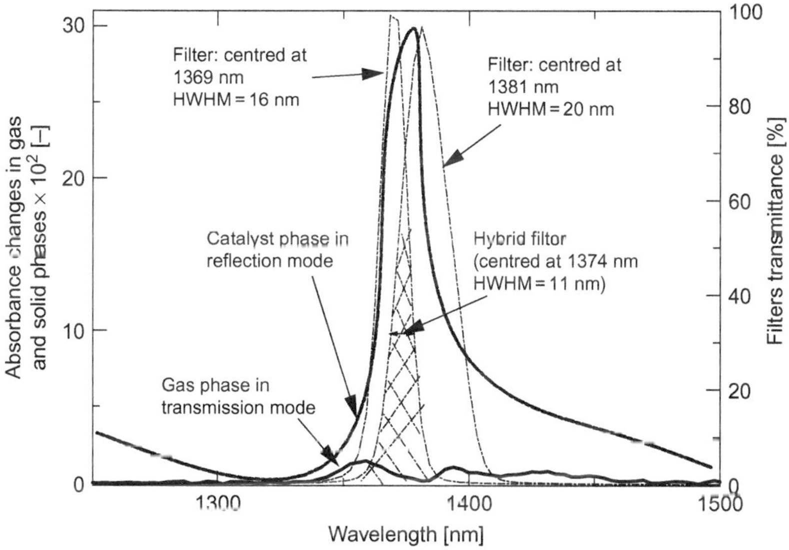


**Fig. 8** NIR spectra of absorbance and transmittance profiles after 15 min reduction at 673 K. $P_{H_2}$ = 10.13 kPa, feed flow rate 1 mL/s (Aiouache et al., 2006). *Reprinted by permission of John Wiley & Sons, Inc.*

temperatures that allowed reduced self-emissions from the samples, and back-reflected images were captured by a frame grabber card (CT3000) and further postprocessed by commercial image-processing software. Fig. 8 shows gas and surface catalyst spectra of a single catalyst well with pure rhodium on ceria

[Rh (1 wt%) and $CeO_2$ (99 wt%)] at 673 K and after 15 min of reduction time by hydrogen. The absorbance of the gaseous phase was much weaker and was caused by radiation of the generated steam, whereas absorbance from the catalyst surface originated from the adsorbed hydroxyl groups. When a full library of 25 catalyst samples was displayed under hydrogen atmosphere and two temperatures of 673 and 823 K as shown in Fig. 9, surface wells of catalyst loaded with pure rhodium displayed significant absorbance values compared with those of bimetallic catalysts of nickel and rhodium, and these results agreed

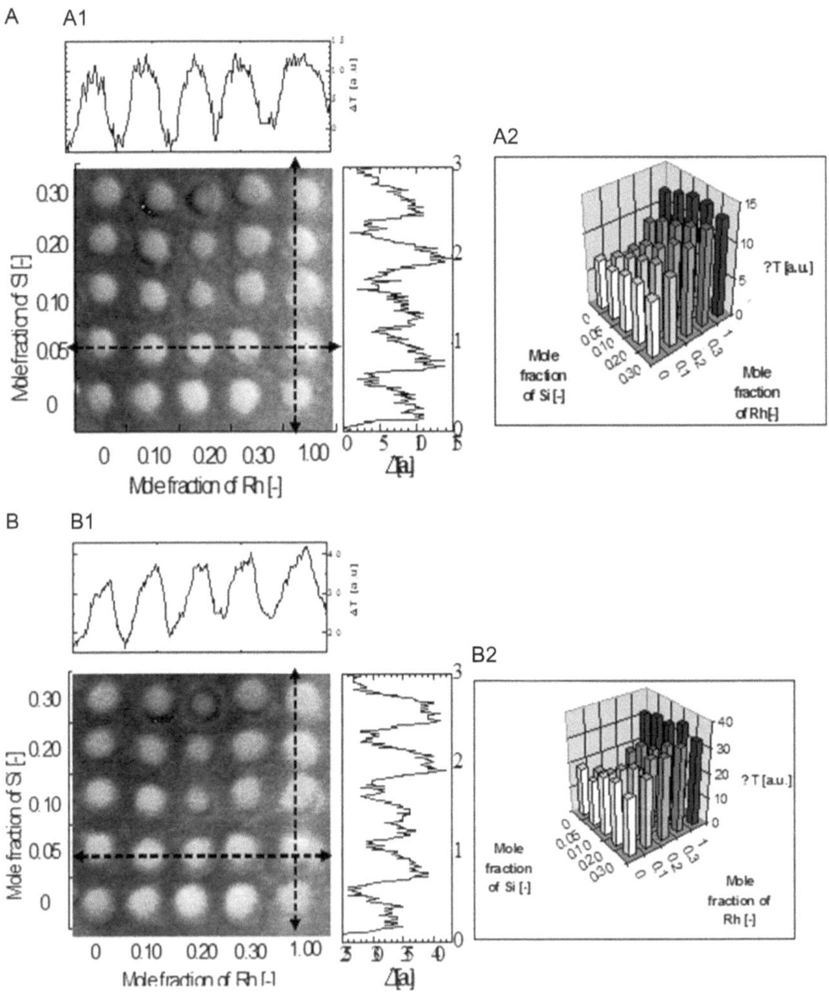


**Fig. 9** NIR self-emission images ($A_1$, $B_1$) and integrated profiles ($A_2$, $B_2$) after 15 min of reduction. $P_{H_2}=10.13$ kPa, feed flow rate 1 mL/s. (A) $T=673$ K, (B) $T=823$ K (Aiouache et al., 2006). *Reprinted by permission of John Wiley & Sons, Inc.*

with those obtained by the thermal imaging in Fig. 10. Absorbance was seen, however, less pronounced at 823 K for all wells, although corresponding temperature changes were found higher than those obtained at 673 K in Fig. 10. It was concluded that an extended reduction to bulky layers of ceria was likely accompanied by surface OH elimination. Screening the coverage of germinal OH bonds on surface of catalysts of various activities toward methane reforming by using NIR imaging was demonstrated in Fig. 11. When temperature was dropped from 823 to 673 K, absorbance values of OH were linear with partial pressure of water vapor pressure, indicating potential mechanism via formation of formates by reaction of water vapor with surface carbonyls, as a relevant step in the overall pathway of $CH_4$ reforming, concluding that the overall reaction proceeded without redox exchanges by carbonate intermediates (Jacobs et al., 2004).

The optical setup that relies on optical filtering of NIR broadband light was extended to planar visualization of gaseous phase flow in a fluidized bed

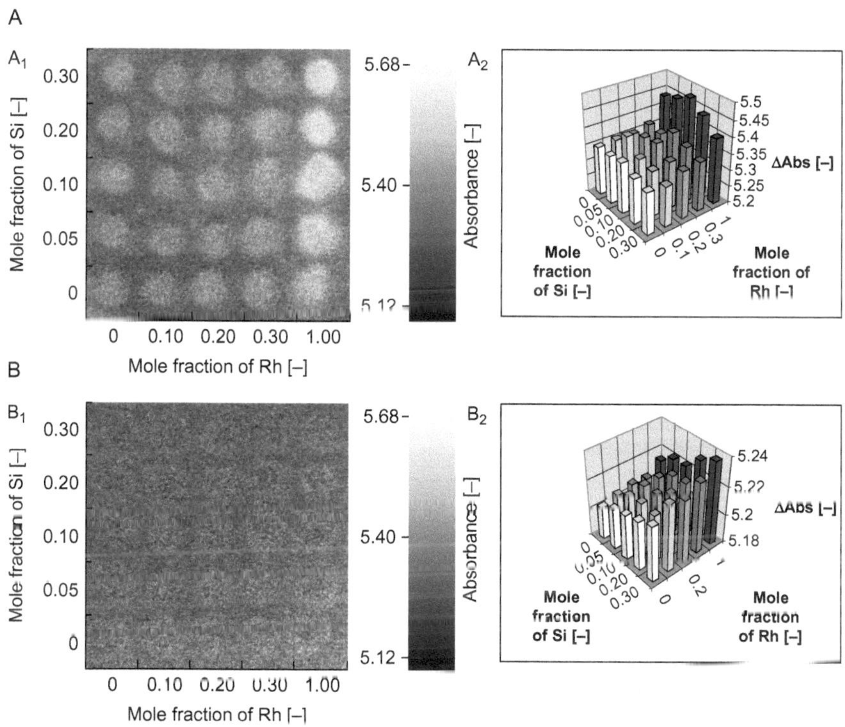


**Fig. 10** Absorbance change images ($A_1$, $B_1$) and integrated value profiles ($A_2$, $B_2$) after 15 min of reduction. $P_{CH_4}=10.13$ kPa, feed flow rate 1 mL/s. (A) temperature 673 K, (B) temperature 823 K (Aiouache et al., 2006). *Reprinted by permission of John Wiley & Sons, Inc.*

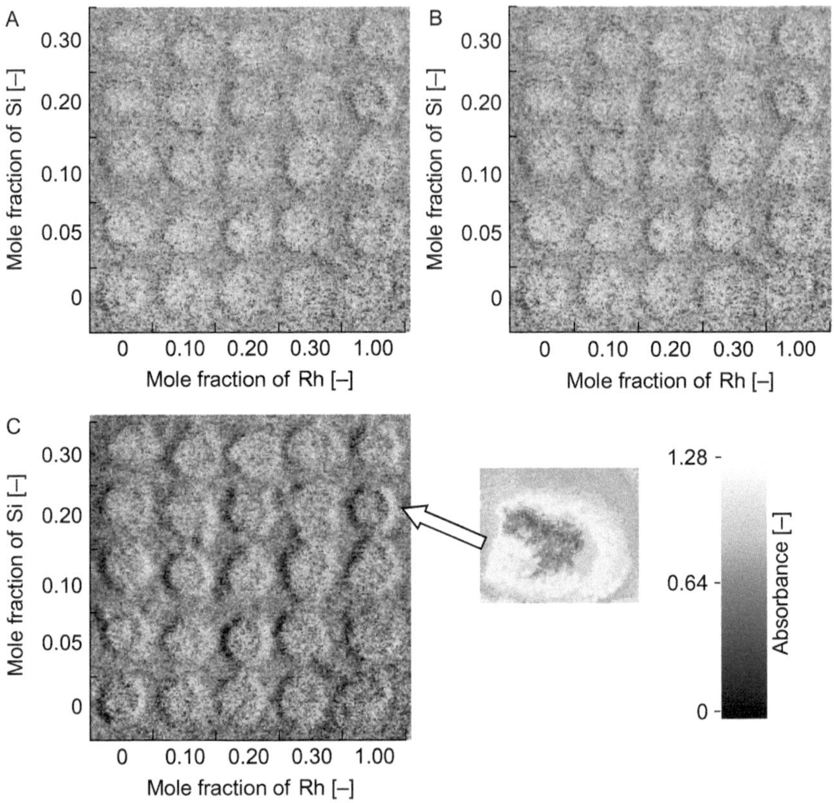


**Fig. 11** Absorbance rate images of catalyst surfaces (reduced at 823 K) at $P_{CH_4} = 10.13$ kPa, $P_{H_2O}/P_{CH_4}$: (A) 1/1, (B) 2/1, (C) 5/1, feed flow rate 1 mL/s, $T = 673$ K (Aiouache et al., 2006). *Reprinted by permission of John Wiley & Sons, Inc.*

of a low AR of tube to particle diameters (Aiouache et al., 2011). The operating conditions were selected to reach both pseudostatic bed conditions and minimum bubbling conditions, ensuring two types of radial porosity profiles. Porosity profiles, and by inference the fluid flow, at the two operating conditions were assessed by observing profiles of concentration gradients along the radial direction of the bed, validating the wall effect on flow dynamics in gas–solid catalytic processes. This work attempted to answer some questions that are associated with low AR fluidized beds. It is well known that large AR fluidized bed reactors are not representative for thin ones. Pressure drop overshoots in low AR fluidized bed reactors were found to be more pronounced, and the physical origins were interpreted by various models. Among these, they were attributed to forces exerted between particles at points of contact and to flow dynamics between the particles and the fluid. A procedure based on 2D NIR imaging was developed to observe gas flow at the exit of a pseudostatic packed reactor and a fluidized bed reactor. The technique validated existing models of thin fixed bed reactors for applications to pseudostatic bed operations.

The experimental approach aimed to observe flow channeling of water vapor which was subject to wall effects inside a gas–solid fluidized bed reactor of a low AR. The technique used equipment partly similar to that cited previously to screen catalyst activity in a high-throughput reactor (Aiouache et al., 2006), including broadband NIR light, a tubular quartz-based reactor and two optical filters centered on NIR absorbing wavelength of water vapor as shown in Fig. 12. The reactor (0.02 m internal diameter (I.D.) and 0.5 m in height) was covered with a carbon-based ribbon, simulating blackbody radiative heating, and two side quartz-based windows ($0.01 \times 0.10\,m^2$) were used for optical observations. NIR imaging of water vapor was demonstrated by loading an inner tube (0.008 mm I.D.) inside the reactor with water vapor at 873 K and recording NIR images at the exit of the inner tube. Absorbance of the directed light was a function of the concentration of water vapor along the optical path as shown in Fig. 13, illustrating distribution of water vapor that exited the inner tube and demonstrating lower concentrations at the center of the tube due to extended optical path.

The inert packing was then replaced by a catalytic ceria–alumina packing for hydrogen reduction. Since the ceria–silica mixture contains oxygen vacancies with diffusion rates that would change during the course of reduction, running experiments at low conversions would lead to kinetics of reduction to be relevant to rapid surface reactions while high conversions would lead to kinetics to be relevant to slow bulk diffusion of oxygen (Villain & Normand, 1994). Effect of the wall on channeling should therefore vary along with the fluid flow-transport interactions, particularly the progress of the reaction. When the inner tube was filled with a catalytic packing of ceria–alumina and reduced by hydrogen, as shown in Eq. (1), at a superficial velocity corresponding to pseudostatic velocity and minimum bubbling velocity conditions; the water vapor produced was then observed by NIR imaging.

$$CeO_2 + \gamma H_2 \Leftrightarrow CeO_{2-\gamma} + \gamma H_2O \tag{1}$$

where $\gamma$ represents oxygen vacancies in the reduced ceria. The dynamic responses that corresponded to step changes in hydrogen step were

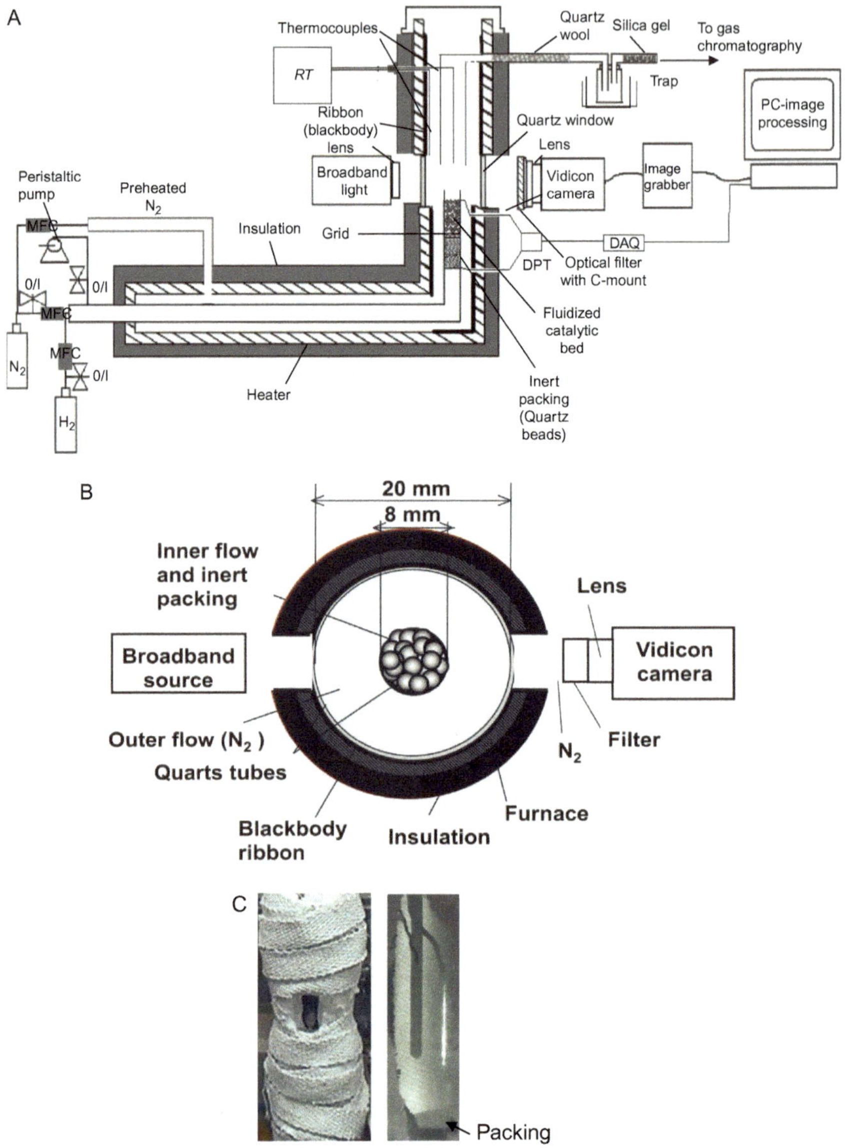


**Fig. 12** Experimental setup. (A) *Front view*, (B) *top view*, (C) outer tube and quartz window photographs. *DAQ*, data acquisition; *DPT*, differential pressure transducer; *I/O*, ON/OFF valves; *MFC*, mass flow controller; *RT*, regulator of temperature (Aiouache et al., 2011). *Reprinted with permission from Elsevier.*

computed by using images of integral data along the radial direction. Cross-sectional concentration profiles of water vapor inside the packed bed as a function of the reduction time and two gas velocities (pseudostatic and minimum bubbling) are illustrated in Fig. 14A and B, respectively. The former

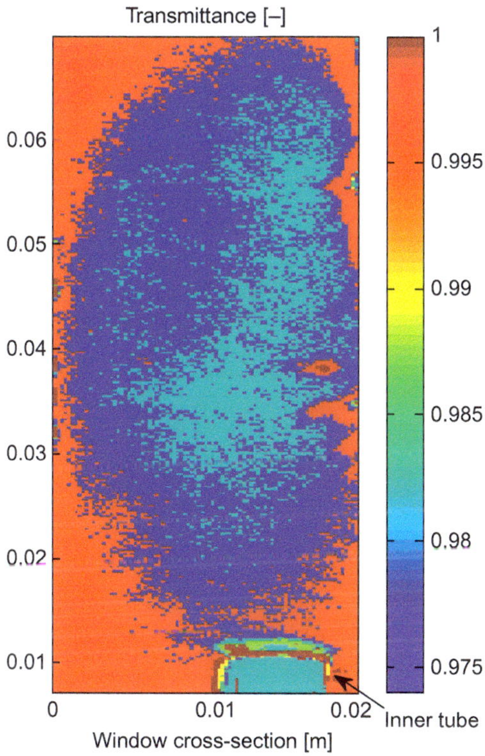


**Fig. 13** Integral transmittance of water vapor at the nozzle exit. Water vapor concentration: 1.35 mol/m$^3$ in $N_2$, temperature: 873 K, and pressure: 101.3 kPa. Inner flow ($N_2$+$H_2O$) 150 mL/min, outer flow ($N_2$): 790 mL/min (Aiouache et al., 2011). *Reprinted with permission from Elsevier.*

figure shows higher concentrations of water vapor in the vicinity of the wall of the inner tube than in the core center of the bed during the period of fast release of water vapor (i.e., 600 s) and where the concentration extended from the wall to the core of the bed, and covered the whole cross-section of the inner reactor. Velocity was then higher near the wall compared with the core of the bed. Flow channeling in the vicinity of the wall occurred at static velocity but was absent at minimum bubbling velocity as seen in Fig. 14B, where apparently the gas velocity served to adjust the flow of the bubbles in the core of the bed, and accordingly induced an increase in the core porosity. Both figures demonstrated faster kinetics at minimum bubbling velocity compared with the static velocity for the first 600 s, where the kinetics was relevant to surface control. Beyond this time of reaction, the kinetics was the same at both velocities, corresponding to relevance of mass transfer inside the catalytic packing. This technique was a proven platform

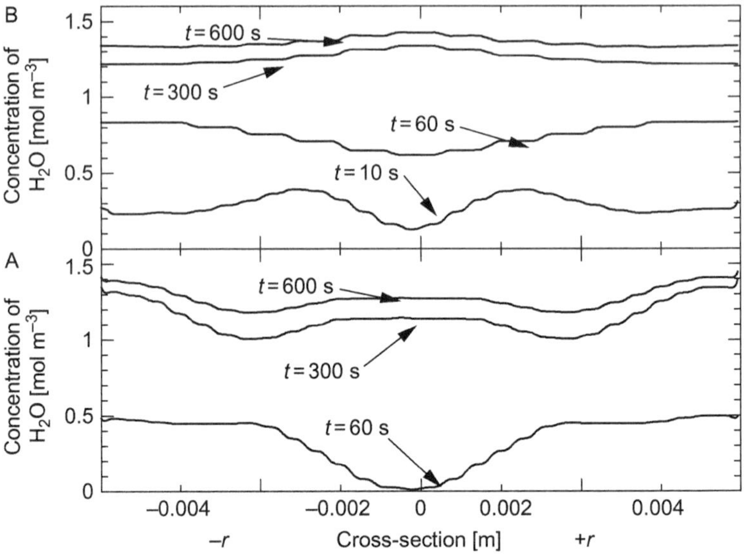


**Fig. 14** Cross-sectional mole fraction reconstructed from the line of sight integral positioned at 2 mm above the nozzle during the catalytic reduction of ceria-based ceria–silica, (A) at $U_c$, (B) at $U_{mb}$, hydrogen concentration: 1.35 mol/m$^3$ in $N_2$, temperature: 873 K, and pressure: 101.3 kPa (Aiouache et al., 2011). *Reprinted with permission from Elsevier.*

which helped validation of NIR imaging and extension to 3D tomographic imaging. By refining the technique and replacing the broadband light source with a fine bandwidth tunable laser, tomographic measurement of gaseous flow could be monitored by diffuse transmittance. Through the use of principles of NIR spectroscopy, a robust system for visualization of cross-sectional profiles of gaseous interactions in packed beds was then developed as illustrated in the following section. The 2D planar NIR imaging was validated by using water vapor exclusively but has potential applications to other gaseous species. The noninvasive in situ measurements were ideal to monitor systems under realistic reaction conditions. These few examples show how important is having access to spatial information that could reduce the search for highly time-resolved techniques for fast processes, as the time-scale can partly be converted into the space-scale. The spatial data which were represented in 2D distributions did not provide, however, local information which often occur because distributions are rarely axisymmetric but change radially and circumferentially in the packed beds, leading recent studies to be focused particularly on tomographic techniques for 3D distributions of flow, mass, and heat.

# 3. SPECIATION IN REACTORS BY OPTICAL 3D IMAGING TECHNIQUE

As discussed in the previous section, the need for a spatially resolved analytical technique which monitors gas-phase chemical reactions is of vital importance. Although the technique in Section 2 provided valuable insight into a gas–solid reactor, it was insufficient to completely describe the process inside a packed bed. A spatially resolved technique which has 3D resolution is required. Subsequent techniques which utilize end-pipe analysis typically miss local variations of the packed bed, leading techniques based on tomography essential for the speciation under *operando* conditions.

## 3.1 3D Distribution In Chemical Systems by Process Tomography

Catalytic reactions take place over extended regions of space within a given reactor with rates that depend on factors, such as the mixture distribution, temperature, and momentum (Fogler, 2013). Process tomography is among very few techniques that are currently available to gather information on the 3D spatial distribution of chemical species, flow, and temperature gradients within a packed bed under *operando* conditions. For years, tomography was mainly the focus of the medical establishment while its application to process engineering has trailed behind. Various alternative challenges relate to application of tomography to process engineering. The conditions of the process with regard to temperature and pressure vary widely to both extremes of the scale, and the varying reactants and products can be in different phases and at different velocities as shown in Fig. 15. The time resolution, which is of utmost importance, varies widely depending on the system being monitored.

Tomography uses a penetrating wave to image sections of an object and has shown large developments over the last decades, particularly in the areas of medical sciences and materials. The first generation of commercial tomographic apparatus relies on a source that emits a beam of radiation and a detector. They rotate opposite sides around the object of interest and measure attenuation of the beam at each position. It is clear that this method is slow and would not be able to capture fast transient operations. Further improvements are found in the scanners of second generation which include a series of sources and detectors that are positioned opposite to one another and continuously rotate around an object to provide a number of projections that are equal to the number of sources or detectors. Other generation of scanners use a source that is directed in such a way that it offers a pattern

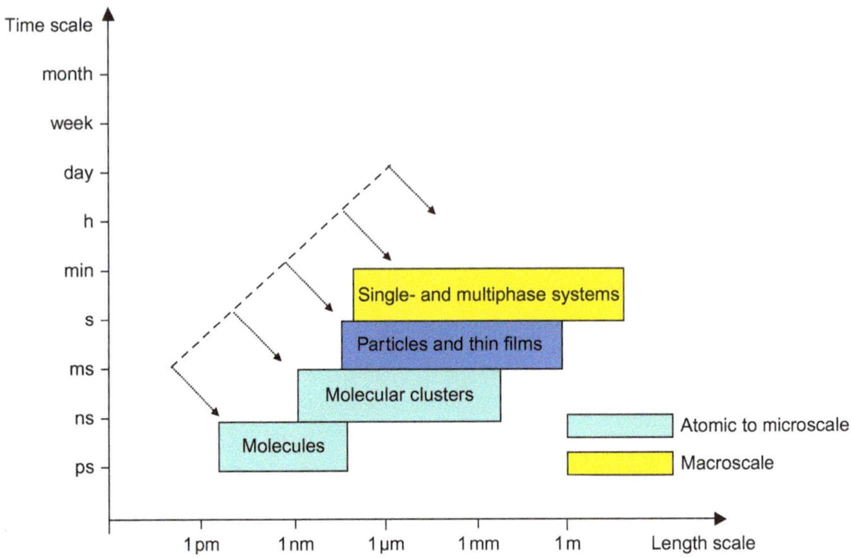


**Fig. 15** Multiscale events in reactor design.

of the beam that has a shape of a fan, reaching a single view for a pattern of detectors. Each tomographic measurement uses one or multiple attenuated beams, which pass through the object, at given spatial and angular orientations and constitute one or multiple projections of the object. Multiple projections are then gathered for the reconstruction method which applies to projections and contains unknown space-dependent image function. Fourier methods such as Radon transform and algebraic reconstruction are the major methods of image reconstruction techniques. The first one is often used in applications in which sufficient projections are available, while algebraic methods are used for reconstructions of limited projections. The latter methods, however, require computational power since they rely on iterative methods and careful filtering of the collected data to produce high-quality reconstructions with excellent computational efficiency. By arranging and accessing parallel projections in a multilevel scheme, the algebraic reconstruction techniques can generate high-quality reconstructions with improved computational efficiency. Adaptive algebraic reconstruction techniques, which introduce adaptive adjustment of the relaxation parameters during the reconstruction procedure, achieve reduced iterations, improved computational efficiency, and thus a quicker convergence.

Tomography techniques for industrial processes can be broken into two main categories based on the source of creating the tomograph:

(1) Nuclear-based imaging techniques: $\gamma$-ray and X-ray transmission tomography, positron emission tomography, X-ray diffraction microtomography, neutron transmission radiography, and nuclear magnetic resonance imaging (MRI) and (2) Non-nuclear-based imaging techniques: tomographic particle imaging velocimetry, electrical capacitance tomography, optical tomography, ultrasonic tomography, microwave tomography, tomographic emission spectroscopy, tomographic deflectometry and tomographic interferometry. Many of these techniques are receiving applications to industrial processes where information on flow and concentration of species are made accessible without the need to invade the process. This information helped industries in monitoring the mixing of multiphase processes, verifying and improving the pipeline conveyors, and processing vessel operations (Pangrle et al., 1992). Among these, MRI was used to measure anisotropic dispersion maps in multiphase reactors and access to scalar data of chemical kinetics and transport in a single experiment. Images of heterogeneities in the porous structure over different scales helped understanding the influence of spatial distribution in pore structure on the mass transfer driven processes, taking place inside the porous medium. A range of diffusivities at different space-scales, which was relevant to the spatial variation of pore size, determined the transport processes occurring within the pore space of commercial porous catalyst pellets.

## 3.2 Transmittance Optical Tomography Through Nonopaque Media

Tomography-based techniques have been successful to investigate processes that hold liquid species for measurements of fluid flow and species distributions, while their number is limited. Planar laser-induced fluorescence-based-tomography involved the replacement of part of the container wall with transparent glass windows, low-noise CCD array, and optical accessories to direct the pulsed laser to the process and fluorescence to the CCD array as shown in Fig. 16. A model of one component along with fluorescent dopant is typically used and data can be retrieved at the repetition rate of the pulsed laser (Reinke et al., 2005). However, planar imaging suffers from the need for spatially continuous optical access over large solid angles in both the excitation and detection paths, limiting its general use in several applications.

Optical tomography is, however, less demanding in terms of optical access because it collects data from sparse volumes. The technique is showing a large interest due to progress in laser technology, spectroscopy, and theoretical developments of nonlinear algebraic reconstructions. The recent advancements in broad bandwidth, frequency-agile lasers, and tunable lasers are

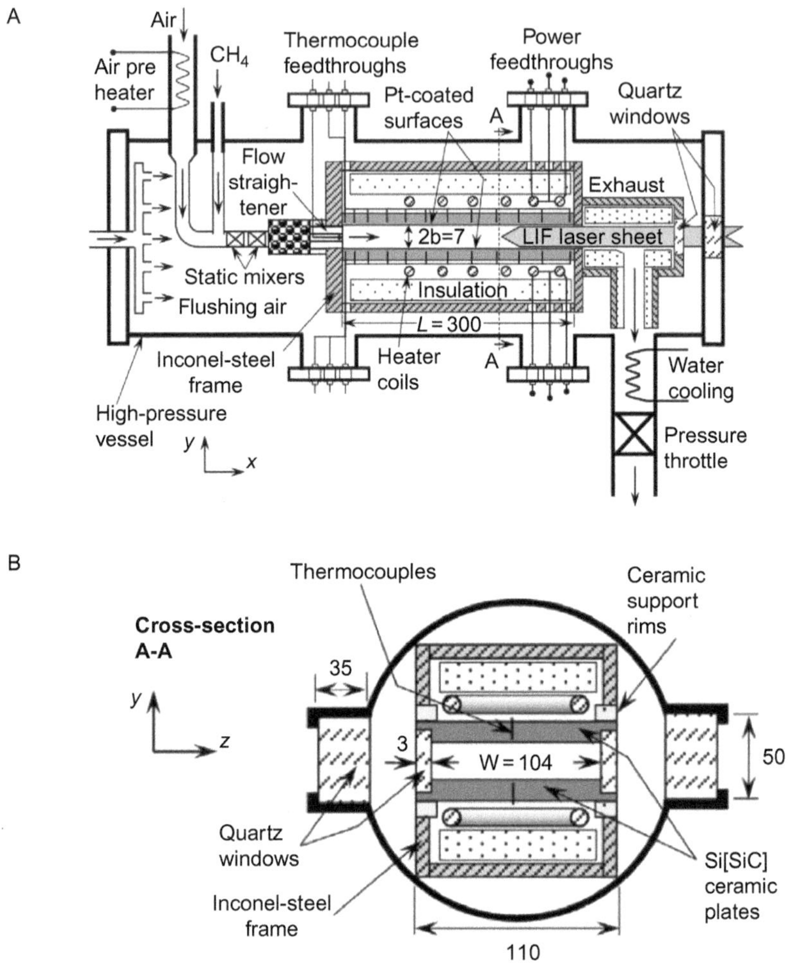


**Fig. 16** Schematic of the high-pressure test rig and the catalytic reactor assembly (Reinke et al., 2005). *Reprinted with permission from Elsevier.*

offering massive information on spectroscopic data from tomography. Increased use of multispectral imaging is reducing impact of noise and allowing access to simultaneous data on species, temperature, and pressure by a single run. As semiconductor technology advances, covering the spectral range of the mid-IR, which is relevant to strong absorption, the peak absorption, shorter optical paths are anticipated. Optical absorption tomography based on NIR spectroscopy has been recurrently used for the measurement of gaseous hydrocarbon distributions. The measurement of distribution of gaseous species is of importance for applications such as pollution emissions monitoring and fuel efficiency in combustion research. The spatial distribution of species, and particularly air–fuel ratio, is a critical determinant of performance of combustion chamber of an internal combustion engine. NIR tomography by using multiple fibers was used to measure distribution of hydrocarbons in a cylinder under transient gas injections as shown in Fig. 17. The light was centered to hydrocarbon absorption wavelength of 1.7 μm, and the measurement domain space was sampled by single- or dual-wavelength fiber-coupled measurement paths. Path integral data of concentration were measured with limited signal-to-noise ratio and spatiotemporal resolution (Hindle et al., 2001).

## 3.3 Diffused Transmittance Optical Tomography Through Semiopaque Media

The behavior of light in nontransparent media is known from the study of wave propagation in turbid media. Light traveling in a nontransparent

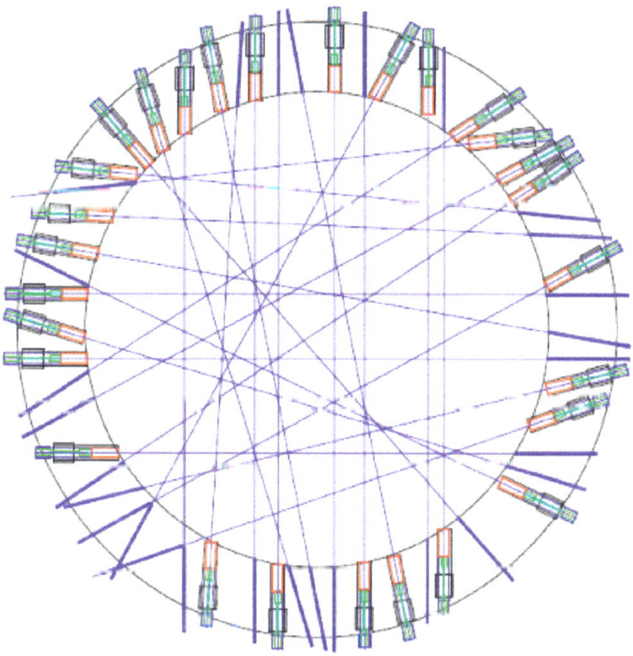

**Fig. 17** Multichannel system NIR fiber-emitters and sensors (Hindle et al., 2001). *Reprinted with permission from Elsevier.*

medium can be broadly classified into three categories: the ballistic light, the quasi-ballistic light, and the diffuse light. The ballistic light is that which either remains nonscattered or undergoes coherent forward scattering in the medium and propagates through the shortest path through the medium. The quasi-ballistic light or *snake photons* is that which undergoes near-forward scattering and follows a path which undulates around the ballistic path. The diffuse light is that which undergoes multiple scattering and totally loses the memory of their initial direction of propagation as shown in Fig. 18 (Yoo & Alfano, 1990). Depending on the turbidity of the medium, the diffuse photons often constitute the largest part of a propagating light. The study of ballistic imaging has largely been concentrated on biomedical applications as noninvasive diagnostic or treatment tool of a number of diseases. During the past two decades, investigations of laser light propagation in skin tissues and the human brain have been particularly extensive (Lee et al., 2005). Process diffuse tomography was used in combustion engineering, and a variety of laser-based diagnostics have been developed over more than three decades in order to determine the physical properties of fuel droplets from atomizing sprays (Yoo & Alfano, 1990). The requirements for the bioimaging systems and those of process monitoring vary mostly through temporal resolution. When imaging a static sample, such as those required in biological tissue sampling, there is no issue of temporal resolution, and the images can be averaged to improve quality. However, in a chemical system, such as the case of monitoring fuel droplets, the temporal resolution is of utmost importance, leading any single image being of high fidelity.

Since ballistic photons pass straight through a nontransparent medium without scattering, they exit within roughly the same solid angle that they entered as shown in Fig. 18. Typically, they constitute a small fraction of the light. Optical imaging inside a nontransparent medium has proven difficult

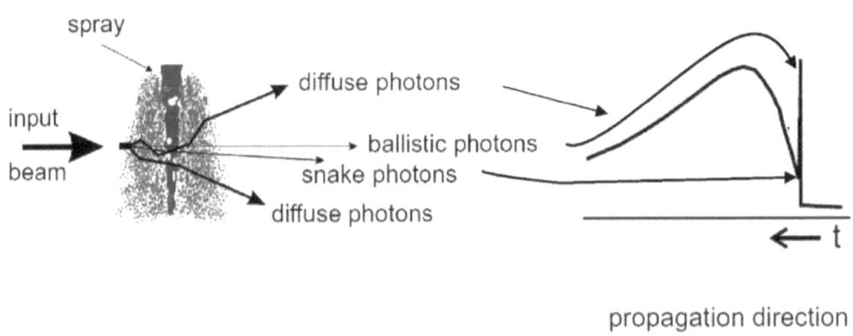


**Fig. 18** Schematic of ballistic, snake, and diffuse photons time dependence.

because most of the photons that would construct such an image are lost by scattering, owing to the presence of the dense field of particles. The attenuation of light that includes both absorption and scattering can be described by the modified Beer–Lambert law.

$$\frac{I_{\text{exit}}}{I_{\text{inlet}}} = e^{-\text{OD}} \tag{2}$$

$$\text{OD} = \varepsilon(T, \lambda) \int C dl \tag{3}$$

where $I_{\text{exit}}$ and $I_{\text{inlet}}$ are light intensities at the exit and the inlet of the packed bed, respectively, $l$ is the optical path length, $C$ is the uneven composition of water vapor across the packing, and $\varepsilon$ is the absorption coefficient. The OD is the optical depth and expresses the number of changes in direction of the photons when it propagates through the packing. The imaging is, therefore, constructed by a combination of ballistic and snake photons (less than one scattering, OD $< 1$) and scattered photons (multiple scattering, OD $> 1$) that pass straight, roughly straight, scattered, or diffusively through the medium (Yoo & Alfano, 1990). For instance, imaging a depth of 1 cm of human tissue would reach an OD of 11, and for sprays, the optical depth is actually a complex function of droplet size, density, refractive index, and the optical wavelength, but most of the conventional techniques operate in the single scattering regime of OD ranging from 1 to 10.

The major limitation of the current existing laser techniques for nonopaque media is commonly related to attenuation and multiple scattering of the source radiation. Under optically thick conditions, such phenomena can introduce substantial errors in the measurements of size and concentration. The magnitude of error varies with position, in a manner dependent on the source, detector, and medium geometries. Imaging scattering media requires therefore the separation of the ballistic and quasi-ballistic light from the scattered light which could be considered as a source of noise. Gating techniques that include time-resolved techniques and heterodyne detection utilizing broad-bandwidth light are efficient to achieve this goal, allowing shot-noise limited imaging in strongly scattering samples with a resolution equivalent to what can be achieved in the presence of weak scattering (Yoo & Alfano, 1990). These techniques, however, require bright light sources of high spatial coherence systems, while the continuous wave lasers remain a relevant task that can be broken into a number of techniques including confocal spatial filtering, Fourier spatial filtering, polarization modulation, and heterodyne detection with narrow-bandwidth light (Kempe et al., 1996; Lee et al., 2005; Yoo & Alfano, 1990).

The transmission of a monochromatic light through porous media consisting of packed glass beads immersed in water was investigated by Yoo and Alfano (Yoo & Alfano, 1990). The light propagation through such strongly scattering random media was characterized by the transport mean-free path and the diffusive absorption length which were determined by combining measurements of total transmission in the strong absorption regime and reference measurements on samples with known mean-free path. This procedure inspired further works on using NIR imaging as tool to observe through weak scattering media such as packed bed adsorbers and reactors and have access to 3D distribution maps of species and temperature under operando conditions.

### 3.3.1 Application of Diffused Transmittance NIR Tomography to Packed bed Adsorbers

Packed beds of high loads and considerable heat releases are commonly arranged by means of multiple tubes of a relatively narrow size. Then, channeling effects occur and consideration of temperature and concentration profiles only with respect to the axial coordinate of the bed or to time is not sufficient; radial profiles across the tube must also be accounted for. Lingg (1995) used measurement probes in the column to sample the flowing gases, and consequently interference with the flow field could not be excluded. Probes at selected positions along axial and radial directions of the packed bed and covering a quarter of the cross-section were used. The authors reported mean concentrations for the wall region and the center of the bed. The technique was then intrusive and affected flow, mass, and heat transfer and was not adequate for transient flow, taking place in adsorbers. The lack of highly temporal and spatial resolutions for concentration measurement data inside adsorption columns led to needs for superior measurement techniques. An alternative way to acquire these data non-intrusively is by means of optical methods. Techniques for fluid concentration measurement by laser absorption spectroscopy yield information along a line of sight, limiting mean concentrations to stand out with the resolution, which is known in the quantitative spectroscopy (Talsky, 1994). Laser absorption spectroscopy was extended by Mewes' group (Salem et al., 2005) to NIR tomography and nonobtrusively observed spatial breakthrough of water vapor at the outlet of a packed bed adsorber. This was realized by prolonging an adsorption column with a hollow glass cylinder that served as area for cross-sectional measurements. This technique relied on an NIR light from a distributed feedback diode laser that was split into three lights as projection sources and was collimated to transilluminate a packed bed adsorber as shown in Fig. 19A. The light absorption by water vapor was measured with three InGaAs photodiode arrays, and the projections between the light sources and the detectors were used for tomographic reconstruction of concentration fields. The wall effect on adsorption breakthrough was demonstrated by tomographic measurements of local water vapor concentrations at exit of the packed bed adsorber of low AR.

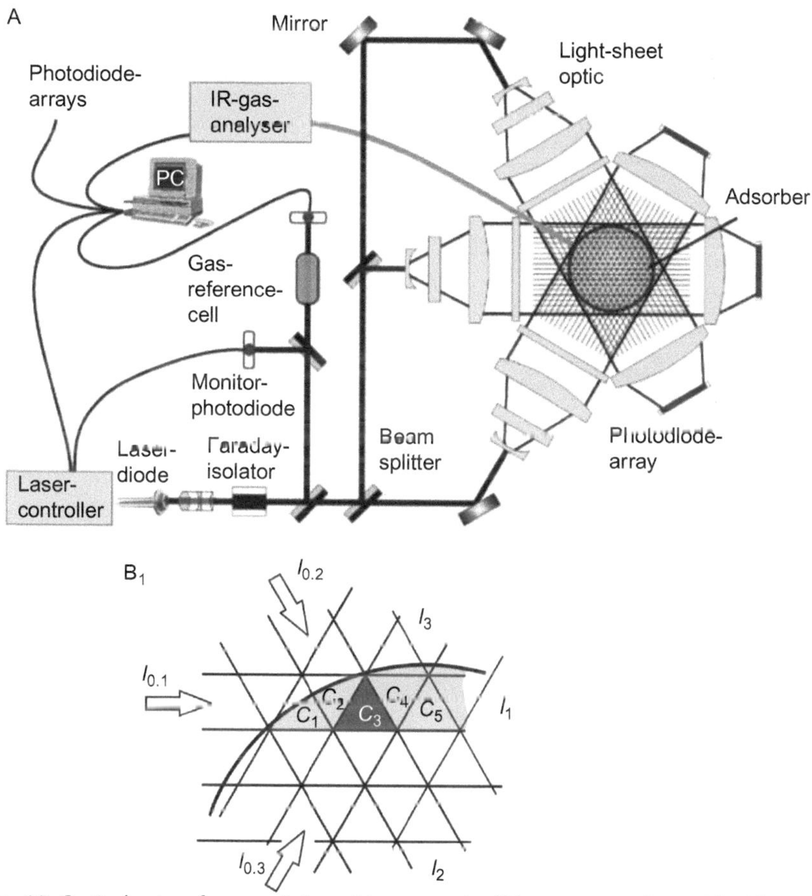


**Fig. 19** Optical setup for near-infrared tomography (A), reconstruction grid with triangular elements ($B_1$ and $B_2$), reconstructed concentration fields during adsorption process (C) (Salem et al., 2005). *Reprinted with permission from Elsevier.*

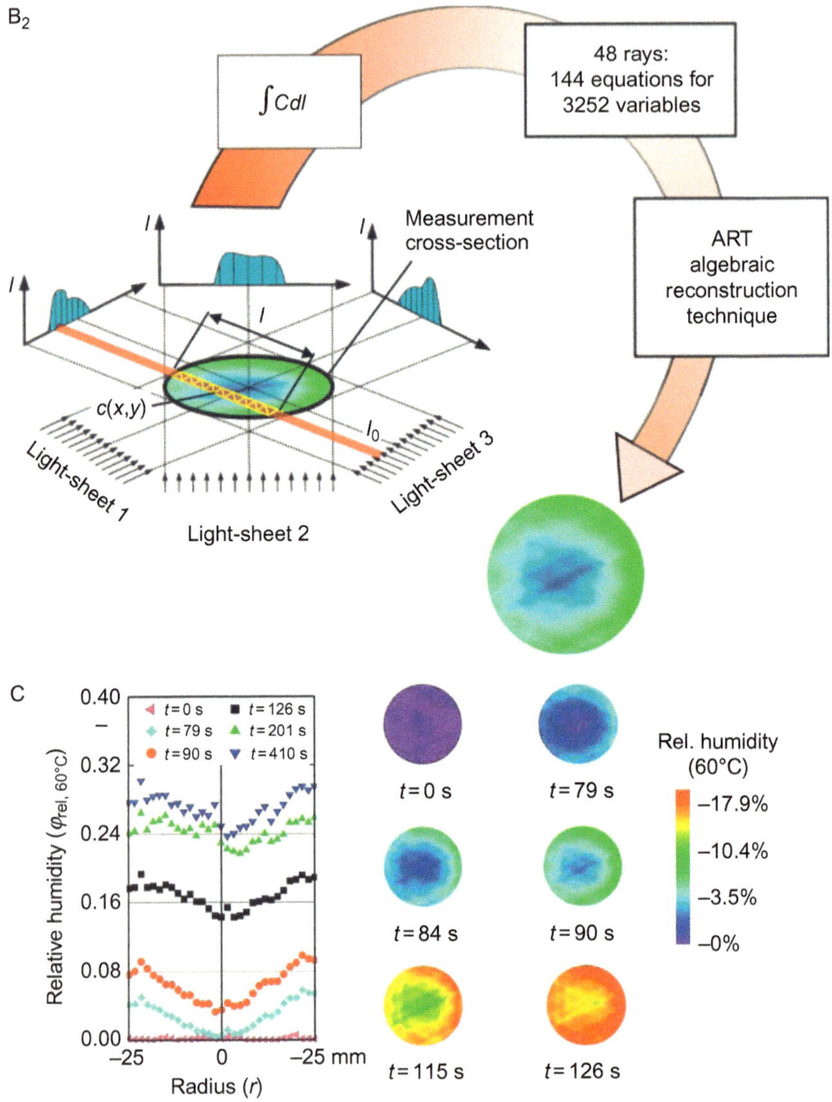


**Fig. 19—Cont'd**

Tomographic reconstruction was used to reconstruct the 2D concentration fields from the measured integral concentration along the ray paths. The procedure of three projections and division of the measurement cross-section into triangular elements, as depicted in Fig. 19B reduced the weighting factors of the algebraic reconstruction, achieving a high speed at convenient convergence. For low water vapor concentrations an error below 8% and a spatial resolution of 2% related to the measurement cross-section were reported. Concentration field measurements during adsorption showed early stage breakthroughs near the column wall due to channeling effects. Some distortion of the crossing light through the cylindrical shape of the adsorber was normalized by comparing the actual ray patterns with the theoretical triangular grids. The reconstructed data in terms of radial profiles and cross-sections are shown in Fig. 19C and indicate disparity in species distribution in the axial direction; however, the measurements were conducted at the exit of the packed bed and therefore did not allow observation of flow along the height of the packed bed. Unfortunately, while the technique did allow access to qualitative values of concentrations at exit of the packed bed, the spatial resolution was limited to three parallel projections, leading experimental parameters related to local mixing such as the dispersion properties not being retrieved. This technique utilized end-pipe analysis of local variations, while those occurring inside the packed bed were missed. It did, however, reiterate importance of considering channeling and uneven distributions inside adsorbers of low AR. A 3D tomography technique was further developed to facilitate this and without affecting the conditions inside the packing.

NIR diffused transmittance tomography was developed by Aiouache et al. (tSaoir et al., 2011) for a semiopaque packed bed adsorber in order to have access to maps of concentration and temperature subject to effects of wall in low AR packed beds and nonisothermal conditions. The 2D projected images from parallel scanning permitted cross-sectional data to be reconstructed from those retrieved from the packing and above the packing sections at spatial resolution of submillimeters. The design could be extended to applications of control of deuterium and by inference tritium in hydrogen isotopic exchange (HIE) process for waste water detritiation and to local deactivation of catalysts in gas–solid catalytic systems. Unlike previous design of optical tomography technique which was adapted to the exit of the packed beds, the diffused transmittance tomography was used to capture cross-sectional information from the interior of the packed bed, and the nonlinearity effects were reduced by a dedicated experimental design and image reconstruction procedure. It was applied to packed beds of low AR which exhibit wall effects on the flow due to variable porosity along radial coordinates. These beds develop regions of high porosities in the vicinity of a wall compared to the core of the bed, causing nonhomogenous flow and maldistribution. The design used a tunable diode laser, an FPA detector, and tomographic reconstruction. A hydrophobic silica gel of reduced light scattering effect was used as packing and adsorbing material. The effects of flow dynamics on composition and temperature fields were discussed.

#### 3.3.1.1 Experimental Setup and Validation

The experimental setup is shown in Fig. 20. The detector chosen was the MOSIR950™ NIR CCD camera which is a NIR image intensified spectroscopy camera of spectral sensitivity ranging from 950 and 1650 nm. It does not require an external controller, holds reduced dark change with a proprietary thermoelectric cooling and vacuum design, and is specifically designed for low light and imaging applications such as the ballistic and semiballistic light that transilluminates weakly opaque media. The combination of high quantum efficiency and low noise gain provided superior sensitivity than conventional CCD or InGaAs arrays as shown in Fig. 21. The MOSIR FPA has a 4:1 AR spectroscopy format (1024 × 256 pixels of 26 × 26-μm pixel pitch) and 16-bit dynamic range. Dark charge was minimized with a proprietary thermoelectric cooling and vacuum design. The camera was orientated on its side to allow the large 1024 pixels to capture the height of the narrow packed bed. The entire experimental setup was encased within a black sealed Perspex box, and dry nitrogen was used to purge the setup and remove any water vapor from the surrounding environment. For added humidity control silica gel was also placed within the rig and could act as an indicator along with the humidity reader of the presence of a leak either within the experimental setup or in the Perspex box itself.

A tunable compact NIR laser (TSL-510, Santec, Japan) of a full 120 nm mode-hop-free tuning range, ranging from 1340 to 1440 nm wavelength was used. The laser has a 100-nm/s rapid wavelength sweep function, which is particularly suitable for fiber optic component testing of transient systems. The laser was connected to a collimator via a monomode fiber optic cable which allowed for a flexible optical setup of light size and direction. Having concluded that the most effective method for this technique would be the alignment of the camera and laser in a parallel fashion, and the correct optics needed to be put in place to facilitate this. A fiber optic cable of 1 m along with a collimator were used to bring the NIR light from the laser to the desired optical path of selected size and shape. Two parallel polarizers placed before and after the packed bed were used to promote proportion of ballistic and pseudoballistic lights being captured by the detector. Cylindrical lenses were also used to expand and collimate the light, creating a light sheet of

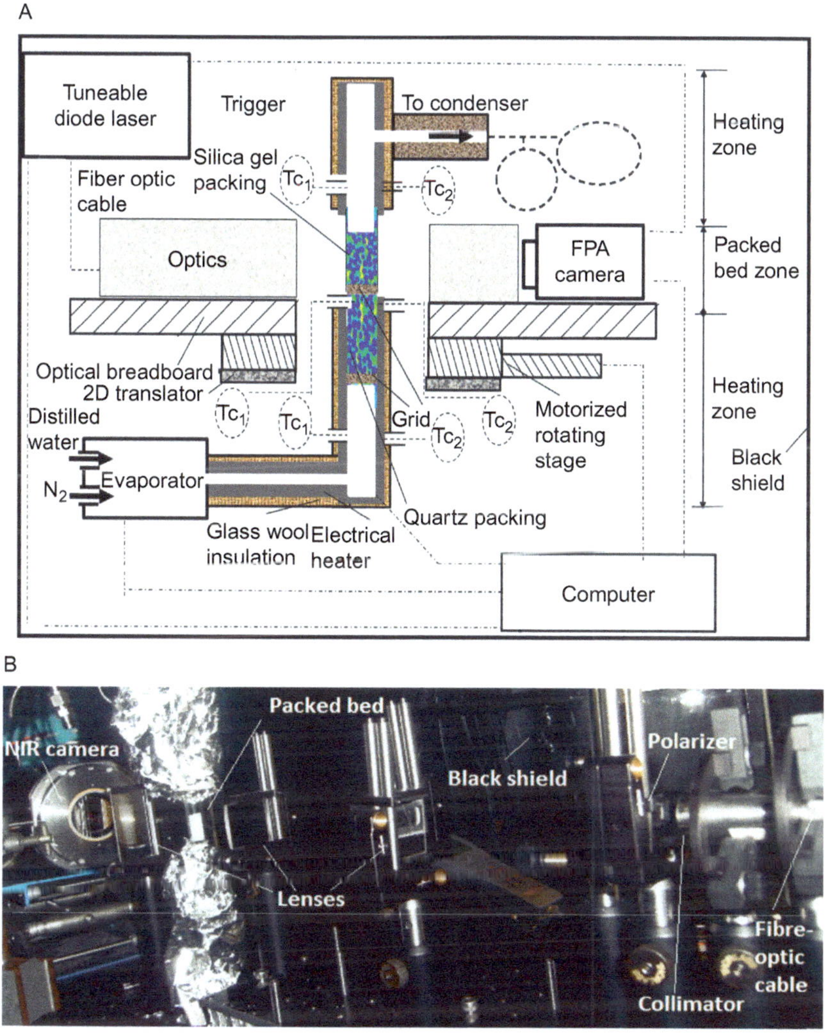


**Fig. 20** Apparatus scheme (A) and photograph of the setup (B): evaporator (Bronkhorst) – mass flow controller ($N_2$), air actuated switching valve, distilled water bath; *H*: humidity sensor; $TC_1$: thermocouples (monitoring); $TC_2$: thermocouples connected to programmable temperature controllers; optics, details of optics are shown in Fig. 3;tunable diode laser: lock-in monomode connected to FPA camera; quartz packing: prepacked bed mixer; silica gel packed bed = height: 14 cm; visible aperture by NIR camera: $1.20 \times 1.82\,cm^2$ (tSaoir et al., 2011). *Reprinted with permission from Elsevier.*

selected size with which the packed bed could be transilluminated. Water vapor was generated by a controlled evaporator (Bronkhorst).

The scattering effect on the transmitted light through the packing required adapted absorption coefficients for the packed bed and the vapor

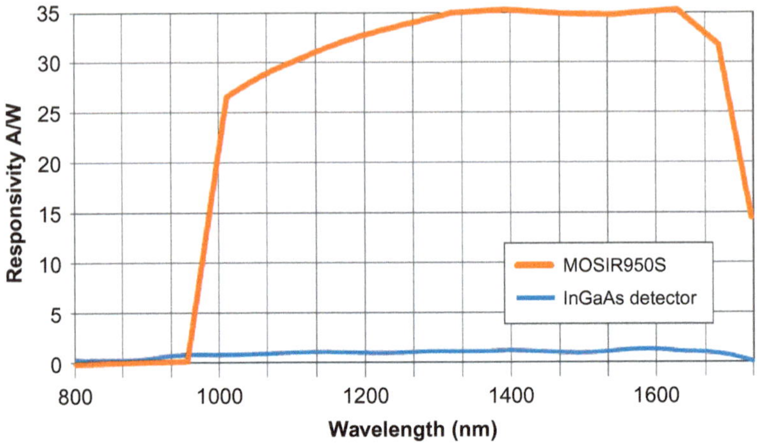


**Fig. 21** Comparative responsivity curves for MOSIR 950 and InGaAs arrays.

phase in the area above the packed bed according to Eqs. (2) and (3). A squared quartz cell of 1 cm I.D., which had been half-filled with the silica, was used. The coefficients were measured at 1380.685 nm across temperatures ranging from 323 to 363 K and fitted to a polynomial model as a function of temperature. The 1380.685 nm wavelength was chosen due to its suitability for the two-line thermometry. In the packing medium, the average optical depth without water vapor was about 0.3, leading to a scattering number of less than unity and to a confirmation of straight propagation of the NIR light with a negligible scattering through the packing. It should be noted that when water vapor was added, the averaged OD increased by 30%.

Spatial profiles of temperature were measured by two-line thermometry technique. The ratio of integral absorbance $Abs_1$ and $Abs_2$ of line strength for two temperature-dependent transitions is given in Eq. (4) (Zhou et al., 2003).

$$\frac{Abs_1(T,\lambda_1)}{Abs_2(T,\lambda_2)}=\frac{S_1(T_0)}{S_2(T_0)}\exp\left[-\frac{hc(E_1-E_2)}{k}\left(\frac{1}{T}-\frac{1}{T_0}\right)\right]\frac{\left[1-\exp\left(-\frac{hc}{kT\lambda_1}\right)\right]}{\left[1-\exp\left(-\frac{hc}{kT\lambda_2}\right)\right]} \tag{4}$$

where $S_i(T_0, \lambda_i)$ is the line strength of the transition centered at line wavelength $\lambda_i$ for a reference temperature $T_0$, often taken at 296 K; $h$ is Planck's constant; $c$ is the speed of light; $k$ is Boltzmann's constant; $E_i$ is the lower state energy; and $T$ is the gas temperature (K). The selection of line pairs was very carefully carried out. The widespread and strong absorption transitions provided many options for measuring $H_2O$ and temperature. The selection was carried out based on criteria set by Zhou et al. (2003) who recommended lines to be strong, not overlapped and to have dissimilar temperature dependence, a search for suitable candidate pair of lines in the $\nu_1 + \nu_3$ combination band, using Hitran database, particularly in the wavelength range of 1340–1440 nm allowed by the tunable diode laser used (Rothman et al., 2009) yielded a pair of lines at 1358.412 and 1380.685 nm. These lines were found to be the good fit in terms of temperature sensitivity, since they offered a difference in lower state energy ($E_2 - E_1$) of 342 $cm^{-1}$, line strength ratio at reference temperature ($S_1(T_0)/S_2(T_0)$) of 1.19 and were close but not overlapping each other.

The adaptative algebraic reconstruction technique (AART) was used for image reconstruction for simultaneous measurements of temperature and concentration as illustrated in Fig. 22. It relied on a discretized domain of an object function, $f(x,y,z)$ and offered flexibility for a limited number of projections, compromising between convergence rate and accuracy. The object function represented an array of discrete unknown voxels, leading to a parallelepiped grid of dimension $m \times m \times N$. Within each voxel, the function $f(x,y,z)$ was assumed to be constant with a value of the voxel, located at $k$th and $j$th pixels, designated $f_{kj}$. For the $i$th hyperplan ray, $p_i$ was the measured value of experimental integral absorbance for concentration

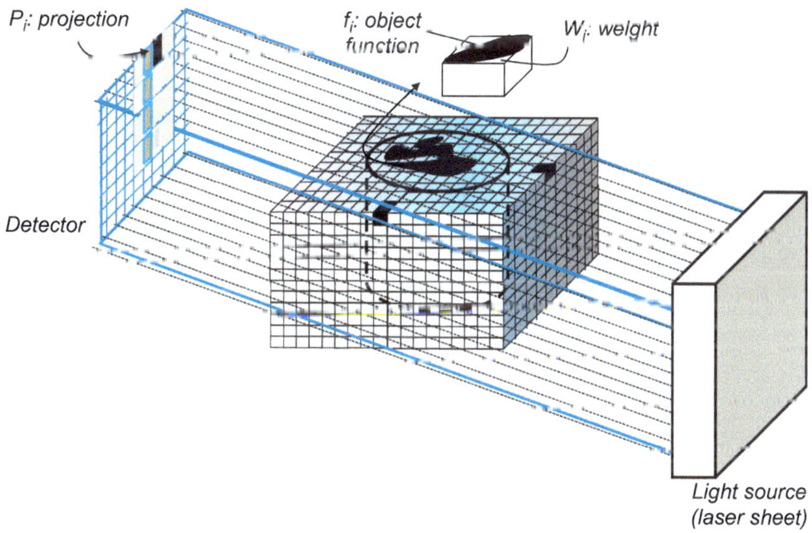


**Fig. 22** Image reconstruction using parallel beam and voxel division (tSaoir et al., 2011). *Reprinted with permission from Elsevier.*

measurements (Abs) or integral absorbance ratio for temperature measurements along the path as expressed by Eqs. (2) and (4), respectively. The algorithm was initiated by giving an arbitrary guess for the voxel vector, projecting onto the first hyperplan equation, and the result was projected onto the next hyperplan and so on until the last one as shown in Eq. (5) and Fig. 23.

$$\sum_{j=1}^{M} w_{ij} f_i = p_i \quad i = 1, 2, \ldots N \tag{5}$$

where $w_{ij}$ is the weighting factor, representing absorbance (i.e., concentration) or absorbance ratio (i.e., temperature) of the contribution of the $j$th voxel to the $i$th measurement, $N$ is the total number of projection rays, and $M$ is the total number of voxels. The set of algebraic equations was solved by the iterative Kaczmarz method expressed in Eq. (6):

$$f_j^l = f_j^{l-1} - \lambda^{ij,l}\left(\sum_{j=1}^{N} w_{ij} f_j^{l-1} - p_i\right) \tag{6}$$

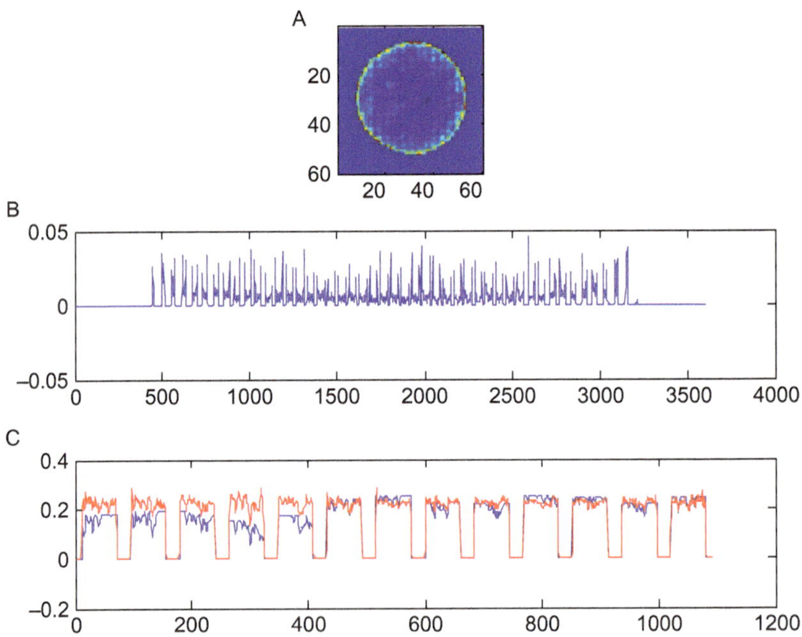


**Fig. 23** Adaptive algebraic reconstruction for 13 projections for a single hyperplan (Matlab® representation) (A) cross-sectional area of a hyperplan, (B) deviation error between assumed and actual projected results for a single projection, and (C) deviation profiles for simultaneous 13 projections.

where $\lambda^{i,1}$ is a relaxation parameter which is defined as follows:

$$\lambda^{ij,l} = \frac{f_j^{l-1} \bullet w_{ij}}{\sum_{j=1}^{N} w_{ij} \bullet \left(f_j^{l-1} \bullet w_{ij}\right)} \tag{7}$$

Herein $f^{k-1} \cdot w_i$ is the dot product, and $k$ is the iteration index. Reconstruction of cross-sectional images or slice matrixes was repeated for each subsequent cross-sectional matrix until 3D integral absorbance ratios for temperature reconstructions and integral absorbance, Abs, for concentration reconstructions were obtained for the packing region and above packing region of the packed cylinder, and were further converted to temperature and concentration using Eqs. (2) and (4), respectively.

In the interests of time resolution, it is desirable to minimize the number of projections yet to maximize the spatial efficiency and accuracy. To find the desired balance, preliminary phantom tests were carried out using a metal tube inserted into an empty quartz tube and 3D reconstructed images of the metal tube were generated. As shown in Fig. 24, the results of six projections clearly show that this is an insufficient number as the cylindrical shape of the pipe cannot be made out. Ten projections were shown to be too few; however, 20 projections were shown to be too time consuming. Thirteen projections,

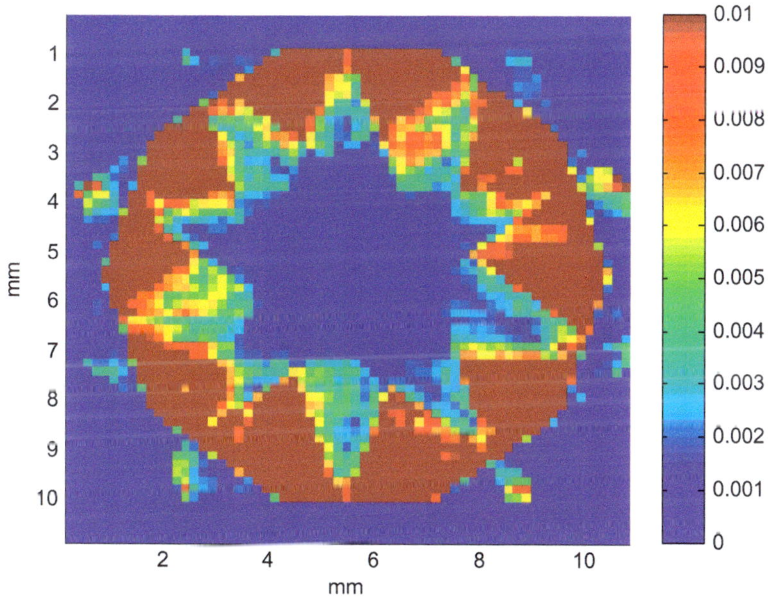


**Fig. 24** Cross-section of the six projection reconstruction of the inner metal pipe (5 mm I.D.).

however, produced accurate reconstruction without an excessive loss of time resolution. The maximum size the voxel array that could be trans'illuminated by the NIR beam was $64 \times 64 \times 700$ voxels. Thirteen projections were found sufficient to run the AART method with volume resolution of $0.150 \times 0.150 \times 0.026\,mm^3$. Projections were taken at fixed angle steps of 15 degrees and 1 cycle of 13 projections lasted approximately 3 min.

The dispersion of water vapor in the hydrophobic silica gel (Davicat® SI 1102 average pore diameter 14.4 nm, pore volume $1.12\,cm^3\,g^{-1}$, BET surface area $331\,m^2\,g^{-1}$), which has been treated with a solution of dimethyldichlorosilane to form a silicone layer around the silica and thus reduced its high hydrophobicity, was carried out by feeding water vapor through a thin tube of 5-mm I.D. The thin tube was inserted at the bottom section of the packed bed as shown in Fig. 25. The bed was fed with $N_2$ by a lateral opening, while the mixture water vapor/$N_2$ (6 mol%) was introduced from the inner tube. A configuration of 13 projections was used to observe water vapor exiting the nozzle as it flowed up through the packing and providing central core regions with higher concentrated water. The accuracy in terms of composition and temperature of the water vapor distribution generated was investigated by varying the water vapor composition from 1 to 6 mol% and temperature from 323 to 353 K.

The original image is presented in Fig. 25, while the reconstruction images of the cylinder before the introduction of water vapor are shown in Fig. 26. The 3D reconstruction of the flow and the temperature, before and after

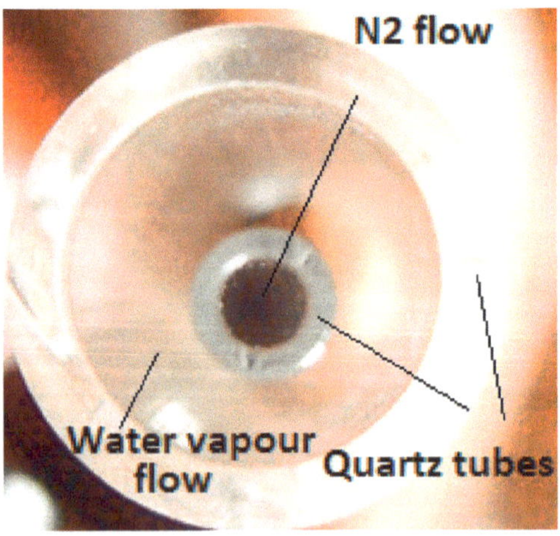


**Fig. 25** Quartz cell for validation tests. Inner tube diameter: 5 mm, quartz cell diameter: 9 mm, thickness of both tubes: 0.5 mm (tSaoir et al., 2011). *Reprinted with permission from Elsevier.*

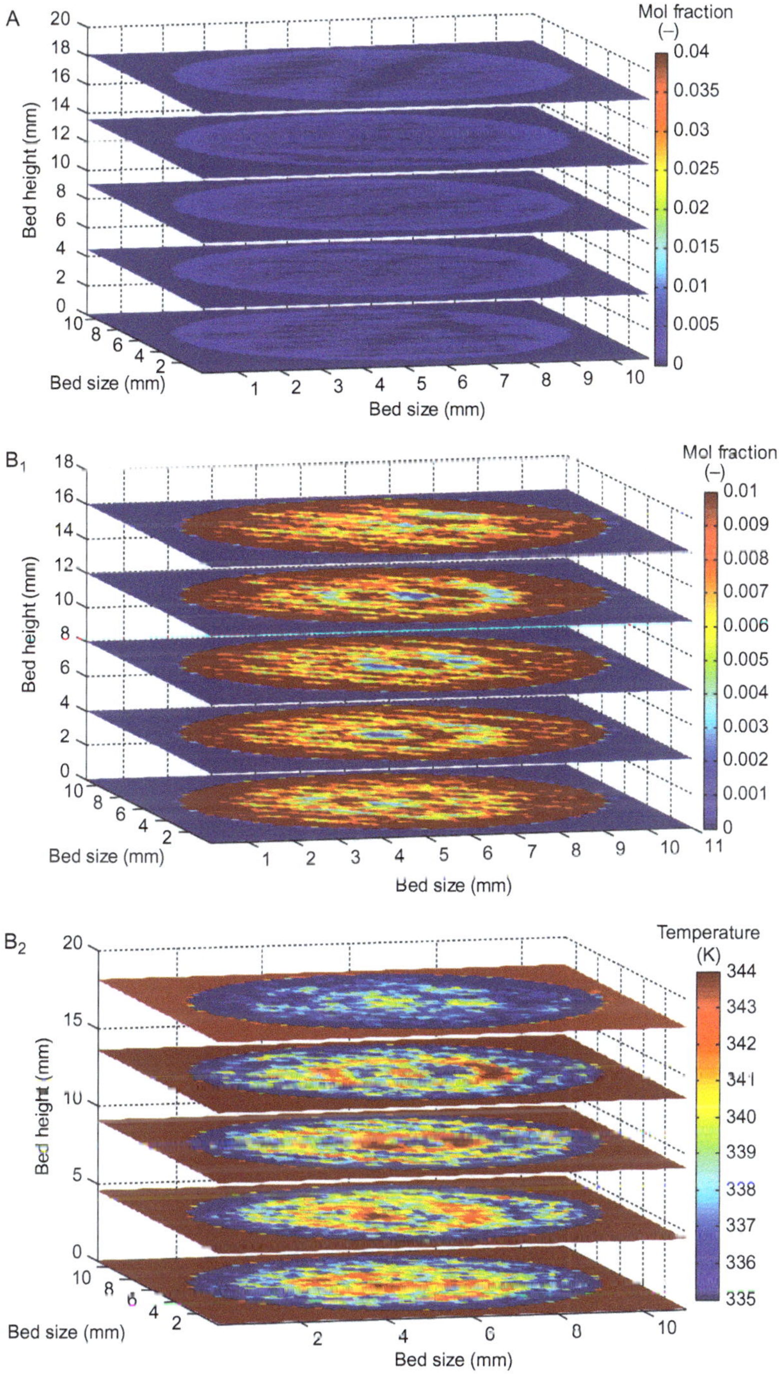


**Fig. 26** 3D reconstruction of cross-sectional composition maps of water vapor (outer tube) and $N_2$ (inner tube) and relevant temperature (A) before water vapor introduction, ($B_1$) and ($B_2$) after water vapor introduction, composition of 1.0 mol% and temperature of 333 K (tSaoir et al., 2011). *Reprinted with permission from Elsevier.*

the introduction of water vapor at a composition of 1.0 mol% and temperature of 333 K, is shown in Fig. 26A, $B_1$, and $B_2$, respectively. For each projection of 64 rays, one can clearly see peaks corresponding to rays of high and low absorbance values of water vapor. There is a noisy region in the decreasing regions of absorbance, particularly with projection number 4, 6, and 11 and noiseless regions with rays at the extremities of projection rays. This feature was not applied to all projections, but the influence of the noise made the distinction of the two separate region peaks more difficult. The noisy area is well displayed in the vicinity of inner wall. Both sides of the inner wall, inner and outer regions show uneven distributions of water vapor while regions far the inner wall show homogeneous distributions. This figure clearly discriminates water vapor flow at concentrations of 6 mol% in the outer flow region from the no water vapor flow region in the inner tube, and the noisy region was likely due to the effect of deflection of the tube. The deflection of NIR light due to inner tube deflection was suppressed by carrying out an experiment similar to procedures used to measure dispersion coefficients in porous media.

#### 3.3.1.2 Flow Visualization and Temperature Measurement in a Packed Bed Adsorber of Low AR

Reconstruction of cross-section profiles of water vapor and temperature was performed for a packed bed adsorber of AR of 13, i.e., particle size of 0.7 mm, inlet and wall temperatures of 333 K, and flow rate of $338\,cm^3\,min^{-1}$, to examine transient profiles of relevant distributions inside the packing and above the packing of the packed bed. Temperature maps after 60 min of adsorption of five selected cross-sectional layers located at equal distances from the bottom of the packing to the top of the vapor phase above the packing are shown in Fig. 27. The four bottom layers belong to the packing section and the top layer to the empty section above the packed bed. Hot and cold zones with changes in temperatures up to 10 K were observed, particularly those where convective heat transfer dominated in contact with the wall and in the center of the packing where heat transfer by conduction and convection occur, leading to efficient crossmixing of the water vapor flow. The effectiveness of wall cooling of the packed bed was assessed by observing along radial direction drops in temperature in the area near the wall, demonstrating inefficient heating from the wall. Hot zones of uniform flow velocity were not spreading sufficiently in the bed, demonstrating an inefficient energy and thus flow maldistribution along both radial and angular coordinates. The vicinity of the wall, as shown in Fig. 28, displays a vertical section temperature map through the center of

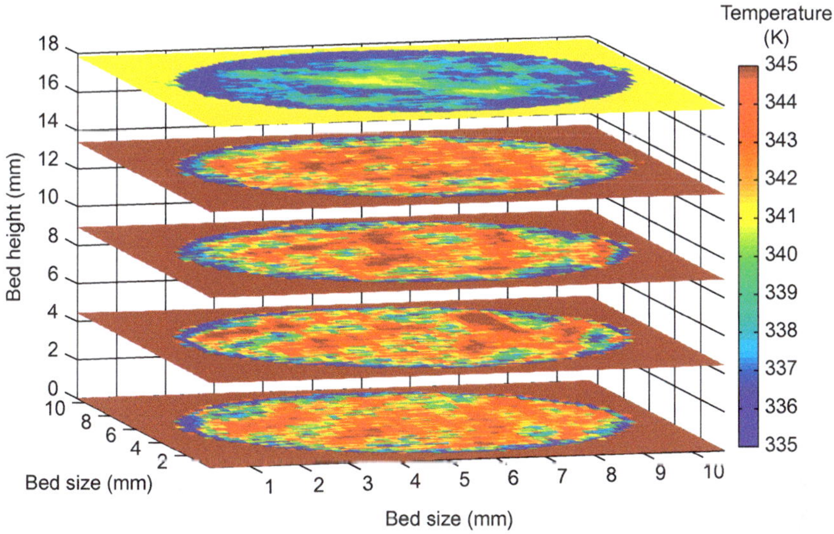


**Fig. 27** Transient temperature maps of five cross-sectional layers located at equal distance from the bottom of the packing to the top of the vapor phase above the packing, time: 60 min, packing height: 14 mm, particle size: 0.7 mm (AR: 13), wall temperature: 333 K, water vapor composition: 6.0 mol%, $N_2$ flow rate: 338 $cm^3\,min^{-1}$ (tSaoir et al., 2011). *Reprinted with permission from Elsevier.*

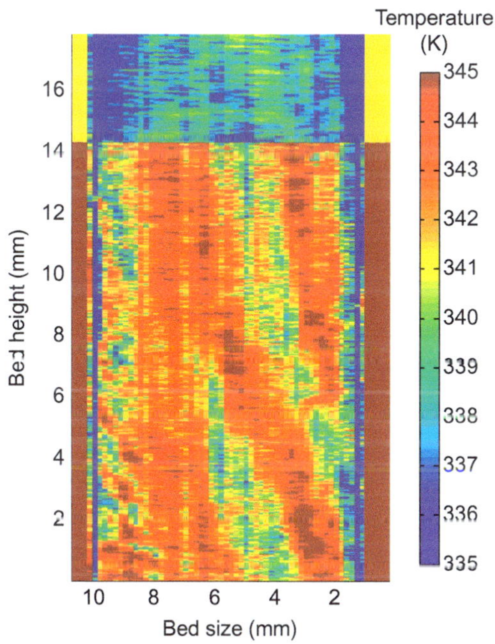


**Fig. 28** Vertical cross-section of transient temperature distribution through the center of the packed bed, time: 60 min, packing height: 14 mm, particle size: 0.7 mm (AR: 13), wall temperature: 333 K, water vapor composition: 6.0 mol%, $N_2$ flow rate: 338 $cm^3\,min^{-1}$ (tSaoir et al., 2011). *Reprinted with permission from Elsevier.*

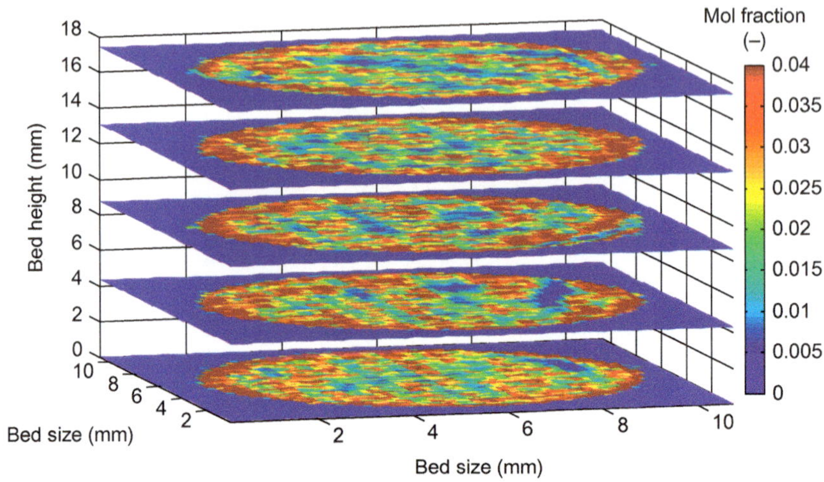


**Fig. 29** Transient water vapor composition maps of five cross-sectional layers located at equal distance from the bottom of the packing to the top of the vapor phase above the packing, time: 60 min, packing height: 14 mm, particle size: 0.7 mm (AR: 13), wall temperature: 333 K, water vapor composition: 6.0 mol%, $N_2$ flow rate: 338 $cm^3 min^{-1}$ (tSaoir et al., 2011). *Reprinted with permission from Elsevier.*

the bed of less uniform temperature distribution in the packed region than in the upper bed region. In addition, successive cold and hot zones in the axial direction of slow flow upward are observed, implying uneven local heat transfer, and thus local flow pattern in the packed bed. The composition profiles of water vapor along the packed bed after an adsorption period of 60 min are shown in Fig. 29. Similar to the trends of temperature, higher values of composition of water vapor are seen in the vicinity of the wall, demonstrating preferential channeling of local flow between the particles and the wall. This reduced mixing was more pronounced at the top of the bed where the flow was less affected by the entrance, and boundary layer near the wall progressed toward more developed regions of laminar flow as shown in Fig. 30. The flow maldistribution was demonstrated to be a complex function of macroscale flow in bed porous media, microscale flow inside the packing particles and adsorption process.

This study was the first to visualize gas flow in packed beds by using NIR optical tomography and measured distribution of composition and temperature in a weakly scattering environment. The design was limited to water vapor due to its high absorption coefficient in NIR. Ray propagation in the weakly scattering packed bed and deflection by the wall were considered by computing dedicated attenuation coefficients in the packing media. Flow

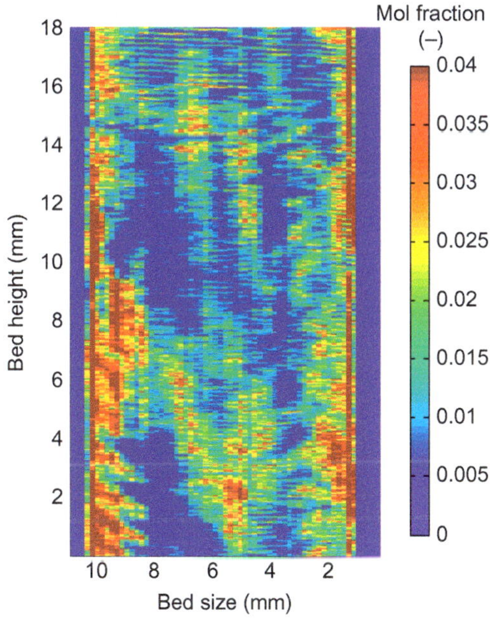


**Fig. 30** Vertical cross-section of transient water vapor composition map through the center of the packed bed, time: 60 min, packing height: 14 mm, particle size: 0.7 mm (AR: 13), wall temperature: 333 K, water vapor composition: 6.0 mol%, $N_2$ flow rate: 338 $cm^3 min^{-1}$ (tSaoir et al., 2011). *Reprinted with permission from Elsevier.*

maldistribution in a packed bed of low AR was imaged by observing temperature and composition maps. The packed bed of a limited size $12.0 \times 12.0 \times 18.2 mm^3$ was accessible at space resolution of submillimeters. The relative deviations in temperature and compositions were 11.1% and 9.3%, respectively. Owing to higher number of projections (i.e., 13 projections), the technique researched deviation in composition at the exit of the packed bed was slightly higher than the one obtained by Mewes' group (Salem et al., 2005) of three projections only.

### 3.3.2 *Spatial Distributions of Composition and Temperature in Gas–Solid Packed Beds Reactor by NIR Tomography*

A further extension of the NIR tomographic technique was the simultaneous observation of flow of reacted water vapor ($H_2O_v$) and produced heavy water vapor ($HDO_v$) during the HIE in a catalytic packed bed reactor. This reaction was selected owing to strong absorption bands of both species in NIR, simultaneous observation of reactants and products, and thermoneutrality, reducing impact of heat exchange on the spectroscopic data.

Owing to its high separation factor and near-ambient operating conditions, the HIE reaction is very promising for detritiation processing of nuclear waste waters (tSaoir et al., 2012). The silica gel of the adsorber of the previous section was replaced by platinum supported on hydrophobic styrene divinyl benzene copolymer, Pt/DSBC, resin to assess flow maldistribution. This reaction is sensitive to liquid water, and hydrophobic materials are preferred but flow channeling and condensed liquid water in liquid phase isotopic exchange reduces both catalyst and overall process efficiency of separation. In this work, the changes in compositions $H_2O_v$, $HDO_v$, and temperature during the reaction course of the vapor phase HIE reaction, between heavy water vapor ($D_2O_v$) and hydrogen gas, were examined by observing the spatial distributions inside and at exit of a packed bed reactor. The main design was similar to the one used for the packed bed adsorber and employed a continuous wavelength (CW) tunable diode laser, FPA detector, and Pt/SDBC (particle size 80–100 μm, average pore size 110–175 Å, surface area 900 $m^2 g^{-1}$, and 2% platinum) resin packing of low scattering effect of NIR light.

$$HDO_v + H_2(g) \Leftrightarrow H_2O_v + HD(g) \tag{8}$$

Herein, water vapor and hydrogen isotopomers were produced by the gas–vapor reactions $(10-12)$ and vapor reaction (13) under kinetically and quasi-equilibrium controlled conditions, respectively (Roland et al., 2006).

$$H_2 + D_2O_v \Leftrightarrow HD + HDO_v \tag{9}$$

$$H_2 + HDO_v \Leftrightarrow HD + H_2O_v \tag{10}$$

$$HD + HDO_v \Leftrightarrow D_2 + H_2O_v \tag{11}$$

$$HD + D_2O \Leftrightarrow HDO + D_2 \tag{12}$$

$$D_2O + H_2O \Leftrightarrow 2HDO_v \tag{13}$$

Typically, water vapor comprised of $D_2Ov$ and $H_2O_v$ mixture was loaded from a controlled evaporator mixer (Bronkhorst), a water composition of 12.0 mol% (relative humidity of 30.5%), and temperature of 333 K, with all pre- and postpacked tube pipes being insulated and heated to the operating temperature. The water vapor was loaded to the system and allowed to stabilize for a period of hours before the commencement of an experiment. A humidity sensor (Exo Terra Digital Hygrometer, accuracy 2% at relative humidity >10%) and a quadrupole mass spectrometer were placed at the exit of the packed bed.

Imaging both $HDO_v$ and $H_2O_v$ by NIR tomography during the HIE reaction required preliminary experiments to quantify relevant extinction coefficients across the packing section and the empty section above the packing. It was achieved by using a squared quartz cell, which was half-filled with the SDBC packing. The pair of transition lines of $H_2O_v$ and $HDO_v$ at 1358.412 and 1380.685 nm, respectively, was found to be a good fit in terms of temperature sensitivity, since it offered a difference in the lower state energy $(E_2 - E_1)$ of 283 $cm^{-1}$, a line strength ratio at reference temperature $(S_1(T_0)/S_2(T_0))$ of 1.19. Spectra obtained above the packing were compared with those available in Hitran database (Rothman et al., 2009) as shown in Fig. 31, illustrating a good fit between the spectra and a deviation in the transmittance of 0.4% at 1380.685 nm for $H_2O_v$ and 0.6% at 1390.24 nm for $HDO_v$. The spectrum of $H_2O_v$ in the packing medium, where both absorption and scattering gathers, presented an OD of 0.31, offering a ray propagation dominated by ballistic light, in agreement with the absorption spectrum profile, and justifying the use of the modified Beer–Lambert law for concentration and temperature measurements. The extinction coefficients in the packing and above the packing for $H_2O_v$ and $HDO_v$ were fitted to a polynomial model as a function of temperature, ranging from 323 to 363 K.

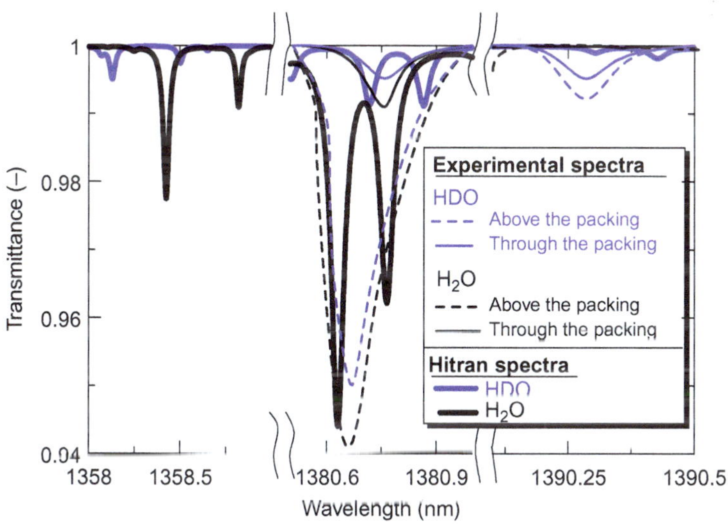


**Fig. 31** Simulated and experimental transmittance spectra of $H_2O_v$ and $HDO_v$ lines, temperature = 333 K; pressure = $1.013 \times 10^5$ Pa, optical path = $10^{-2}$ m, composition of $H_2O_v$ in $N_2$ = 6 mol%, and composition of $HDO_v$ in $N_2$ = 6 mol% (tSaoir et al., 2012). *Reprinted with permission from Elsevier.*

Tests of simultaneous visualization of spatial profiles of dynamic compositions of reactant, $D_2O_v$, product, $HDO_v$, and relevant temperature during the HIE reaction in Pt/SDBC resin-packed bed reactor of low AR were performed on a packed bed of an AR of 13—i.e., particle size of 0.7 mm, inlet and wall temperatures of 333 K, and flow rate of 250 $cm^3\,min^{-1}$. The steady-state compositions were achieved after 50 min. Fig. 32 shows transient composition profiles of the water and hydrogen isotopomers $H_2$, $H_2O_v$, $D_2O_v$, and $HDO_v$ at the exit of the packed bed. The inlet $D_2O_v$ was subject to $H_2O_v$ intrusion from the surrounding environment, ca. 1.7–1.8 mol%, in all experiments. After 2 min, as shown in Fig. 33$A_1$ and $B_1$, the composition maps of both species $H_2O_v$ and $HDO_v$ display weakly pseudohomogeneous distribution but compositions increased, particularly at 60 min, in the vicinity of the wall of the packed bed. $HDO_v$ compositions are more visible than $H_2O_v$ compositions, which agree with the progressive production $HDO_v$ formation obtained by mass spectrometry in Fig. 32. These two figures clearly demonstrate the effect of the wall on flow maldistribution in small AR packed beds and follows well trends of respective porosity profiles. Temperature maps, as shown in Fig. 34, display an uneven distribution as well, indicating an ineffective heat transfer, but colder zones were displayed closer to the wall, at 60 min, where convective heat transfer dominated due to wider porosity, and hotter zones were in the center of packing where heat transfer by conduction and convection occurred, achieving efficient cross-mixing. Hot zones reflected a flow of uniform

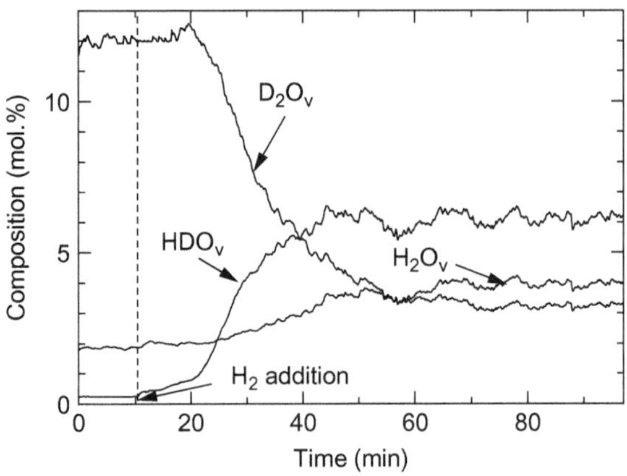


**Fig. 32** HIE dynamics of $HDO_v$, $H_2O_v$, and $D_2O_v$ compositions at the exit of the packed bed reactor obtained by mass spectrometry, packing height: 14 mm, particle size: 0.7 mm (AR: 13), wall temperature: 333 K, $D_2O_v$ (inlet): 12 mol%, $H_2$ (inlet): 25 mol%, and $N_2$ 63 mol%, $N_2$ flow rate: 250 $cm^3\,min^{-1}$ (tSaoir et al., 2012). *Reprinted with permission from Elsevier.*

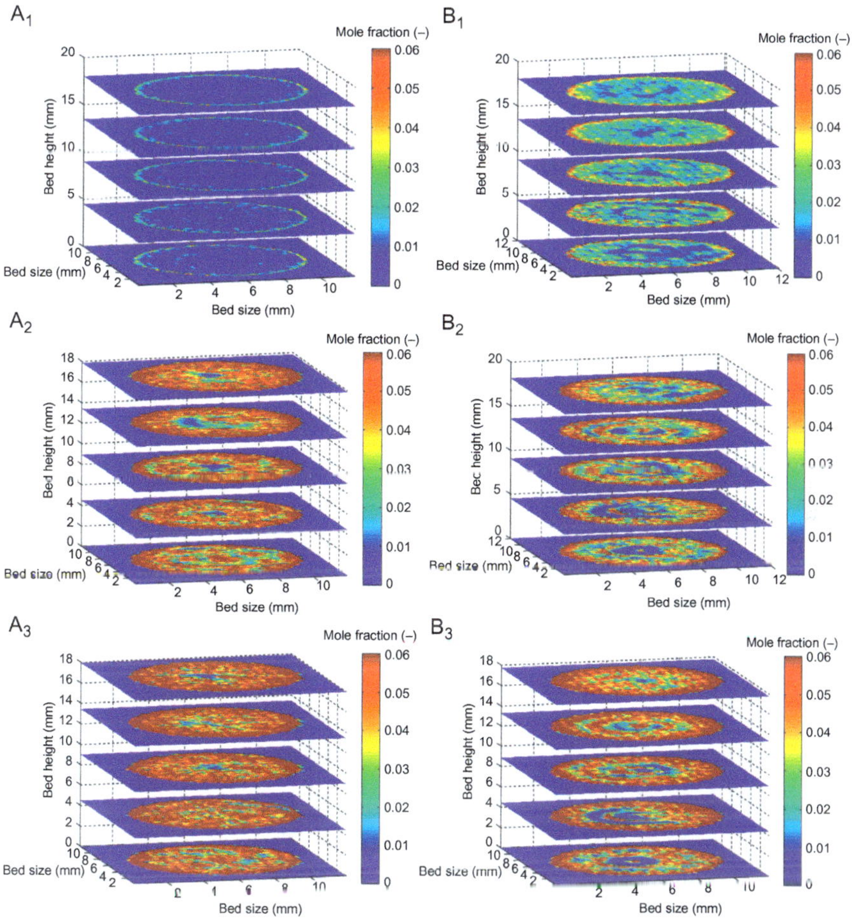


**Fig. 33** Hydrogen isotopic exchange dynamics by spatial distributions of $HDO_v$ and $H_2O_v$ compositions of five cross-sectional layers located at equal distance from the bottom of the packing to the top of the vapor phase above the packing, ($A_1$–$A_3$): $HDO_v$ at 2, 30, and 60 min, respectively, ($B_1$–$B_3$): $H_2O_v$ at 2, 30, and 60 min, respectively, packing height: 14 mm, particle size: 0.7 mm (AR: 13), wall temperature: 333 K, D2O (inlet): 12 mol%, $H_2$ (inlet): 25 mol%, and $N_2$ 63 mol%, $N_2$ flow rate: 250 $cm^3\,min^{-1}$ (Sauir et al., 2012). *Reprinted with permission from Elsevier.*

distribution along the axial and radial directions, while cold zones reflected a minimal flow velocity and resulted in an inefficient transfer of heat between the wall and the packing. Successive cold and hot zones along the bed suggest spatial changes in the flow in these zones through passages between the neighboring layers of particle in the radial direction. This study confirmed potential application of NIR tomography to investigate behavior of flowing gases in packed bed reactors by measuring the 3D distribution of

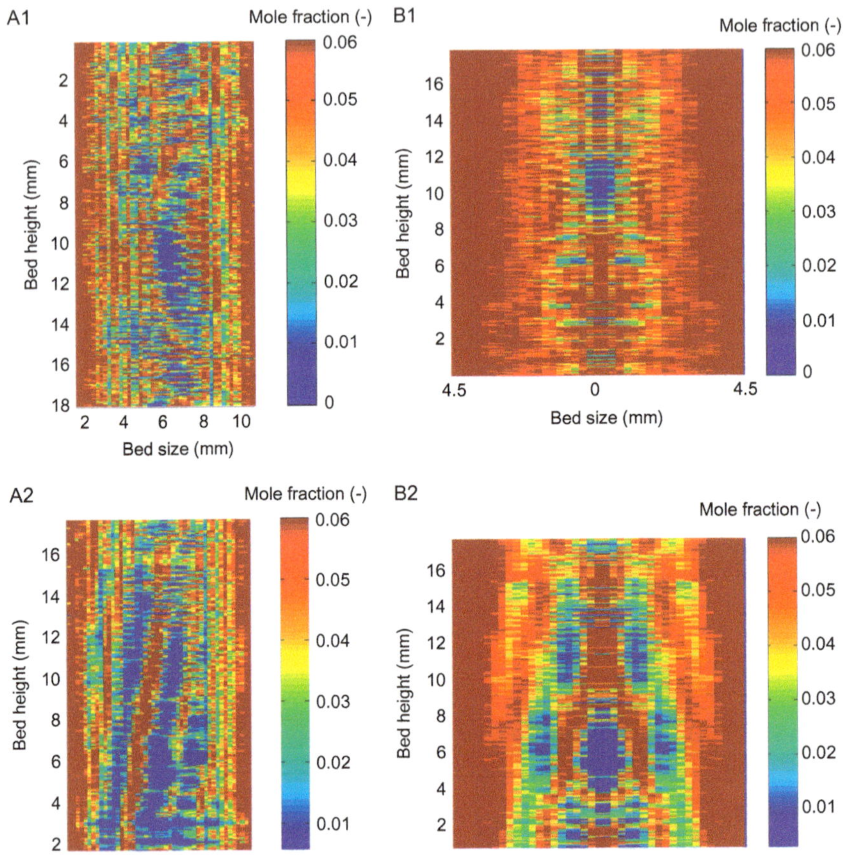


**Fig. 34** Hydrogen isotopic exchange dynamics by spatial distributions of composition by vertical cross-section of the packed bed ($A_1$): $HDO_v$, ($A_2$): $H_2O_v$ and by 2D spatial distributions ($B_1$): $HDO_v$, ($B_2$): $H_2O_v$ at 60 min of reaction time, (packing height: 14 mm, particle size: 0.7 mm (AR: 13), wall temperature: 333 K, $D_2O_v$ (inlet): 12 mol%, $H_2$ (inlet): 25 mol%, and $N_2$ 63 mol%, $N_2$ flow rate: 250 $cm^3\,min^{-1}$ (tSaoir et al., 2012). *Reprinted with permission from Elsevier.*

composition and temperature maps. Flow maldistribution could be observed along the bed by imaging uneven distribution of temperature between the core of the packed bed and near the wall and the dynamic lag between the near-wall $H_2O_v$ and $HDO_v$ compositions.

### *3.3.3 Dispersion Profiles in Gas–Solid Packed Beds by NIR Tomography*

A further application of the NIR imaging was the investigation of flow heterogeneity by the mass dispersion trends in packed bed reactors of low AR under diffusion-dominated flow regime. The diffusion-dominated flow regime is among the five mass dispersion regimes in packed beds depending on Peclet number ($Pe$), including: dispersion beyond the Darcy range ($Pe > 100{,}000$) where the effects of inertia and turbulence are large; pure mechanical dispersion ($300 < Pe < 100{,}000$) where the effect of molecular diffusion is limited; predominant mechanical dispersion ($5 < Pe < 300$); superposition of mechanical dispersion and diffusion ($0.3 < Pe < 5$) where the effect of molecular diffusion and dispersion is comparable; and pure molecular diffusion regime ($Pe < 0.3$) where the effects of inertia and turbulence are negligible) (Delgado, 2007). The trends of dispersion are often expressed in terms of the radial component ($D_{rad}$) and axial component ($D_{ax}$) and are typically accessible by running experiments which rely on a continuous tracing of chemical concentration or a discontinuous tracing of particles and fitting relevant dispersion constants under steady-state or dynamic conditions to pseudohomogenous models. The second method takes advantage of the discrete element method (DEM) approach where motion of particles is followed instead of species concentration (Freund et al., 2005). Results of NIR tomography on dispersion trends were used to validate those obtained by 3D and 2D modeling. The 3D modeling of the flow field and transport using pellet shapes that are from computer-generated methods, such as the DEM, provides spatiotemporal resolution that replicates very closely the one achieved by 3D tomographic techniques (i.e., X-ray computer tomography, MRI, laser Doppler anemometry, etc.). Spatial data of internals such as profiles of porosity, velocity, temperature, and concentration are accessible by using a limited number of tests. The approaches using 3D modeling are still, however, limited by requirements of large memories and computational power. An alternative option to consider local structures of packed beds is to couple the fluid flow with the radial variation of porosity by means of 2D modeling. Approaches by these two types of modeling and NIR tomography were compared for the visualization of the spatial distributions of gaseous species in packed beds of randomly distributed particles, subject to wall effects. Packed beds of low AR have been shown in Sections 3.3.1 and 3.3.2 being heterogeneous, of insufficient mixing and uneven porosity distribution, making dispersion profiles unexpected. NIR imaging was used, as alternative approach to conventional multiexperiment one to reduce the number of tests for dispersion studies. The design of the experiments was similar to the one used for HIE reactor, but the packed bed tube was a slightly larger (12 mm I.D. and 39 mm length) and was filled with hydrophobic quartz glass beads instead of a hydrophobic resin (Alzahrani et al., 2015). Water vapor was introduced to the system from a controlled evaporator mixer (Bronkhorst) at lower flow rate than previously used (0.1–0.5 $cm^3 min^{-1}$) of $N_2$. After a period of stabilization to reach packing saturation, the feed was switched to deuterated $H_2O_v$ and $D_2O_v$.

#### 3.3.3.1 Mass Dispersion Profiles by Spatially Resolved NIR Imaging

NIR imaging was carried out by using a packed bed of AR of 4 due to design limits of the field of view (39 mm), while a minimum length of the packing to particle ratio of 12 was required to avoid length effects. The tests for radial dispersion were carried out at low velocity (i.e., 0.00002–0.00011 $ms^{-1}$) to meet the residence time of the flow with the time resolution of a single tomographic test. The reconstructed images, as shown in Fig. 35$A_1$ and $A_2$, of cross-sectional and vertical compositions of $H_2O_v$ displayed higher values in the vicinity of the wall (outer flow) and clear limits between the $H_2O_v$ flow region and non-$H_2O_v$. In addition, the 3D data, once 2D angularly averaged as shown in Fig. 35$C_1$, show a clear distinction between $H_2O_v$ and non-$H_2O_v$ regions, which extends radially from the bottom to the top of the packed bed. A comparison with the results from the mass spectrometer at the exit of the packing indicated relative deviations in $H_2O_v$ composition of 9.1%. These results are seen as satisfactory given inherent errors associated with the experimental design (acquisition time per tomography cycle, light diffusion through packing, diffraction by the packing tube, convenient signal-to-noise ratio, and both entrance and exit effects).

The axial dispersion tests followed the same approach used for the radial dispersion but were carried out under transient operations, and the full inlet flow was switched to $D_2O_v$ to begin the experiments. Axial dispersion, as shown by cross-sectional and vertical compositions of $H_2O_v$ in Fig. 35$A_2$ and $B_2$, is not observed through the composition changes of $H_2O_v$ along the radial coordinates, but a heterogeneous distribution is discerned well, indicating a preferential channeling between the glass beads. Insufficient mixing along the bed length suggested that the flow was dominated by the axial direction. The 2D circumferential distribution (Fig. 35$C_2$) confirms the flow maldistribution associated with the high composition of water vapor near the wall and a damped one toward the core of the packed bed, correlating well with profiles of porosity profiles of low AR packed beds.

The tomographic technique offered qualitative knowledge of spatial distribution of $H_2O_V$ and showed promising results relevant to local dispersion coefficients along the radial and axial coordinates, but quantitative data were not accessible because the maps of $H_2O_v$ were not of sufficient

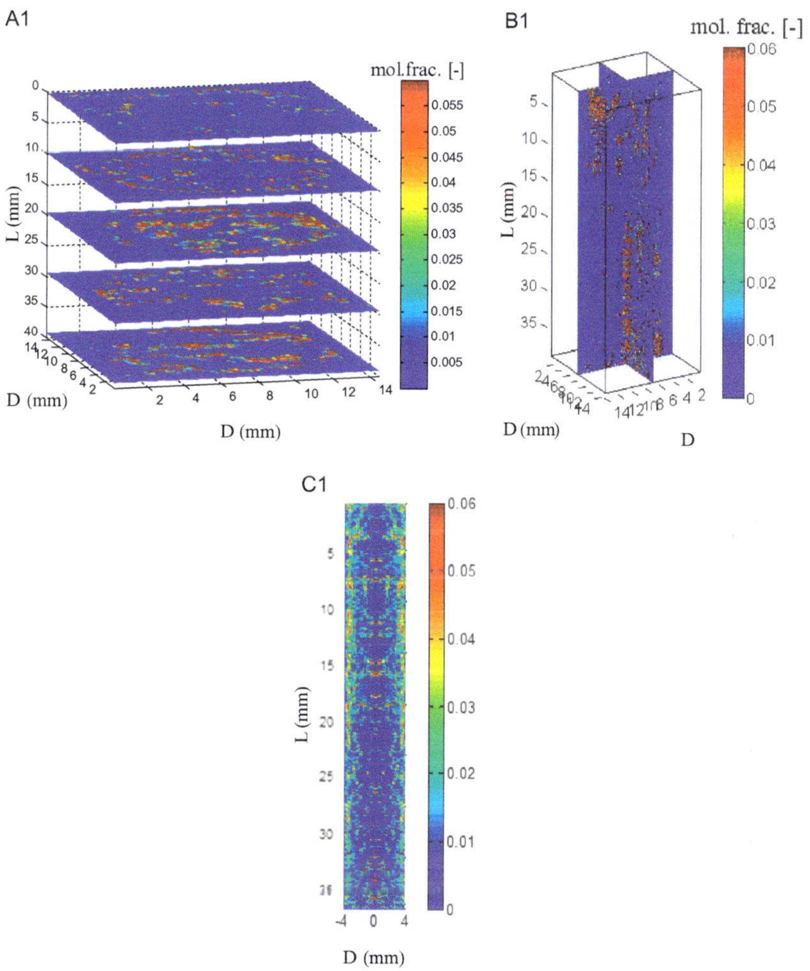


**Fig. 35** Visualization of $H_2O_v$ composition by NIR imaging. ($A_1$–$C_1$) Radial dispersion experiments and ($A_2$–D) axial dispersion experiments. Inner nozzle diameter: 5 mm, quartz tube diameter: 12 mm, thickness of both tubes: 0.5 mm. ($A_1$, $A_2$) Equidistant cross-sectional maps of $H_2O_v$, ($B_1$, $B_2$) vertically sectional maps, ($C_1$, $C_2$) circumferentially averaged maps, and (D) circumferentially and radially averaged maps (*straight line model*), inlet velocity: 0.00011 m $s^{-1}$. *Reprinted with permission from Alzahrani F, Aldehani M, Rusi H, et al.: Gas flow visualization in low aspect ratio packed beds by three-dimensional modeling and near-infrared tomography, Ind Eng Chem Res 54: 12714–12729, 2015. American Chemical Society.*

spatiotemporal resolutions. The overall trends of $D_{rad}$ and $D_{ax}$ for the full packed bed as a function of flow profiles were, however, obtained by fitting experimental data of $H_2O_v$ by NIR tomography to the analytical solutions of 2D diffusive and convective models of mass transfer in packed beds.

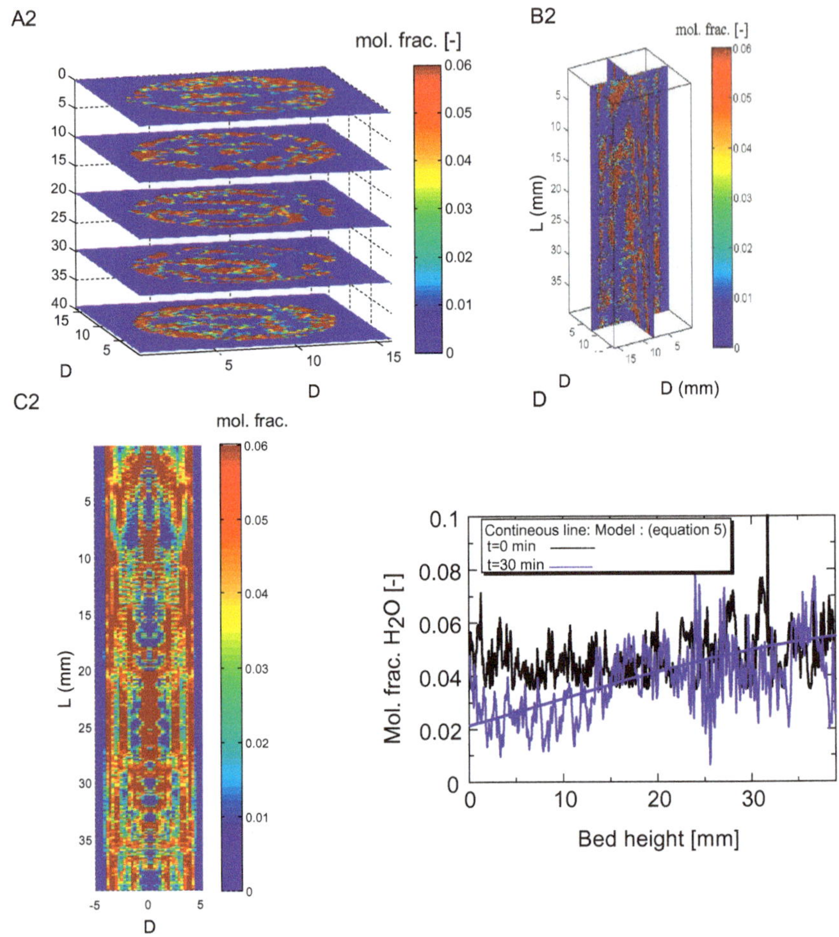


**Fig. 35—Cont'd**

Among these, those that rely on continuous species tracing under steady-state transversal and dynamic radial mass dispersion experiments, respectively, are expressed in Eq. (14), assuming constant velocity $u$ and constant dispersion coefficients.

$$\frac{\partial C}{\partial t} = D_{\mathrm{ax}} \frac{\partial^2 C}{\partial z^2} + \frac{1}{r}\frac{\partial}{\partial r}\left(D_{\mathrm{rad}} r \frac{\partial C}{\partial r}\right) - u\frac{\partial C}{\partial z} \tag{14}$$

The analytical solutions, by assuming a maximum concentration at the center of the packed bed, as expressed by Eqs. (15) and (16), follow continuous tracing of species under steady-state radial dispersion and dynamic axial dispersion, respectively.

$$\frac{C}{C_o} = \frac{1}{2} erfc\left(\frac{r}{2\sqrt{(zD_{rad}/u)}}\right) \tag{15}$$

$$\frac{C}{C_o} = \frac{1}{2} erfc\left(\frac{z-ut}{2\sqrt{D_{ax}t}}\right) + \frac{1}{2}\exp\left(\frac{uz}{D_{ax}}\right) erfc\left(\frac{z+ut}{2\sqrt{D_{ax}t}}\right) \tag{16}$$

where $C_0$ is the inlet concentration. The 3D maps were first circumferentially and radially averaged into 1D composition profiles along the bed height as shown in Fig. 35D and were fitted to the model of Eq. (16) to obtain unknown $D_{ax}$ values. On the other hand, the radial dispersion study used the full 2D maps of $H_2O_v$ obtained at steady-state conditions. Fig. 35D shows a good fit of experimental data with model, and a small slight deviation in the regions close to the inlet, which would originate from the effect of the inlet on the flow, is known to be significant in laminar regimes.

The 2D fits of data, along axial and radial coordinates with Eq. (14) for $D_{rad}$ and $D_{ax}$, normalized to molecular diffusion $D_m$ are shown in Fig. 36B$_1$ and B$_2$. Deviations of radial dispersion results are observed between the experimental data of $H_2O_v$ and those obtained from the solution of the

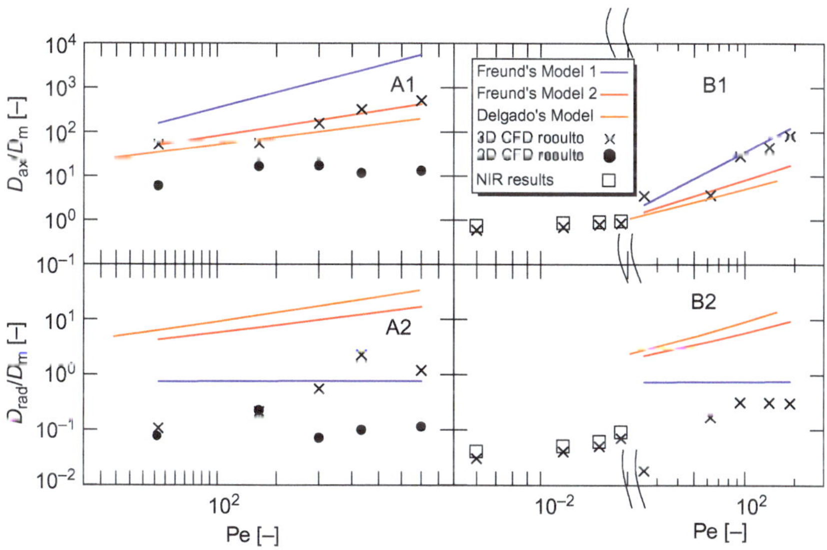


Fig. 36 Axial and radial dispersion coefficients along with flow dynamics, respectively: (A$_1$, A$_2$) for AR of 2, (B$_1$, B$_2$) for AR of 4. *Reprinted with permission from Alzahrani F, Aldehani M, Rusi H, et al.: Gas flow visualization in low aspect ratio packed beds by three-dimensional modeling and near-infrared tomography, Ind Eng Chem Res 54: 12714–12729, 2015. American Chemical Society.*

model (Eq. 15) or popular literature models. These deviations are less pronounced for the axial dispersion tests, although these two models are not straightforwardly applicable to low convective flow regimes. The results were broadly consistent with the range of dispersion coefficients obtained for mass transfer regimes under diffusion control, where $D_m/D_{ax}$ approaches the tortuosity of the bed (i.e., a theoretical value of 1.4).

#### 3.3.3.2 Mass Dispersion Profiles by DEM Modeling

Since the computation by diffusive and convective 2D modeling was not consistent enough with experimental data from NIR tomography, DEM approach was used to validate mass dispersion by NIR tomography. DEM is a valid tool which has been extensively used for mass dispersion studies (Freund et al., 2005) and relies on a discontinuous Lagrangian DEM method, by tracking particles that travel along a trajectory in a steady-state velocity field, as expressed in Eq. (17).

$$D_{ax/rad} = \frac{1}{2}\frac{d\sigma^2_{ax/rad}}{dt} \tag{17}$$

where $\sigma_{ax/rad}$ is the second moment or mean square deviation, as defined by Eqs. (18) and (19):

$$\sigma^2_{ax}(t) = \frac{1}{NP}\sum_{n=1}^{NP}\left(x_n(t) - \overline{x}(t)\right)^2 \tag{18}$$

$$\sigma^2_{rad}(t) = \frac{1}{NP}\sum_{n=1}^{NP}\left(r_n(t) - \overline{r}(t)\right)^2 \tag{19}$$

Here, $n$ is the particle index, $NP$ is the total number of tracked particles, $x$ and $r$ are displacements of particles along the axial and radial coordinates, respectively. $\overline{x}$ and $\overline{r}$ are the averaged displacements of all particles along the axial and radial coordinates, respectively.

The particle tracking module of Comsol was coupled with a 3D gas flow model (Eqs. 20 and 21) to give trajectories to individual particles of similar trends to those developed by the carrying fluid. The model is governed by both the mass conservation (Eq. 20) and the Navier–Stokes momentum equations (Eq. 21), where the inertial forces include both the pressure and viscous forces.

$$\rho \nabla \cdot u = 0 \tag{20}$$

$$\rho(u \cdot \nabla u) = -\nabla pI + \nabla \cdot \left(\mu\left(\nabla u + (\nabla u)^T - \frac{2}{3}\mu(\nabla \cdot u)I\right)\right) \tag{21}$$

The particles were assumed to be of negligible mass and subject to bounce conditions at the fluid–solid interface. The dispersion coefficients were computed by accessing first the transient positions of individual particles and computing the second moment or mean square deviation of relevant positions of these particles along axial and radial directions according to Eqs. (18) and (19), respectively, until asymptotic values of dispersion coefficients were achieved. The profiles of axial dispersion normalized to molecular diffusion, as shown in Fig. 36, were in good agreement with those obtained by NIR tomography and literature models. The radial dispersion by DEM showed some discrepancies with literature models and NIR-based results, and this behavior could originate from the size of the packed bed, which was not large enough to allow asymptotic profiles of velocity to be reached by the moving particles. The wall was found to be more effective toward the axial dispersion compared to the radial one and confirmed sensitivity of the mixing to overall geometry of low ARs packed beds. Finally, the results achieved by the 3D DEM were compared with those obtained from the 2D DEM, which used 2D angularly averaged porosity, as shown in Fig. 36. Unlike radial dispersion, axial dispersion values by 2D modeling were in good agreement with those from 3D modeling, and thus validated the benefit of using 2D averaged maps of porosity for large-scale packed beds for mass dispersion studies.

### 3.3.4 Local Catalyst Activity Profiles in Gas–Solid Packed Beds by NIR Tomography

Planar NIR imaging in Section 2.2.1 was a proven stage to validate its use for catalyst activity under realistic operating conditions. 2D integral data of catalyst surface at the exit of a multichannel reactor were accessible by observing dynamics of surface species. Extension of the technique to 3D observation permitted access to interactions of catalyst activity along with transport phenomena. By replacing the broadband light source with a fine bandwidth tunable laser, tomographic measurements of gaseous flow were monitored by diffuse transmittance. Through the use of principles of NIR spectroscopy, a robust system for visualization of cross-sectional profiles of gaseous interactions in packed beds was developed. NIR tomography was then applied to investigate local catalyst deactivation in packed bed reactors of low AR, and the results were compared with those achieved by 2D and

3D modeling. The strongly exothermic reaction of preferential oxidation (PROX) of CO was selected as it is typically carried out in packed beds of low AR, allowing gradients of temperature and species, and uneven distribution of flow, heat, and mass to take place. Since this reaction is also subject to catalyst deactivation by the reactants, the products, or both, dynamic profiles of local flow, mass, and heat transport and deactivation were anticipated to occur, leading to critical effects on performance and stability. This is because exothermic reactions with catalyst deactivation are known to develop rate differences between heat and mass during their excursions along the packed bed, and thus to end with wrong-way behavior and instability (Jaree et al., 2003). The effect of local deactivation on thermal instability in packed bed reactors was demonstrated experimentally for the reversible self-poisoning of CO oxidation on $Pt/Al_2O_3$ catalyst, where temperature waves were seen amplified under selected operating conditions. The differential flow (DIF) of mass and heat typically occurs because heat being subject to absorption by the thermal capacity of packing slows down during its excursion compared with propagation of species. As a result, packed bed reactors tend to amplify perturbations driven by uneven distribution of activity. For instance, when a drop in activity in a particular location of the bed takes place, a concentration wave is induced by the decrease in rate of reactant consumption prior to the reaction zone, travels at stream velocity, and leaves behind the thermal wave, partially retained by thermal inertia of the packing, and any clash between the two waves would result in increased rates and high-amplitude temperature excursions (Jaree et al., 2003).

It is therefore important to look at interaction of deactivation and transport phenomena. Conventional research into catalyst deactivation takes place at laboratory scale, free of transport phenomena, and receives less attention in open literature than the discovery of new catalysts. This is partly due to complex phenomena taking place, requiring multidisciplinary skills, suitable equipment, detailed knowledge of process conditions, and feedstock, and in most cases, time, particularly for processes of slow deactivation (Forzatti & Lietti, 1999). Catalyst deactivation, if often cited in works carried out at laboratory scale, is little quoted in industrial research. It should be recognized, however, that research at industrial scale is more interested in the process of deactivation along the packed bed than in the detailed surface mechanism that is generated in the laboratories. Insufficient information on detailed description of insights into the reactor is, however, frequent, leading dynamics of deactivation along the packed bed to be known only approximately. This is true since any extension to reactor design would not require the exact shape of the deactivation trend. A well-known example is nonuniform distribution of catalyst in packed beds. If the deactivation process is especially severe in one region of the bed, then it may be better to use a special, deactivation-resistant catalyst formulation in that region, and a different formulation in the rest of the bed. This approach has been adopted for the steam reforming of hydrocarbons; an alkalized $Ni/Al_2O_3$ catalyst is used in the upper bed to provide better resistance to carbon lay-down, and a nonalkalized formulation is used in the rest of the bed. Multilayered packed bed reactors with zones of catalyst activity and other of reduced or non-activity are becoming popular with exothermic reactions with successful improvement in catalyst life, reaction yield, and product selectivity (Jaree et al., 2001).

The conventional approach to test uniformity of activity in catalytic beds follows testing samples from multiple locations. This approach is time and effort consuming as the charging procedure should be carefully monitored to avoid causing broken pellets, excessive dust, or change in packing structural activity. It is therefore important to observe deactivation profiles by in situ monitoring of process parameters, including flow, compositions, and temperature profiles under *operando* conditions. Existing provision includes measurement devices such as temperature sensors, in situ analytical probes, and catalyst sampling devices (Petersen, 1997). Cases of reactions with strong heat exchange need more careful consideration and periodic return to standard operating conditions as changes in activity with position in the bed are anticipated to occur, leading to a dramatic consumption of catalyst in particular zones of the bed. Techniques of detection based spatially resolved probes are promising options to conventional sampling techniques to screen the first signs of local catalyst deactivation. In fact, phenomena of axial variation of catalyst activity in packed bed reactors are frequently observed and can be monitored by changing temperature, concentrations, arranging uneven distribution of catalyst activity along the bed or through on-stream deactivation. Yakhnin and Menzinger (1999) demonstrated in case of catalytic process of CO oxidation, under kinetically controlled operations and ignition behavior, that it could form a zone of deactivated catalyst downstream of the entrance to the packed bed, but also well forward of the end of the bed. Thus, a zone of active catalyst followed by a zone of completely deactivated catalyst, followed in turn by active catalyst, was observed. Deactivation was more pronounced in case of spatially variable (heterogeneous) activity in a packed bed and was illustrated by a segregated catalyst bed into alternate layers of active and inactive materials. Layers of different catalysts have been popular for improving selectivity for some classes of complex reactions under nonisothermal conditions (Jaree et al., 2001; Szépe & Levenspiel, 1968, 1971), for providing higher conversion with reactors exposed to composition modulation or for extending the region of stability with highly exothermic reactions. Because of concentration change and temperature differences, catalyst in these packed beds does not deactivate uniformly. It has been shown that abrupt variations in catalyst activity would result in temperature excursions that were much larger than those in beds of uniform activity when disturbances occur upstream of a reactor.

Understanding the details of local deactivation is an essential prerequisite for performance of reliable predictive simulations. During the recent decades, models associated with packed bed reactors subject to deactivation have been essentially been based on simplified pseudohomogenous or heterogeneous 1D and 2D models with inclusion of axial dispersion, radial dispersion, or both axial and radial dispersions along with local porosity. These models, however, are still relying on simplifying assumptions, and there are problems when trying to simulate nonsymmetric geometries inside reactors. This is especially important when it comes to low AR reactors, where the significance of the wall effect cannot be overlooked. The requirement for a more detailed level of knowledge has seen the emergence of direct 3D modeling of the flow field and transport processes using beds composed of either orderly or randomly packed particles. This approach, though limited by the large memory and computational time requirements, has become more and more feasible due to the continual increase in computing power (Dixon et al., 2010). Currently, simulations consisting of a few hundred to over a thousand particles still require large parallel computers. This means that nowadays simulations can only be done for small or laboratory scale reactors. Simulations of full-scale industrial reactors are still a long way off due to all kinds of realistic reaction conditions that are very hard to achieve. This approach is of comparable spatiotemporal resolution to much more expensive methods such as X-ray computer tomography, MRI or laser Doppler anemometry. From the 3D simulation results, more reliable profiles of velocity, temperature, concentration, and local deactivation can be obtained, allowing for correlations to be obtained based on large amounts of data for multiple rather than single experiments. With ongoing development of powerful computational tools, the CFD models are gaining substantial confidence to address a range of problems associated with 3D modeling. Packed beds of 3D geometry and discrete particles are being used to mimic actual laboratory experiments. It is therefore possible to simulate the flow, heat, and mass transports, catalytic reactions, and deactivation in a packing consisting of a few thousands of particles but is still not possible to reach a 3D simulation of a packed bed of industrial scale, leading this approach to be convenient exclusively for investigation of complex interaction between the flow, heat, and mass transports on a formerly unreachable level of detail, helping thus to establish cross-links of knowledge between this approach and conventional simplified modeling approaches. It is therefore primordial to screen insights in local information of flow, mass, and heat and deactivation by experimental techniques based on spatially resolved methods, and then reach fundamental understanding of how these phenomena are taking place by 3D models.

#### 3.3.4.1 PROX of Carbon Monoxide

The PROX of carbon monoxide on $Pt/Al_2O_3$ was selected because, as exothermic reaction, the oxidation of CO was demonstrated to exhibit local deactivation owing to DIF of heat and mass (Jaree et al., 2003; Yakhnin & Menzinger, 1999). Also, this reaction has been extensively studied for packed bed dynamics, and enormous amount of information is available, owing to its role in sectors of catalytic control of automotive emissions and fuel cells. Special attention was paid exclusively to ignition (or light-off) associated with the carbon monoxide oxidation of the PROX. Physically, the ignition is related to instability and self-acceleration of exothermic reactions due to the heat release. Before ignition, the carbon monoxide oxidation is in a low-reactive state and the mass-transfer limitation is typically negligible. After ignition, the carbon oxidation is in a high-reactive state, and the reaction rate is controlled by transport. With changing local deactivation, one can observe extinction of carbon monoxide oxidation, i.e., the transition from a high-reactive state to a low-reactive state. This biinstability and its effect on catalyst deactivation of carbon monoxide oxidation have been investigated experimentally and by 1D pseudohomogeneous modeling (Jaree et al., 2008; Menzinger et al., 2004; Silveston & Hudgins, 2003).

#### 3.3.4.2 Local Deactivation Profiles by NIR Tomography

The design of experiments was similar to the one used for the HIE reactor, but the inner tube of the packed bed was slightly larger (12 mm I.D. and 39 mm length) and was filled with Pt/alumina catalyst. The two concentric tubes, which were previously used to insure isothermal heating, were managed to insure a near-adiabatic conditions. The annulus between the two tubes of 5 mm thickness was evacuated by a diffusion pump to a pressure in the range of $10^{-5}$ mmHg. The tests of observing insights of the packed bed reactor by NIR tomography were performed on a packed bed of an AR of 13—i.e., particle size of 0.7 mm. A typical test was carried out at inlet temperature of 413 K and feed flow rate of 333 $cm^3\,min^{-1}$ and composition was 0.01% CO, 0.01 $O_2$, 0.2 $H_2$ and $N_2$ was used as carrier and exit gases were analyzed by mass spectrometry. Similar to Sections 3.3.1–3.3.3, the data retrieved from the tomographic tests were processed by using the AART and relevant absorption coefficients which were obtained by preliminary calibration tests.

### 3.3.4.3 Results

**3.3.4.3.1 Kinetics of PROX of CO** Kinetics study was first investigated, and the resulting models were implemented in the 2D and 3D CFD models and then were validated by NIR tomography. The mechanism of PROX of CO is essentially driven by the CO oxidation, particularly at low temperatures. At the low-reactive state, the surface of catalyst is mainly covered by CO, and the reaction rate is controlled by $O_2$ dissociation in contrast to the high-reactive state, where the surface of catalyst is mainly covered by oxygen, and the reaction rate is proportional to CO pressure.

$$CO + 1/2\,O_2 \rightarrow CO_2 \quad \Delta H^0_{298} = -283\,\text{kJ/mol} \tag{22}$$

Different kinetic models were proposed to express changes in reaction rates along with the reactive states, including first-order kinetics, power law expressions, and surface Langmuir–Hinshelwood mechanism. According to the latter mechanism, carbon monoxide and oxygen are both adsorbed at the surface of the catalyst such as the rate expression suggested by Venderbosch et al. (1998):

$$R = \frac{k_{\exp}\left(-\frac{E_a}{RT}\right)\cdot c_{CO}\cdot c_{O_2}}{\left(1 + k_{\exp}\left(\frac{\Delta H_{ads}}{RT}\right)\cdot c_{CO}\right)^2} \tag{23}$$

where $c_{CO}$ and $c_{O2}$ represent the CO and $O_2$ concentrations, respectively, $k$ is the reaction rate constant, $E_a$ is the activation energy, $K$ is the equilibrium adsorption constant of CO, and $\Delta H_{ads}$ the enthalpy of adsorption of CO. Eq. (23) expresses first-order kinetics in CO at low concentrations of CO and apparent negative order at high concentrations of CO. Self-inhibition of CO oxidation by CO is a characteristic of this reaction. In addition, upon arriving in the mass transfer controlled regime, the apparent activation decreases to a value in line with the temperature dependency of diffusion transport. Power law models have also been recurrently reported, particularly for precious metal catalysts. These models are easy to implement in CFD simulations as they reduce nonlinearity issues when they are combined with transport phenomena.

Another simplified model of first-order rate expression for CO oxidation on $Pt/Al_2O_3$ catalyst is the one suggested by Amphlett et al. (1996).

$$R = kc_{CO} \tag{24}$$

where

$$k = 0.026\exp\left(-\frac{1000}{T}\right)\left(m^3\,s^{-1}\,kg^{-1}\right) \tag{25}$$

when CO oxidation (Eq. 22) is associated with hydrogen oxidation (Eq. 26) though the PROX, the reverse water–gas shift reaction (Eq. 27) takes place, reducing the amount of CO to be converted.

$$H_2 + 1/2\,O_2 \rightarrow H_2O \quad \Delta H^0_{298} = -288\,kJ/mol \tag{26}$$

$$H_2 + CO_2 \rightarrow CO + H_2O \quad \Delta H^0_{298} = 41\,kJ/mol \tag{27}$$

Similar to the kinetics of CO oxidation, the one of PROX of CO has been generally modeled by the surface Langmuir–Hinshelwood, Eley–Rideal, or linear adsorption mechanisms, where CO and $O_2$ molecules are adsorbed at competitive active sites. The concentration of $H_2$, along with relevant adsorption capacities of CO at low concentrations, leads the rate for each surface reaction to be assumed of the first-order in coverage for each species. Following the first-order mechanism by Amphlett et al. (1996) and the one by Choi and Stenger (2004) for PROX of CO, a simplified kinetic model associated with the three reactions and deactivation was proposed as follows:

$$r_{21} = ak_1c_{CO} \tag{28}$$

$$r_2 = ak_2c_{H_2} \tag{29}$$

$$r_3 = ak_3\left(c_{CO}c_{H_2O} - \frac{c_{CO_2}c_{H_2}}{K_c}\right) \tag{30}$$

where

$$k_i = A_i e^{-E_i/RT} \quad i=1-3$$

$$\frac{da}{dt} = k_d c_{CO} \tag{31}$$

$$K_c = \exp\left(\frac{4577.8}{T} - 4.33\right), \tag{32}$$

where $a$ is the activity and $k_d$ is the deactivation rate constant.

The reverse water gas shift reaction has a chemical equilibrium constant $K_c$ of 0.001–0.004 at temperature ranging from 413 to 573 K, reaching quasi-equilibrium state when the feed contains sufficient amounts of $H_2$, and leading the overall rate of the PROX to be dominated by CO oxidation.

A plug flow packed bed gas–solid model was developed, where dispersions inside the packing were assumed negligible. The minimization of the sum of squares of residuals was performed by the nonlinear least squares method, using the Marquardt method to adjust the kinetic parameters. The validity of the kinetic model was verified by calculating the relative deviation between experimental data and predicted results from the kinetic model. The Arrhenius plots of the kinetic constants are given in Fig. 37 along with the activation energies and preexponential factors of each reaction. The activation energy values are within the range of reported values (Amphlett et al., 1996; Choi & Stenger, 2004), taking into account the weakening effect of platinum/alumina interactions due to hydroxyl

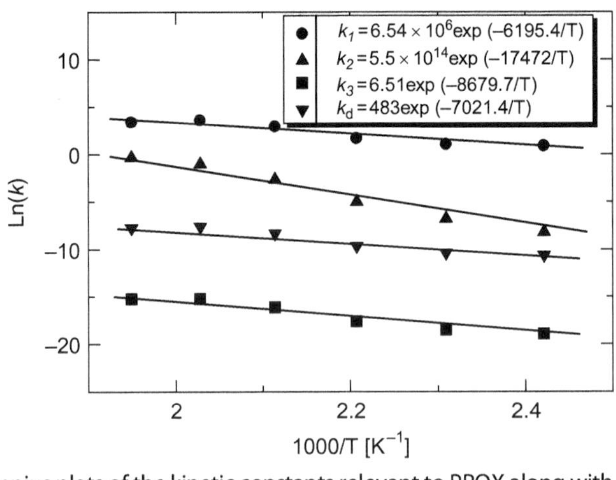


**Fig. 37** Arrhenius plots of the kinetic constants relevant to PROX along with deactivation.

formation via reverse water–gas shift reaction. This result clearly anticipated the potential merit of using water vapor as tracing species to assess catalyst deactivation by CO oxidation since the rate of this latter is the relevant step and well reflected by water vapor production.

The dynamic changes of the PROX of CO were followed by observing the production of water vapor produced by $CO_2$ conversion through the reverse water gas shift reaction. NIR tomography was convenient to observe the trends of interaction of chemical kinetics of PROX of CO with mass and heat transports and deactivation in a packed bed reactor. NIR tomography was demonstrated as a useful technique to trace spatial distribution of water vapor but would not be capable of tracing other species such as CO and $CO_2$ which hold absorption lines two to three magnitudes lower than those exhibited by water vapor in the NIR spectrum. The study of catalyst deactivation of $Pt/Al_2O_3$ by CO was therefore investigated through PROX of CO by following the dynamic maps of water vapor along the length of the packed bed by NIR tomography.

Fig. 38A and B show profiles cross-sectional area of water vapor and temperature from the bottom to the top of the packed bed reactor at 60 min of reaction time, respectively. The feed contained a mixture of 1% CO, 1% $O_2$, and 20% $H_2$. The generation of water vapor in Fig. 38A was not homogeneous in the packed bed owing to flow maldistribution, and NIR tomography was able to discriminate the flow of high and low concentration of water vapor where limited channeling is anticipated. This heterogeneous distribution helped understanding complex interplay between the fluid flow, transport phenomena of heat and mass at gas–solid bound aries, and chemical kinetics. Temperature distribution (Fig. 38B) followed those of concentration of water vapor by presenting uneven distribution and pronounced hot regions at some local regions in the center of the packed bed. This was partly due to exothermicity of the PROX process along with inefficient transport of the generated heat by flowing fluid. The ignition took place at the second cross section, corresponding to about 12 mm of reactor height and would anticipate the oxidation of the remaining CO in the gas stream to $CO_2$. Further reaction of this later with hydrogen into CO and $H_2O_v$, would lead to an incomplete CO oxidation. It is deduced therefore that desorption of CO was sufficient to allow some oxygen to adsorb, dissociate, and react with the adsorbed CO. The heat produced from the increased rate of reaction was adequate to further increase desorption of CO and thus the reaction rate for water vapor production. NIR tomography allowed discrimination of colder zones, particularly those closer to the wall

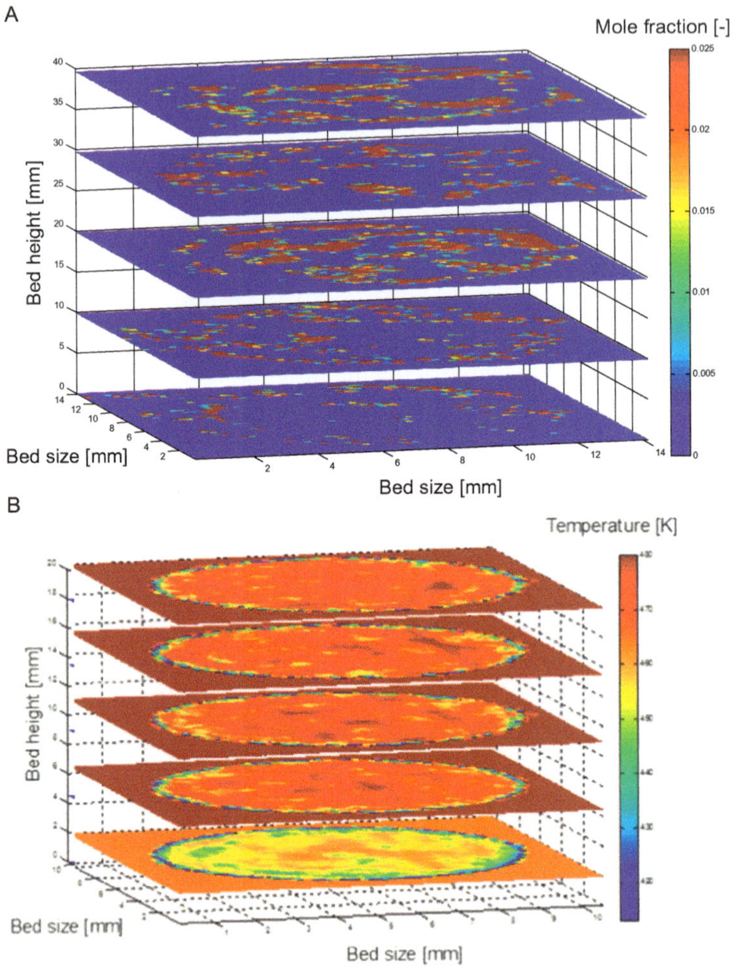


**Fig. 38** Cross-sectional profiles of water vapor (A) and temperature (B) at $t = 60$ min. Feed flow rate $= 333$ mL/min and composition: 1% CO, 1% $O_2$, and 20% $H_2$.

where convective heat transfer dominates owing to a larger porosity in contact with the wall. The hot zones are in the core packing area and start from catalyst particles, which miss contact with the wall, to the center of the packing, where heat transfer by conduction and convection occur, leading to efficient cross-mixing. Cold zones, therefore, correspond to the situation, where the flow velocity was uniform along the axial and radial directions, while hot zones reflect a minimal flow velocity, resulting in a reduced energy transfer between the particles. In addition, an inefficient energy release is seen in some areas by an increase in temperature, which demonstrates that the maldistribution was present along both radial and angular coordinates, and again, the complex interplay between the fluid flow, heat transports at gas–solid boundaries, and chemical kinetics of the PROX of CO occurred. Fig. 39A and B presents the results of circumferentially and radially averaged concentration of water vapor along the packing height

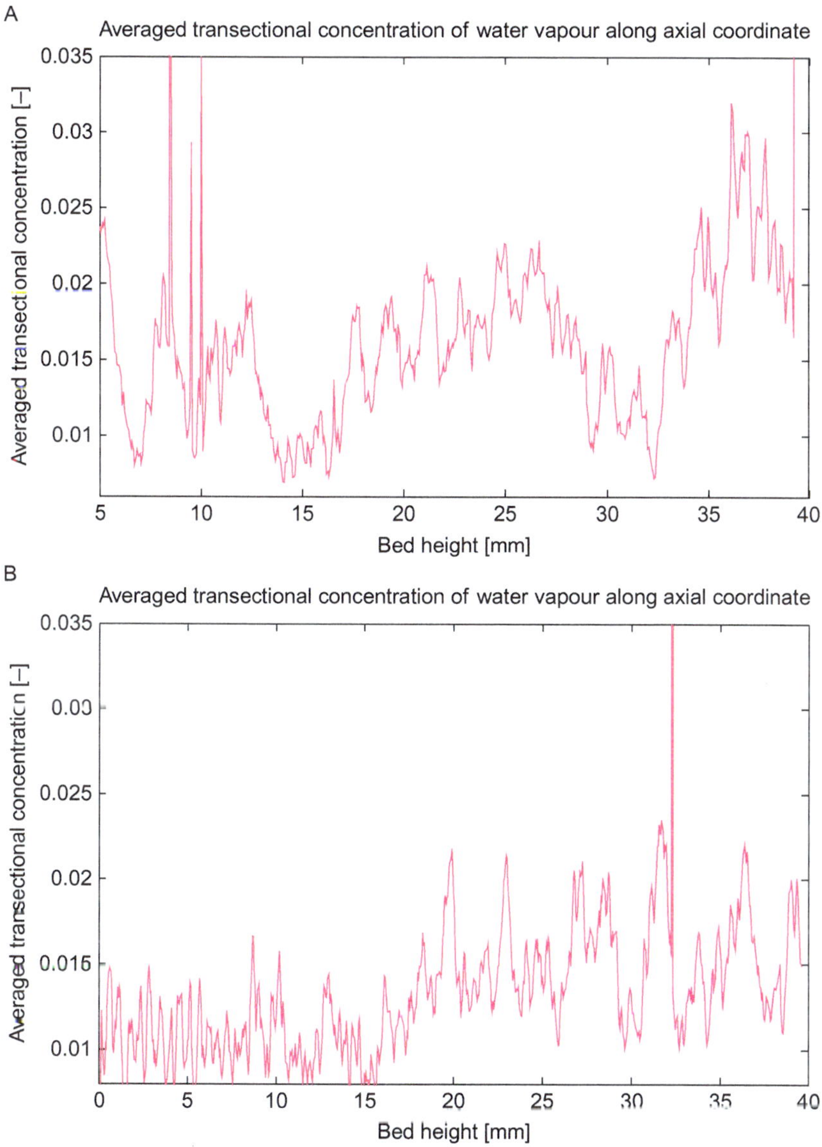


**Fig. 39** Circumferentially and radially averaged profiles of water vapor and temperature along the packed bed height: (A) $t=60$ min and (B) $t=180$ min, feed flow rate $=333$ mL/min, and composition: 1% CO, 1% $O_2$, and 20% $H_2$.

at 60 and 180 min, respectively. The two figures clearly show profiles of the decrease of water vapor production with time and raise importance of local catalyst deactivation. Spatiotemporal profiles of the produced water vapor are seen therefore to be of valuable importance to get access to local deactivation in the packed beds and thus provide significance of temporal insights in the packed beds of composition and temperature.

#### 3.3.4.3.2 Local Deactivation Profiles by 2D and 3D Modeling

Results of NIR tomography on dispersion trends were used to validate those obtained by 3D and 2D modeling. Similar to the previous section 3.3.3.2 on modeling the dispersion, the 2D and 3D modeling of the flow field, transport, and deactivation used randomly distributed packing that was built by the DEM. The latter provided spatial resolution between the particles and inside the pellets as well. A combination of heat, mass, and momentum balances along with the kinetic model of the PROX of CO and catalyst deactivation was used for screening local catalyst activity. The program code numerically solved the Navier–Stokes equations and associated differential equations for energy and mass balances over a number of approximate differential controlled surfaces (2D) or volumes (3D). These controlled elements were considered small within the flow domain that as a whole described the domain geometry. The four differential equations that were used to describe the dynamic behavior of the packed bed reactor included equations of motion for momentum conservation, continuity equation for conservation of mass, thermal energy equation for conservation of energy and mass balance for conservation of chemical species. This latter included the kinetic model, and the deactivation model associated with the PROX of CO.

The catalyst activity was studied by simulation using COMSOL Multiphysics and a parametric study on effects of some important properties like: (1) deactivation rate expressed in terms of dimensionless activation energy of catalyst deactivation and (2) the Damköhler number (ratio of residence time and reaction time) on the deactivation profiles on local catalyst deactivation was carried out by 2D and 3D CFD modeling. As stated earlier, most of the works have been carried out on catalyst deactivation in packed bed reactors using 1D modeling (Jaree et al., 2008; Silveston & Hudgins, 2003), but a limited number of works developed 3D models. Works based on 1D modeling allowed understanding of physical phenomena inside packed reactors but were not able to reflect local information in terms of velocity, mass, and heat transfers and catalyst deactivation. As to the works by means of 3D modeling and deactivation, they were limited to operations of slow deactivation, allowing the modeling of the packed bed to be assumed under successive steady-state operation (Behnam et al., 2010). Herein, both 2D and 3D models were investigated in order to get a comprehensive investigation of detailed catalyst deactivation in packed bed reactors under both simplified and complex modeling approaches.

*3.3.4.3.2.1 Analysis of Catalyst Deactivation in 2D Packed Bed Reactors of Low AR* 2D pseudohomogeneous model was first used to simulate an adiabatic packed bed reactor, operating the PROX reaction of CO. The 2D modeling was carried out because it did not require large memory and computation time, allowing a simplified approach to validate, prior to 3D modeling, with experimental data by NIR imaging and those achieved in literature (Menzinger et al., 2004; Silveston & Hudgins, 2003). The 1D pseudohomogeneous mass and energy balances developed by Jensen and Ray (Jensen & Ray, 1982) and Jaree et al. (Menzinger et al., 2004) for single oxidation of CO were extended to 2D modeling of PROX which included momentum, heat, and mass balances as expressed by Eqs. (33)–(35). The effect of catalyst deactivation on the thermal behavior was extended to two ARs (i.e., 3 and 4) of the packed bed to investigate impact of the wall, and by inference flow heterogeneity.

- Momentum balance in a homogenous porous medium

$$\varepsilon\rho_g u(\nabla u) = -\varepsilon\nabla P + \mu\nabla\bullet\nabla u - \frac{2}{3}\varepsilon\mu\nabla\bullet\nabla u + \varepsilon F \tag{33}$$

$$\nabla\bullet\left(\rho_g u\right) = 0 \tag{34a}$$

The drag force per unit of volume $F$ was expressed using Ergun's relation as:

$$F = -\frac{\mu u}{K} - \beta\rho u^2 \tag{34b}$$

The permeability of the porous media $K$ and the non-Darcy term $\beta$ were calculated from Ergun's model.

$$K = \frac{\varepsilon^3 dp^2}{A(1-\varepsilon)^2},\ \beta = \frac{B(1-\varepsilon)}{\varepsilon^3 dp} \tag{34c}$$

$$\text{where } A = 150,\quad B = 1.75$$

$$u|_{\tau=0} = u_0,\ \nabla u|_{\tau=1} = 0, \tag{34d}$$

$$\nabla u|_{r=0} = 0 \tag{34e}$$

- Heat balance (both gas and solid phases)

$$\left(\varepsilon_{p_g} C_{p_g} + (1-\varepsilon) p_s C_{p_s}\right) \frac{\partial T}{\partial t} + \varepsilon_{p_g} C_{p_g} u \cdot \nabla T = \nabla(k \nabla T) + Q \tag{35}$$

$$k \nabla T|_{r=R} = 0 \tag{35a}$$

$$Q = \Delta H_R R \tag{35b}$$

$$\nabla T|_{r=0} = 0 \tag{35c}$$

- Mass balance (both gas and solid phases)

$$\frac{\partial \varepsilon C_i}{\partial t} + \nabla(-D_i \nabla_{C_i}) + \varepsilon u \nabla C_i = R_i \tag{36}$$

$$\nabla c|_{r=0} = 0$$

where $\mu$ is the viscosity, $C_p$ is the heat capacity, $k$ is the thermal conductivity, $D_i$ is the diffusion coefficient, $R_i$ is the stoichiometric chemical reaction rate (Eqs. 28–32), $Q$ is the heat generated by the chemical reaction, and subscripts s and g refer to the solid (catalytic) phase and gas phases, respectively. A linear first-order kinetic model was assumed (Eqs. 28–32) along with separate deactivation model as it decreased nonlinearity associated with effects of reactant or product activation or inhibition of surface kinetic models and facilitated model convergence. A simplified power law deactivation rate model (Eq. 31) relevant to self-deactivation by CO was assumed. The deactivation rate constant was considered to follow Arrhenius trend: $k_d = A_D e^{-\frac{E_D}{RT}}$, $E_D$ is the activation energy of deactivation and $A_D$ is the pre-exponential Arrhenius factor of deactivation. The density and thermal capacity of the solid and gaseous phases were taken from literature handbook as follows: $\rho_s = 1230\,\text{kg/m}^3$, $\rho_g = 1\,\text{kg/m}^3$, $Cp_s = 900\,\text{J/kgK}$, $C_{p_g} = 1045\,\text{J/kgK}$, respectively.

The effect of activation energy of catalyst deactivation ($\Upsilon_D$) on deactivation trends was first investigated. An increase in the activation energy of deactivation, $\Upsilon_D$ promoted the PROX rate by decreasing the rate of deactivation. Fig. 40 shows activity coefficient profiles along the dimensionless axial position ($\xi = z/L$, where $z$ is the axial position and $L$ is the length of the

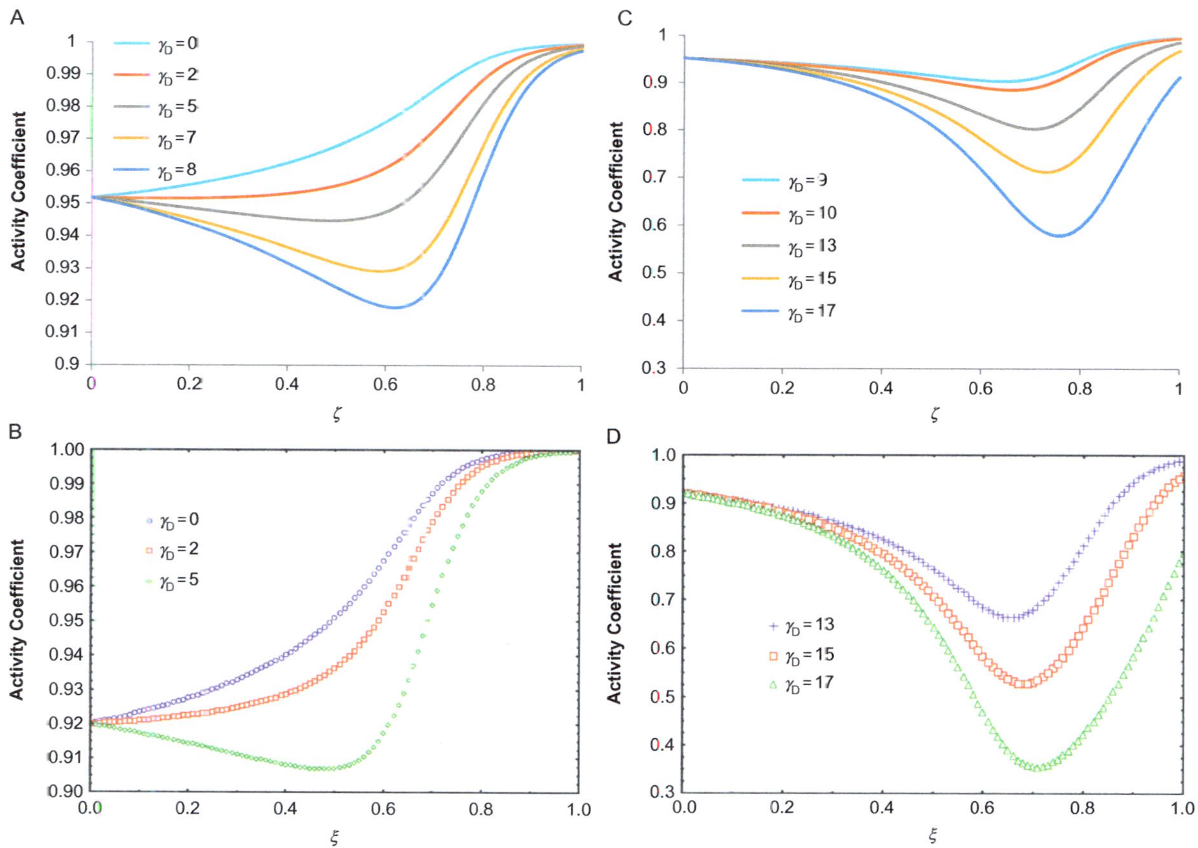


**Fig. 40** Effect of deactivation energy ($\Upsilon_D$) on the catalytic activity profiles: (A) and (C) 2D CFD simulation of PROX of CO; (B) and (D) Jaree's 1D model of CO oxidation (Jaree et al., 2008) ($A_D = 4 \times 10^{-6}\,m^3$/mol.s, $D = 0.000167\,m^2$/s, $k = 1.3$ W/mK, $A = 1.6 \times 10^6\,s^{-1}$, $E = 51.5$ kJ/mol K, and at $\tau = 50{,}000$ s).

packed bed) of the packed reactor for an activation energy of deactivation $\Upsilon_D$ ranging from 0 to 17. Accelerating the rate of PROX allowed release of thermal energy, which in turn accelerated the deactivation rate. Decreasing $\Upsilon_D$ caused the zone of catalyst deactivation to shift forward to the front section of the reactor, while the reaction zone was still further down in the catalyst bed. These results follow the trends presented by Jaree et al. (2008) for a single CO oxidation. In our simulation, a catalytic activity for different $\gamma_D$ (ranging from 0 to 17) started from 0.95 with a deviation of 0.03 from Jaree's starting values. Activity coefficient curves exhibited similar changing trends; i.e., an increase of the $\Upsilon_D$ moved the deactivation zone forward to the front of the packed bed. However, at $\Upsilon_D=0$, the curve went up slowly, while it shows a relatively bigger slope in the one presented by Jaree et al. (2008); at $\Upsilon_D=2$, the curve kept constant values of activity coefficient till 0.4 of length $\zeta$ of the packed bed, and then increased progressively; at $\Upsilon_D=5$, both curves went down; at $\Upsilon_D=13$, 15, or 17, the deactivation zone appears between $0.6\xi$ and $0.8\xi$, while the drop in our simulation is much smaller (i.e., 0.15, 0.25, and 0.35) compared with 0.27, 0.40, and 0.57 in the literature. As we can see, the drop increased gradually with the growing value of $\Upsilon_D$. The deviations observed in our work were driven by the kinetics of CO oxidation which was slightly higher in the PROX owing to water gas shift reaction by water inhibition as discussed in the kinetic study. It is interesting to see that the rate of deactivation was the fastest at the entrance of the catalytic bed where the concentration was the highest. This suggests that locating the reaction zone and knowing the deactivation energy could potentially help reduce its impact on process efficiency.

Following the activation energy of deactivation, effect of the Damköhler number of deactivation ($Da_D$) on deactivation trends was investigated. The Damköhler number is associated with the concentration of catalyst and extent of reaction zone and expressed in terms of the dimensionless deactivation rate constant to velocity rate or expressed in terms of residence time of the gas in the packed bed to deactivation time. Fig. 41 illustrates the catalyst activity coefficient profiles vs dimensionless axial position $\xi$ and the Damköhler number of deactivation ($Da_D=(A_D/u)e(-E_D/RT)$). The Damköhler number affected catalyst activity almost in the same way as the activation energy of deactivation ($\Upsilon_D$). For high values of $Da_D$, catalyst activity decreased, particularly in regions toward the exit, i.e., about $0.7\xi$–$0.8\xi$. It is clear, therefore, that local deactivation can be driven to the exit of the packed bed by operating the packed bed at high temperatures or low velocity rates.

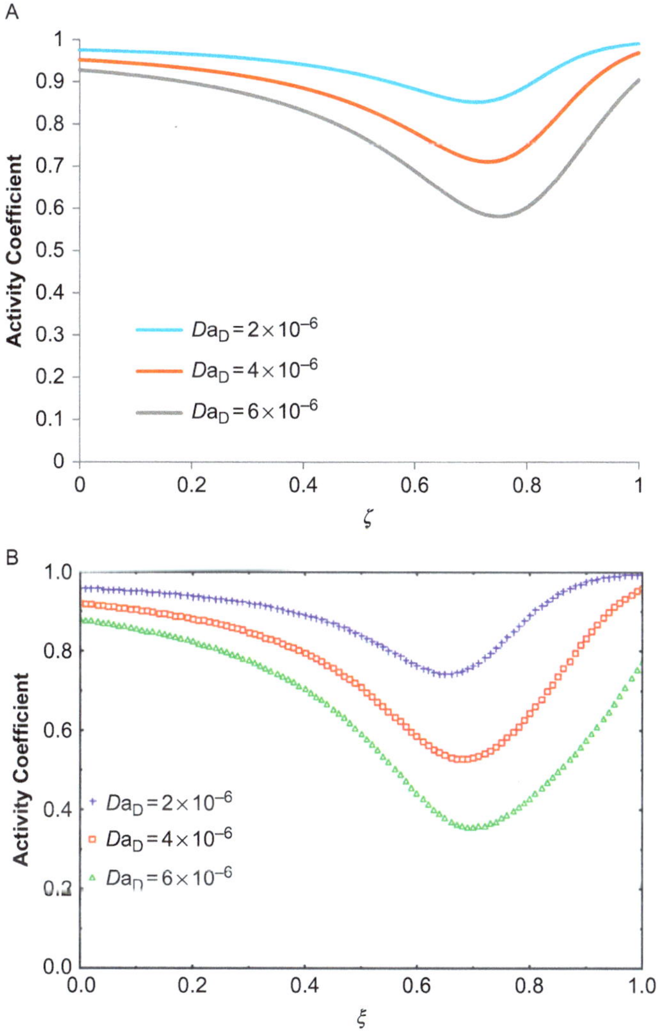


**Fig. 41** Effect of Damköhler number of deactivation $Da_D$ on the catalytic activity profiles (A) 2D CFD simulation of PROX of CO (B) Jaree's 1D model of CO oxidation (Jaree et al., 2008) ($A_D = 4 \times 10^{-6}\,m^3/mol\,s$, $E_D = 51.5\,kJ/mol\,K$, $D = 0.000167\,m^2/s$, $k = 1.3\,W/mK$, $A = 1.6 \times 10^6\,s^{-1}$, $E = 51.5\,kJ/mol\,K$, and at $\tau = 50{,}000\,s$).

*3.3.4.3.2.2 Analysis of Catalyst Deactivation in 3D Packed Bed Reactors of Low AR* The simulation by 3D modeling was carried out under similar operating conditions used in the 2D modeling. The effect of deactivation rates and mass transfer on local deactivation under transient operations was investigated. Governing equations which couple balances of momentum, heat, and mass transfer in a 3D randomly packed bed reactor are illustrated as follows.

- Momentum balance

Navier–Stokes and continuity equations for momentum and mass conservations, respectively, were used. A steady-state velocity profile was assumed at the inlet, whereas the pressure at the exit was assumed at local atmospheric pressure. In addition, no-slip boundary condition at the entire solid (packing or wall)–fluid interface was considered.

$$\rho_g u \nabla u = -\nabla P + \mu \nabla . \nabla u - \frac{2}{3}\mu \nabla . \nabla u \tag{37}$$

$$\nabla \cdot \left(\rho_g u\right) = 0 \tag{38}$$

$$u\big|_{\tau=0} = u_0,\ P\big|_{\tau=1} = P_0 \tag{39}$$

- Mass balance in the gas phase

$$\frac{\partial C_i}{\partial t} + \nabla \cdot (-D_i \nabla_{C_i}) + u \cdot \nabla C_i = 0 \tag{40}$$

The diffusion of reaction species in the inert nitrogen gas was considered as molecular (Fickian) due to low values of flow rates and concentrations used.

- Mass balance in the catalytic phase

Mass transfer in the catalytic phase was assumed to be driven by diffusion and free of convection (negligible velocity) and the chemical reaction occurs in the catalytic phase only.

$$\frac{\partial C_i}{\partial t} + \nabla \cdot (-D_{ie} \nabla_{C_i}) = R_i \tag{41}$$

where $D_{ie}$ is the effective diffusion coefficient which was approximated based on catalyst particle properties (porosity $\varepsilon_c$ and tortuosity $\tau_c$):

$$D_{ie} = \frac{\varepsilon_c D_i}{\tau_c} \tag{42}$$

The diffusion coefficients were assumed to be unaffected by catalyst deactivation and were estimated by using appropriate models of prediction of diffusivity in gaseous phase. These values, however, were assumed not affected by the solid phase, when the 3D model was compared with the 2D, reaching similar operating conditions.

- Heat balance in the gas phase

$$\left(\rho_g Cp_g \frac{\partial T}{\partial t} + \rho_g Cp_g u \cdot \nabla T = \nabla \cdot \left(k_g \nabla T\right) \tag{43}$$

$$k_g \nabla T\big|_{r=R} = 0$$

- Heat balance in the catalytic phase

$$\rho_s s Cp_s \frac{\partial T}{\partial t} = \nabla \cdot (k_s \nabla T) + Q \tag{44}$$

$$k_s \nabla T\big|_{r=R} = 0 \tag{45}$$

The thermal conductivities of gaseous phase $k_g$ and catalytic phase $k_s$ were considered as those of air and $Al_2O_3$ support, respectively.

Effect of activation energy of deactivation reaction ($\Upsilon_D$) on local deactivation is illustrated in Fig. 42, where a cut line at the center of the packed bed reflects profile of catalyst activity coefficient along dimensionless axial position $\zeta$, in a packed bed of AR of 3 and AR of 4, respectively. Relevant 3D and cross-sectional profiles of activity coefficient along the bed are shown in Fig. 43A, $B_1$, $B_2$. It can be seen that for a specific AR, the activity coefficient started at the same value at any value of activation energy of the deactivation reaction $\Upsilon_D$, and with the increase of $\Upsilon_D$, the value of activity coefficient decreased with a moving wave toward the front of the packed bed. Similar trends were observed in Fig. 40 using 2D modeling. Increasing $\Upsilon_D$ speeded up the PROX reaction along with thermal release, and thus caused more deactivation, particularly with AR of 4. Decreasing $\Upsilon_D$ caused the zone of catalyst deactivation to shift forward to the front section of the reactor, while the reaction zone was still further down in the catalyst bed. Fig. 43 shows that deactivation affects first the surrounding shell of the spherical particle as distinguished regions between the same or different catalyst particles are observed. In addition, more deactivation occurs in regions close to the wall and regions surrounding particles where high velocities dominate.

The effect of the Damköhler number ($Da_D$) was next investigated. Fig. 44 illustrates activity coefficient profiles along the length of the 3D

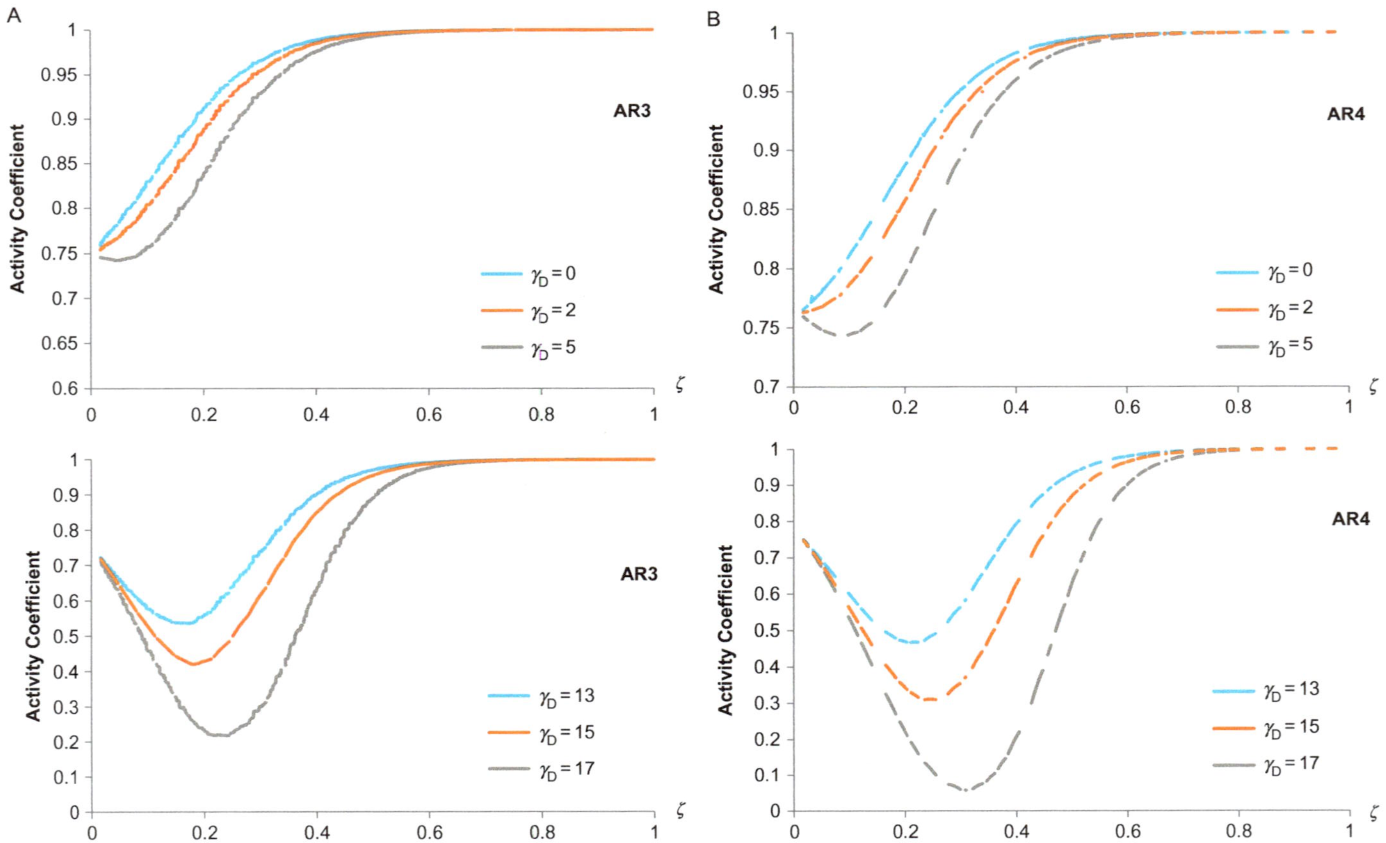


**Fig. 42** Effect of $\Upsilon_D$ on catalyst activity profile ($D = 0.000167$, $k = 1.30625$, $Da_D = 4 \times 10^{-6}$, $Da = 0.4$, $\Upsilon = 15$ and at $\tau = 50{,}000$ s), (A) AR = 3 and (B) AR = 4.

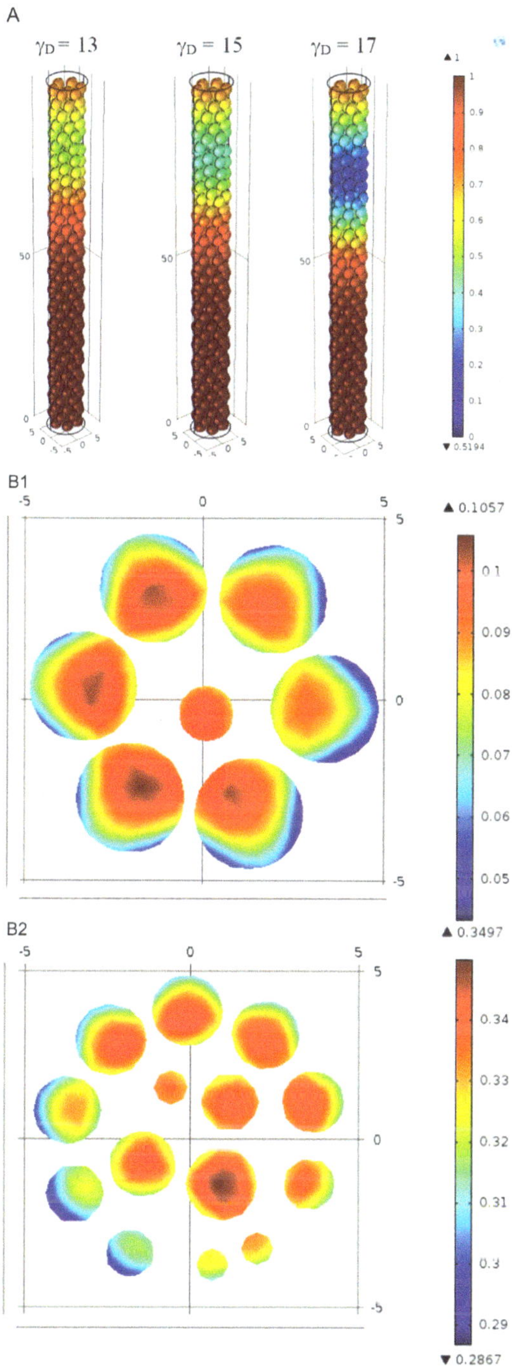


**Fig. 43** Spatial profiles of activity coefficient at $\Upsilon_D=17$, (A) 3D profiles; (B1) and (B2) cross-section profiles of activity deactivation of AR=3 and AR=4, respectively; $D=0.000167$, $k=1.30625$, $Da_D=4\times10^{-6}$, $Da=0.4$, $\Upsilon=15$, and at $\tau=50{,}000$ s.

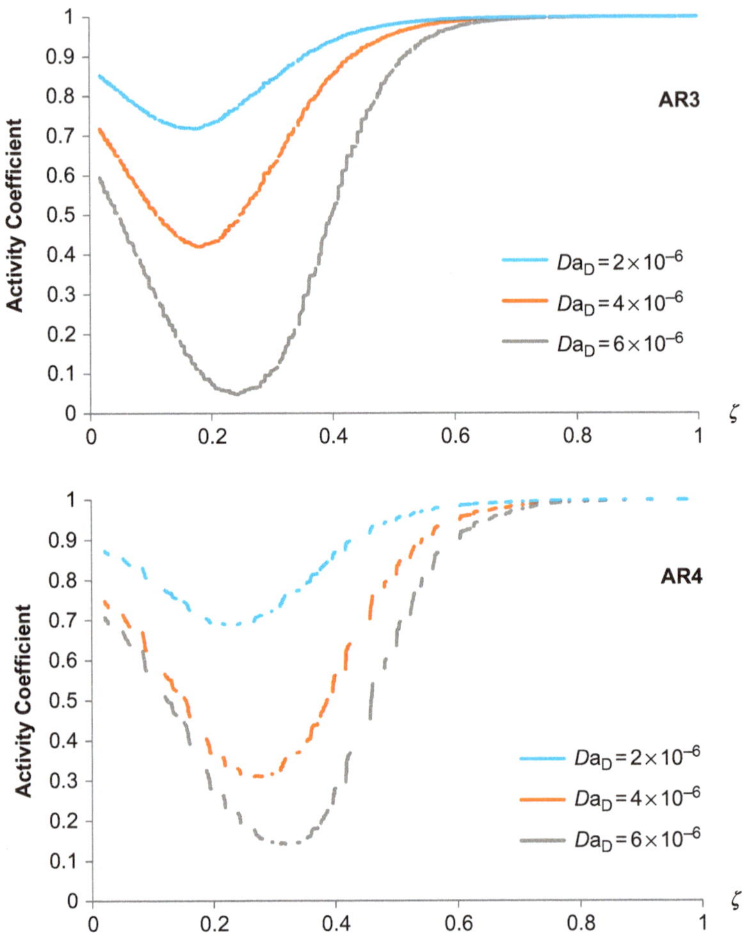


**Fig. 44** Effect of $Da_D$ on the catalytic activity profile (3D) ($D$=0.000167, $k$=1.30625, $Da$=0.4, $\Upsilon$=15, $\Upsilon_D$=15, and at $\tau$=50,000 s), AR=3 and 4.

packed bed at various values of the Damköhler number ($Da_D$). Compared with 2D model, activity coefficients fall to lower values and profile of waves move more toward the front of the packed bed reactor, particularly with AR of 3.

The 2D and 3D models have shown similar trends of deactivation profiles when deactivation rate, mass, and heat transfer were varied. The extent of catalyst deactivation in the 3D models was larger than in the 2D models. Furthermore, the deactivation waves moved more toward the front of the packed bed in the 3D models. The 3D models, however, allowed local profiles of deactivation to be observed at packing pore level. The deactivation was quite asymmetrical along axial and radial directions of the particle surface leading to uneven rates of thermal expansion and contraction, which have the potential to break the particle, or to cause local deactivation associated with temperature runaways at longer times.

Results of experimental kinetic studies and NIR imaging were extended to CFD modeling in order to get a comprehensive investigation of nonuniform catalyst deactivation in packed bed reactors. A parametric study that includes the effects of parameters such as activation energy of deactivation and Damköhler number on local deactivation profiles was investigated. A highly sensitive deactivation to temperature promoted the reaction rate by releasing additional thermal energy, and thus inhibiting the deactivation. The 3D modeling allowed observation of local catalyst deactivation at packing pore level which was not accessible by the 2D modeling. Deactivation occurred more in regions surrounding particles (shells), particularly parts of these regions facing high velocities or convective rates. The extent of catalyst deactivation in the 3D model was larger than in the 2D model, and the deactivation waves moved more toward the front of the packed bed in the 3D model. The deactivation was quite asymmetrical along axial and radial directions of the particle surface, leading potentially to uneven rates of thermal expansion and contraction, which have the potential to damage the particles. The 3D model has shown that catalysts in packed beds with low AR get more deactivated than large packed beds at low activation energy of deactivation and high Damköhler number. Since low AR packed beds were demonstrated to disperse less mass and heat along the radial coordinates, they exhibited less mixing, particularly at low fluid velocity (high Damköhler number), allowing the heat released by CO oxidation to accelerate the deactivation reaction along the axial coordinate.

## 4. CONCLUSIONS

Application of NIR imaging of 2D and 3D spatial resolution permitted access to behavior of flowing gases and catalyst activity by measuring spatial distribution of composition and temperature in a weakly scattering packed bed reactor, subject to wall effects, intraparticle rate resistance, and nonisothermal conditions. Water vapor was used owing to its high absorption coefficient in NIR. Heterogeneous distribution of activity, flow maldistribution, and dynamic lags between the local mixing zones were observed and quantified. The uneven distribution of concentration and temperature highlights the importance of mass transport and heat transport at gas–solid boundaries. The spatial resolution of NIR imaging was of importance to give reliable profiles of physical data, and dispersion coefficients were cited as example of acceptable validation. Spatially resolved data were cited of critical significance to reach insights in some packing states at a given stage of chemical process. Local catalyst activity was reported as example and ongoing works are attempting to convert these data (i.e., water vapor profiles of the PROX of CO) into local catalyst deactivation for validation of 2D and 3D models. The next logical step for this technique is the expansion from the detection of water to other species such as the commonly used gases of CO, $CO_2$, and $NO_x$. These gases are known to exhibit low absorption coefficients compared with water vapor, but at present, new developments are being designed to make this technique more sensitive to these species. With the development of IR technology, the principles of this technique are being progressively transferred to an IR laser setup. However, key to that would be the development of affordable tunable lasers, optical accessories and FPA sensors of similar sensitivities to those available in the NIR spectrum, allowing the technique to be applied to packed beds and other non-transparent reactive media.

## REFERENCES

Aiouache F, Oyama H, Kitagawa K: Spatial near-infrared imaging of hydroxyl band coverage on ceria-based catalysts, *AIChE J* 52:4, 2006.

Aiouache F, tSaoir MN, Kitagawa K: Screening wall effects of a thin fluidized bed by near-infrared imaging, *Chem Eng J* 167:288–296, 2011.

Alzahrani F, Aldehani M, Rusi H, et al: Gas flow visualization in low aspect ratio packed beds by three-dimensional modeling and near-infrared tomography, *Ind Eng Chem Res* 54:12714–12729, 2015.

Amphlett JC, Mann RF, Peppley BA: On board hydrogen purification for steam reformation/PEM fuel cell vehicle power plants, *Int J Hydrogen Energy* 21:673–678, 1996.

ASTM: *Standard practice for near infrared qualitative analysis,* USA, 2016, ASTME1790–04.

Behnam M, Dixon AG, Nijemeisland M, Stitt EH: Catalyst deactivation in 3D CFD resolved particle simulations of propane dehydrogenation, *Ind Eng Chem Res* 49:10641–10650, 2010.

Choi Y, Stenger HG: Kinetics, simulation and insights for CO selective oxidation in fuel cell applications, *J Power Sources* 129:246–254, 2004.

Delgado JMPQ: Longitudinal and transverse dispersion in porous media, *Chem Eng Res Des* 85:1245–1252, 2007.

Dixon AG, et al: CFD method to couple three-dimensional transport and reaction inside catalyst particles to the fixed bed flow field, *Ind Eng Chem Res* 49:9012–9025, 2010.

Fogler SH: *Elements of chemical reaction engineering,* 2013, University of Michigan.

Forzatti P, Lietti L: Catalyst deactivation, *Catal Today* 52:165–181, 1999.

Freund H, Bauer J, Zeiser T, Emig G: Detailed simulation of transport processes in fixed-beds, *Ind Eng Chem Res* 44:6423–6434, 2005.

Hindle F, Carey S, Ozanyan K, Winterbone D, Clough E, McCann H: Measurement of gaseous hydrocarbon distribution by a near infra-red absorption tomography system, *J Electron Imaging* 10:593–600, 2001.

Horn R, Korup O, Geske M, Zavyalova U, Oprea I, Schloegl R: Reactor for in situ measurements of spatially resolved kinetic data in heterogeneous catalysis, *Rev Sci Instrum* 81:064102, 2010.

Jacobs G, Patterson PM, Williams L, Sparks D, Davis BH: Low temperature water-gas shift: role of pretreatment on formation of surface carbonates and formats, *Catal Lett* 96:97–105, 2004.

Jaree A, et al: Temperature excursions in packed bed reactors with an axial variation of catalyst activity, *Catal Today* 69:137–146, 2001.

Jaree A, Hudgins RR, Budman HM, Silveston PL: Hysteresis and extinction waves in catalytic CO oxidation caused by reactant concentration perturbations in a packed-bed reactor, *Ind Eng Chem Res* 42:1662–1673, 2003.

Jaree A, Boonsomlanjit B, Limtrakul J: On the dynamical instability of packed-bed reactors in the presence of catalyst deactivation, *Comput Chem Eng* 32:2897–2902, 2008.

Jensen KF, Ray WH: The bifurcation behavior of tubular reactors, *Chem Eng Sci* 37:199–222, 1982.

Kempe M, Rudolph W, Welsch E: Comparative study of confocal and heterodyne microscopy for imaging through scattering media, *J Opt Soc Am A* 13:46–52, 1996.

Korup O, Mavlyankariev S, Geske M, Goldsmith CF, Horn R: Measurement and analysis of spatial reactor profiles in high temperature catalysis research, *Chem Eng Process* 50(10):998–1009, 2011.

Lee CK, Sun CW, Lee PL, et al: Study of photon migration with various source-detector separations in near-infrared spectroscopic brain imaging based on three-dimensional Monte Carlo modelling, *Opt Express* 13:8339–8348, 2005.

Lingg G: *Die Modellierung gasdurchströmter Festbettadsorber unter Beachtung der ungleichmäßigen Strömungsverteilung und äquivalenter Einphasenmodelle,* Dissertation, 1995, Technische Universität München.

Menzinger M, Yakhnin V, Jaree A, Silveston PL, Hudgins RR: Dynamic responses of packed bed reactors, *Chem Eng Sci* 59:4011–4022, 2004.

Nadeau P, Berk D, Munz RJ: Measurement of residence time distribution by laser absorption spectroscopy, *Chem Eng Sci* 51:2607–2612, 1996.

Niemantsverdriet JW, Engelen AFP, de Jong AM, Wieldraaijer W, Kramer GJ. Realistic surface science models of industrial catalysts, *Appl Surf Sci* 366:144–145, 1999.

Pangrle BJ, Walsch EG, Moore SC, DiBiasio D: Magnetic resonance imaging of laminar flow in porous tubes and shell systems, *Chem Eng Sci* 47:517, 1992.

Partridge WP, Storey JME, Lewis SA, et al: Time-resolved measurements of emission transients by mass spectrometry, *SAE Int J Fuels Lubr* 109:2992–2999, 2000.

Petersen EE: Catalyst deactivation: opportunity amidst woe, *Stud Surf Sci Catal* 111:87–98, 1997.

Qin F, Wolf EE: Spatially resolved infrared spectroscopy: a novel technique for in situ study of spatial surface coverage during CO, oxidation on supported catalysts, *Catal Lett* 39:19–25, 1996.

Reich G: Near-infrared spectroscopy and imaging: basic principles and pharmaceutical applications, *Adv Drug Deliv Rev* 57:1109–1143, 2005.

Reinke M, et al: Gas phase chemistry in catalytic combustion of methane/air mixtures over platinum at pressures of 1 to 16 bars, *Combust Flame* 141:448–468, 2005.

Roland TF, Borysow J, Fink M: Surface mediated isotope exchange reactions between water and gaseous deuterium, *J Nucl Mater* 353:193–201, 2006.

Rothman LS, Gordon IE, Barbe A, et al: The HITRAN 2008 molecular spectroscopic database, *J Quant Spectrosc Radiat Transfer* 110:533–572, 2009.

Sa J, Fernandes DLA, Aiouache F, et al: SpaciMS: spatial and temporal operando resolution of reactions within catalytic monoliths, *Analyst* 135:2260–2272, 2010.
Salem K, Tsotsas E, Mewes D: Tomographic measurement of breakthrough in a packed-bed adsorber, *Chem Eng Sci* 60:517–522, 2005.
Silveston PL, Hudgins RR: Periodic temperature forcing of catalytic reactions, *Chem Eng Sci* 59:4043–4053, 2003.
Snively CM, Oskarsdottir G, Lauterbach J: Chemically sensitive parallel analysis of combinatorial catalyst libraries, *Catal Today* 67:357–368, 2000.
Szépe S, Levenspiel O: Optimal temperature policies for reactors subject to catalyst deactivation-I batch reactor, *Chem Eng Sci* 23:881–894, 1968.
Szépe, S. and O. Levenspiel, Proceedings of the fourth Europrean symposium on chemical reaction engineering, Pergamon Press, Brussels. 1971.
Talsky G: *Derivative spectrophotometry: low and higher order,* Weinheim, 1994, VCH.
tSaoir MN, Fernandes DLA, Sá J, et al: Visualization of water vapour flow in a packed bed adsorber by near-infrared diffused transmittance tomography, *Chem Eng Sci* 66:6407–6423, 2011.
tSaoir MN, Fernandes DLA, Sá J, et al: Transient distributions of composition and temperature in a gas–solid packed bed reactor by near-infrared tomography, *Chem Eng J* 189–190:383–392, 2012.
Venderbosch RH, Prins W, Swaaij WPM: Platinum catalyzed oxidation of carbon monoxide as a model reaction in mass transfer measurements, *Chem Eng Sci* 53(19):3355–3366, 1998.
Villain F, Normand F: Redox processes on pure ceria and on Rh/CeOz catalyst monitored by X-ray absorption, *J Phys Chem* 98:5522–5533, 1994.
Yakhnin VZ, Menzinger M: Convective instability and its suppression in packed-bed- and monolith reactors, *Chem Eng Sci* 54:4547–4557, 1999.
Yoo KM, Alfano RR: Time-resolved coherent and incoherent components of forward light scattering in random media, *Opt Lett* 15:320–322, 1990.
Zhou X, Liu X, Jeffries JB, Hanson RK: Development of a sensor for temperature and water concentration in combustion gases using a single tunable diode laser, *Meas Sci Technol* 14:1459–1468, 2003.

# INDEX

Note: Page numbers followed by "*f*" indicate figures, and "*t*" indicate tables.

## T

## W



## X

FSC
www.fsc.org
MIX
Papier | Fördert gute Waldnutzung
FSC® C083411

Kolibri 360 GmbH
Ferdinand-Jühlke-Straße 7
99095 Erfurt, Deutschland
produktsicherheit@kolibri360.de